FREE Study Skills D

Dear Customer,

Thank you for your purchase from Mometrix! We consider it an honor and privilege that you have purchased our product and want to ensure your satisfaction.

As a way of showing our appreciation and to help us better serve you, we have developed a Study Skills DVD that we would like to give you for FREE. **This DVD covers our "best practices" for studying for your exam, from using our study materials to preparing for the day of the test.**

All that we ask is that you email us your feedback that would describe your experience so far with our product. Good, bad or indifferent, we want to know what you think!

To get your **FREE Study Skills DVD**, email freedvd@mometrix.com with "FREE STUDY SKILLS DVD" in the subject line and the following information in the body of the email:

a. The name of the product you purchased.

b. Your product rating on a scale of 1-5, with 5 being the highest rating.

c. Your feedback. It can be long, short, or anything in-between, just your impressions and experience so far with our product. Good feedback might include how our study material met your needs and will highlight features of the product that you found helpful.

d. Your full name and shipping address where you would like us to send your free DVD.

If you have any questions or concerns, please don't hesitate to contact me directly.

Thanks again!

Sincerely,

Jay Willis
Vice President
jay.willis@mometrix.com
1-800-673-8175

Officer Candidate School Test SECRETS

Study Guide
Your Key to Exam Success

OCS Test Review
for the ASVAB, ASTB (OAR), and AFOQT

Published by
Mometrix Test Preparation
OCS Exam Secrets Test Prep Team

Copyright © 2017 by Mometrix Media LLC

All rights reserved. This product, or parts thereof, may not be reproduced, stored in a retrieval system, or transmitted in any form or by any means—electronic, mechanical, photocopy, recording, scanning, or other—except for brief quotations in critical reviews or articles, without the prior written permission of the publisher.

Written and edited by the OCS Exam Secrets Test Prep Staff

Printed in the United States of America

This paper meets the requirements of ANSI/NISO Z39.48-1992 (Permanence of Paper).

Mometrix offers volume discount pricing to institutions. For more information or a price quote, please contact our sales department at sales@mometrix.com or 888-248-1219.

Mometrix Test Preparation is not affiliated with or endorsed by any official testing organization. All organizational and test names are trademarks of their respective owners.

ISBN 13: 978-1-5167-0227-5
ISBN 10: 1-5167-0227-1

Dear Future Exam Success Story:

Congratulations on your purchase of our study guide. Our goal in writing our study guide was to cover the content on the test, as well as provide insight into typical test taking mistakes and how to overcome them.

Standardized tests are a key component of being successful, which only increases the importance of doing well in the high-pressure high-stakes environment of test day. How well you do on this test will have a significant impact on your future, and we have the research and practical advice to help you execute on test day.

The product you're reading now is designed to exploit weaknesses in the test itself, and help you avoid the most common errors test takers frequently make.

How to use this study guide

We don't want to waste your time. Our study guide is fast-paced and fluff-free. We suggest going through it a number of times, as repetition is an important part of learning new information and concepts.

First, read through the study guide completely to get a feel for the content and organization. Read the general success strategies first, and then proceed to the content sections. Each tip has been carefully selected for its effectiveness.

Second, read through the study guide again, and take notes in the margins and highlight those sections where you may have a particular weakness.

Finally, bring the manual with you on test day and study it before the exam begins.

Your success is our success

We would be delighted to hear about your success. Send us an email and tell us your story. Thanks for your business and we wish you continued success.

Sincerely,

Mometrix Test Preparation Team

Need more help? Check out our flashcards at: http://MometrixFlashcards.com/ASVAB

TABLE OF CONTENTS

TOP 20 TEST TAKING TIPS 1

COMMON SUBJECT: WORD KNOWLEDGE 2
   DETERMINING WORD MEANING 2
   TESTING TIPS 10

COMMON SUBJECT: READING COMPREHENSION 13
   COMPREHENSION SKILLS 13
   CRITICAL THINKING SKILLS 24
   TESTING TIPS 30

COMMON SUBJECT: MATHEMATICS 38
   MATH BASICS 38
   GEOMETRY BASICS 47
   PROBABILITY BASICS 54
   STATISTICS BASICS 57
   FINAL NOTES 60

COMMON SUBJECT: SCIENCE 62
   EARTH AND SPACE SCIENCE 62
   BIOLOGY 68
   CHEMISTRY 81
   PHYSICS 91

COMMON SUBJECT: MECHANICAL COMPREHENSION 102
   KINEMATICS 102
   KINETICS 109
   WORK/ENERGY 120
   MACHINES 126
   MOMENTUM/IMPULSE 140
   FLUIDS 141
   HEAT TRANSFER 147
   OPTICS 149
   ELECTRICITY 149
   MAGNETISM 158

ASVAB: ASSEMBLING OBJECTS 159

AFOQT: VERBAL ANALOGIES 161

AFOQT: SITUATIONAL JUDGMENT 168

AFOQT: PHYSICAL SCIENCE 169

AFOQT: TABLE READING 173

AFOQT: INSTRUMENT COMPREHENSION 175

AFOQT: BLOCK COUNTING 179

AFOQT: AVIATION INFORMATION 181

ASVAB PRACTICE TEST 192
   ARITHMETIC REASONING 192
   WORD KNOWLEDGE 197
   PARAGRAPH COMPREHENSION 202
   MATHEMATICS KNOWLEDGE 206

- General Science 210
- Assembling Objects 214
- Mechanical Comprehension 220

ANSWER KEY 230

- Arithmetic Reasoning 230
- Word Knowledge 232
- Paragraph Comprehension 234
- Mathematics Knowledge 235
- General Science 238
- Assembling Objects and Mechanical Comprehension 243

OAR PRACTICE TEST 245

- Math Skills Test 245
- Reading Comprehension Test 250
- Mechanical Comprehension Test 262

ANSWER KEY 273

- Math Skills Test 273
- Reading Comprehension Test 277
- Mechanical Comprehension Test 279

AFOQT PRACTICE TEST 282

- Verbal Analogies 282
- Arithmetic Reasoning 286
- Word Knowledge 291
- Math Knowledge 295
- Reading Comprehension 299
- Situational Judgment 310
- Physical Science 323
- Table Reading 326
- Instrument Comprehension 334
- Block Counting 338
- Aviation Information 340

ANSWER KEY 344

- Verbal Analogies 344
- Arithmetic Reasoning 346
- Word Knowledge 348
- Math Knowledge 350
- Reading Comprehension 354
- Situational Judgment 357
- Physical Science 358
- Table Reading 360
- Instrument Comprehension 361
- Block Counting 362
- Aviation Information 363

SECRET KEY #1 - TIME IS YOUR GREATEST ENEMY 366

- Pace Yourself 366

SECRET KEY #2 - GUESSING IS NOT GUESSWORK 367

- Monkeys Take the Test 367
- $5 Challenge 368

SECRET KEY #3 - PRACTICE SMARTER, NOT HARDER	369
Success Strategy	369
SECRET KEY #4 - PREPARE, DON'T PROCRASTINATE	370
SECRET KEY #5 - TEST YOURSELF	371
GENERAL STRATEGIES	372
APPENDIX: AREA, VOLUME, SURFACE AREA FORMULAS	377
ADDITIONAL BONUS MATERIAL	379

Top 20 Test Taking Tips

1. Carefully follow all the test registration procedures
2. Know the test directions, duration, topics, question types, how many questions
3. Setup a flexible study schedule at least 3-4 weeks before test day
4. Study during the time of day you are most alert, relaxed, and stress free
5. Maximize your learning style; visual learner use visual study aids, auditory learner use auditory study aids
6. Focus on your weakest knowledge base
7. Find a study partner to review with and help clarify questions
8. Practice, practice, practice
9. Get a good night's sleep; don't try to cram the night before the test
10. Eat a well balanced meal
11. Know the exact physical location of the testing site; drive the route to the site prior to test day
12. Bring a set of ear plugs; the testing center could be noisy
13. Wear comfortable, loose fitting, layered clothing to the testing center; prepare for it to be either cold or hot during the test
14. Bring at least 2 current forms of ID to the testing center
15. Arrive to the test early; be prepared to wait and be patient
16. Eliminate the obviously wrong answer choices, then guess the first remaining choice
17. Pace yourself; don't rush, but keep working and move on if you get stuck
18. Maintain a positive attitude even if the test is going poorly
19. Keep your first answer unless you are positive it is wrong
20. Check your work, don't make a careless mistake

Common Subject: Word Knowledge

Determining Word Meaning

An understanding of the basics of language is helpful, and often vital, to understanding what you read. The term *structural analysis* refers to looking at the parts of a word and breaking it down into its different components to determine the word's meaning. Parts of a word include prefixes, suffixes, and the root word. By learning the meanings of prefixes, suffixes, and other word fundamentals, you can decipher the meaning of words which may not yet be in your vocabulary. Prefixes are common letter combinations at the beginning of words, while suffixes are common letter combinations at the end. The main part of the word is known as the root. Visually, it would look like this: prefix + root word + suffix. Look first at the individual meanings of the root word, prefix and/or suffix. Using knowledge of the meaning(s) of the prefix and/or suffix to see what information it adds to the root. Even if the meaning of the root is unknown, one can use knowledge of the prefix's and/or suffix's meaning(s) to determine an approximate meaning of the word. For example, if one sees the word *uninspired* and does not know what it means, they can use the knowledge that *un-* means 'not' to know that the full word means "not inspired." Understanding the common prefixes and suffixes can illuminate at least part of the meaning of an unfamiliar word.

> **Review Video: Determining Word Meanings**
> Visit *mometrix.com/academy* and enter *Code:* **894894**

The following is a list of common prefixes and their meanings:

Prefix	Definition	Examples
a	in, on, of, up, to	abed, afoot
a-	without, lacking	atheist, agnostic
ab-	from, away, off	abdicate, abjure
ad-	to, toward	advance
am-	friend, love	amicable, amatory
ante-	before, previous	antecedent, antedate
anti-	against, opposing	antipathy, antidote
auto-	self	autonomy, autobiography
belli-	war, warlike	bellicose
bene-	well, good	benefit, benefactor
bi-	two	bisect, biennial
bio-	life	biology, biosphere
cata-	down, away, thoroughly	catastrophe, cataclysm
chron-	time	chronometer, synchronize
circum-	around	circumspect, circumference
com-	with, together, very	commotion, complicate
contra-	against, opposing	contradict, contravene
cred-	belief, trust	credible, credit

Prefix	Meaning	Example
de-	from	depart
dem-	people	demographics, democracy
dia-	through, across, apart	diameter, diagnose
dis-	away, off, down, not	dissent, disappear
epi-	upon	epilogue
equi-	equal, equally	equivalent
ex-	out	extract
for-	away, off, from	forget, forswear
fore-	before, previous	foretell, forefathers
homo-	same, equal	homogenized
hyper-	excessive, over	hypercritical, hypertension
hypo-	under, beneath	hypodermic, hypothesis
in-	in, into	intrude, invade
in-	not, opposing	incapable, ineligible
inter-	among, between	intercede, interrupt
intra-	within	intramural, intrastate
magn-	large	magnitude, magnify
mal-	bad, poorly, not	malfunction
micr-	small	microbe, microscope
mis-	bad, poorly, not	misspell, misfire
mono-	one, single	monogamy, monologue
mor-	die, death	mortality, mortuary
neo-	new	neolithic, neoconservative
non-	not	nonentity, nonsense
ob-	against, opposing	objection
omni-	all, everywhere	omniscient
ortho-	right, straight	orthogonal, orthodox
over-	above	overbearing
pan-	all, entire	panorama, pandemonium
para-	beside, beyond	parallel, paradox
per-	through	perceive, permit
peri-	around	periscope, perimeter
phil-	love, like	philosophy, philanthropic
poly-	many	polymorphous, polygamous
post-	after, following	postpone, postscript
pre-	before, previous	prevent, preclude
prim-	first, early	primitive, primary
pro-	forward, in place of	propel, pronoun
re-	back, backward, again	revoke, recur
retro-	back, backward	retrospect, retrograde
semi-	half, partly	semicircle, semicolon
sub-	under, beneath	subjugate, substitute
super-	above, extra	supersede, supernumerary
sym-	with, together	sympathy, symphony
trans-	across, beyond, over	transact, transport

ultra-	beyond, excessively	ultramodern, ultrasonic, ultraviolet
un-	not, reverse of	unhappy, unlock
uni-	one	uniform, unity
vis-	to see	visage, visible

The following is a list of common suffixes and their meanings:

Suffix	Definition	Examples
-able	able to, likely	capable, tolerable
-age	process, state, rank	passage, bondage
-ance	act, condition, fact	acceptance, vigilance
-arch	to rule	monarch
-ard	one that does excessively	drunkard, wizard
-ate	having, showing	separate, desolate
-ation	action, state, result	occupation, starvation
-cy	state, condition	accuracy, captaincy
-dom	state, rank, condition	serfdom, wisdom
-en	cause to be, become	deepen, strengthen
-er	one who does	teacher
-esce	become, grow, continue	convalesce, acquiesce
-esque	in the style of, like	picturesque, grotesque
-ess	feminine	waitress, lioness
-fic	making, causing	terrific, beatific
-ful	full of, marked by	thankful, zestful
-fy	make, cause, cause to have	glorify, fortify
-hood	state, condition	manhood, statehood
-ible	able, likely, fit	edible, possible, divisible
-ion	action, result, state	union, fusion
-ish	suggesting, like	churlish, childish
-ism	act, manner, doctrine	barbarism, socialism
-ist	doer, believer	monopolist, socialist
-ition	action, state, result	sedition, expedition
-ity	state, quality, condition	acidity, civility
-ize	make, cause to be, treat with	sterilize, mechanize, criticize
-less	lacking, without	hopeless, countless
-like	like, similar	childlike, dreamlike
-logue	type of written/spoken language	prologue
-ly	like, of the nature of	friendly, positively
-ment	means, result, action	refreshment, disappointment
-ness	quality, state	greatness, tallness
-or	doer, office, action	juror, elevator, honor
-ous	marked by, given to	religious, riotous
-ship	the art or skill of	statesmanship
-some	apt to, showing	tiresome, lonesome
-th	act, state, quality	warmth, width

-tude	quality, state, result	magnitude, fortitude
-ty	quality, state	enmity, activity
-ward	in the direction of	backward, homeward

When defining words in a text, words often have a meaning that is more than the dictionary definition. The **denotative** meaning of a word is the literal meaning. The **connotative** meaning goes beyond the denotative meaning to include the emotional reaction a word may invoke. The connotative meaning often takes the denotative meaning a step further due to associations which the reader makes with the denotative meaning. The reader can differentiate between the denotative and connotative meanings by first recognizing when authors use each meaning. Most non-fiction, for example, is fact-based, the authors not using flowery, figurative language. The reader can assume that the writer is using the denotative, or literal, meaning of words. In fiction, on the other hand, the author may be using the connotative meaning. Connotation is one form of figurative language. The reader should use context clues to determine if the author is using the denotative or connotative meaning of a word.

> **Review Video: Denotation and Connotation**
> Visit **mometrix.com/academy** and enter **Code: 310092**

Readers of all levels will encounter words with which they are somewhat unfamiliar. The best way to define a word in **context** is to look for nearby words that can help. For instance, unfamiliar nouns are often accompanied by examples that furnish a definition. Consider the following sentence: "Dave arrived at the party in hilarious garb: a leopard-print shirt, buckskin trousers, and high heels." If a reader was unfamiliar with the meaning of garb, he could read the examples and quickly determine that the word means "clothing." Examples will not always be this obvious. For instance, consider this sentence: "Parsley, lemon, and flowers were just a few of items he used as garnishes." Here, the possibly unfamiliar word *garnishes* is exemplified by parsley, lemon, and flowers. Readers who have eaten in a few restaurants will probably be able to identify a garnish as something used to decorate a plate.

In addition to looking at the context of a passage, readers can often use contrasts to define an unfamiliar word in context. In many sentences, the author will not describe the unfamiliar word directly, but will instead describe the opposite of the unfamiliar word. Of course, this provides information about the word the reader needs to define. Consider the following example: "Despite his intelligence, Hector's low brow and bad posture made him look obtuse." The author suggests that Hector's appearance was opposite to his actual intelligence. Therefore, *obtuse* must mean unintelligent or stupid. Here is another example: "Despite the horrible weather, we were beatific about our trip to Alaska." The word *despite* indicates that the speaker's feelings were at odds with the weather. Since the weather is described as "horrible," *beatific* must mean something good.

In some cases, there will be very few contextual clues to help a reader define the meaning of an unfamiliar word. When this happens, one strategy the reader may employ is substitution. A good reader will brainstorm some possible synonyms for the given word, and then substitute these words into the sentence. If the sentence and the surrounding passage continue to make sense, the substitution has revealed at least some information about the unfamiliar word. Consider the sentence, "Frank's admonition rang in her ears as she climbed the mountain." A reader unfamiliar with *admonition* might come up with some substitutions like "vow," "promise," "advice," "complaint," or "compliment." All of these words make general sense of the sentence, though their meanings are diverse. The process has suggested, however, that an admonition is some sort of message. The substitution strategy is rarely able to pinpoint a precise definition, but can be effective as a last resort.

It is sometimes possible to define an unfamiliar word by looking at the descriptive words in the context. Consider the following sentence: "Fred dragged the recalcitrant boy kicking and screaming up the stairs." *Dragged*, *kicking*, and *screaming* all suggest that the boy does not want to go up the stairs. The reader may assume that *recalcitrant* means something like unwilling or protesting. In that example, an unfamiliar adjective was identified. It is perhaps more typical to use description to define an unfamiliar noun, as in this sentence: "Don's wrinkled frown and constantly shaking fist identified him as a curmudgeon of the first order." Don is described as having a "wrinkled frown and constantly shaking fist," suggesting that a *curmudgeon* must be a grumpy old man. Contrasts do not always provide detailed information about the unfamiliar word, but they at least give the reader some clues.

When a word has more than one meaning, it can be tricky to determine how it is being used in a given sentence. Consider the verb *cleave*, which bizarrely can mean either "join" or "separate." When a reader comes upon this word, she will have to select the definition that makes the most sense. So, take as an example the following sentence: "The birds cleaved together as they flew from the oak tree." Immediately, the presence of the word *together* should suggest that in this sentence *cleave* is being used to mean "*join.*" A slightly more difficult example would be the sentence, "Hermione's knife cleaved the bread cleanly." It doesn't make sense for a knife to join bread together, so the word must be meant to indicate separation. Discovering the meaning of a word with multiple meanings requires the same tricks as defining an unknown word: looking for contextual clues and evaluating substituted words.

Literary Devices

Understanding how words relate to each other can often add meaning to a passage. This is explained by understanding **synonyms** (words that mean the same thing) and **antonyms** (words that mean the opposite of one another). As an example, *dry* and *arid* are synonyms, and *dry* and *wet* are antonyms. There are many pairs of words in English that can be considered synonyms, despite having slightly different definitions. For instance, the words *friendly* and *collegial* can both be used to describe a warm interpersonal relationship, so it would be correct to call them synonyms. However, *collegial* (kin to *colleague*) is more often used in reference to professional or academic relationships, while *friendly* has no such connotation. Nevertheless, it would be appropriate to call these words synonyms. If the difference between the two words is too great, however, they may not be called synonyms. *Hot* and *warm* are not synonyms, for instance, because their meanings are too distinct. A good way to determine whether two words are synonyms is to substitute one for the other and see if the sentence means the same thing. Substituting *warm* for *hot* in a sentence would convey a different meaning.

> ➤ **Review Video: <u>Synonyms and Antonyms</u>**
> *Visit **mometrix.com/academy** and enter **Code: 105612***

Antonyms are opposites. *Light* and *dark*, *up* and *down*, *right* and *left*, *good* and *bad*: these are all sets of antonyms. It is important to distinguish between antonyms and pairs of words that are simply different. *Black* and *gray*, for instance, are not antonyms because gray is not the opposite of black. *Black* and *white*, on the other hand, are antonyms. Not every word has an antonym. For instance, many nouns do not. What would be the antonym of *chair*, after all? On a standardized test, the questions related to antonyms are more likely to concern adjectives. Remember that adjectives are words that describe a noun. Some common adjectives include *red*, *fast*, *skinny*, and *sweet*. Of these four examples, only *red* lacks a group of obvious antonyms.

> ➢ **Review Video:** <u>Synonyms and Antonyms Continued</u>
> Visit **mometrix.com/academy** and enter *Code:* **440473**

There are many types of language devices that authors use to convey their meaning in a more descriptive or interesting way. Understanding these concepts will help you understand what you read. These types of devices are called *figurative language* – language that goes beyond the literal meaning of the words. **Descriptive language** that evokes imagery in the reader's mind is one type of figurative language. **Exaggeration** is also one type of figurative language. Also, when you compare two things, you are using figurative language. **Similes** and **metaphors** are ways of comparing things, and both are types of figurative language commonly found in poetry. An example of figurative language (a simile in this case) is: "The child howled like a coyote when her mother told her to pick up the toys." In this example, the child's howling is compared to that of a coyote. Figurative language is descriptive in nature and helps the reader understand the sound being made in this sentence.

Alliteration is a stylistic device, or literary technique, in which successive words (more strictly, stressed syllables) begin with the same sound or letter. Alliteration is a frequent tool in poetry but it is also common in prose, particularly to highlight short phrases. Especially in poetry, it contributes to euphony of the passage, lending it a musical air. It may act to humorous effect. Alliteration draws attention to itself, which may be a good or a bad thing. Authors should be conscious of the character of the sound to be repeated. In the above example, a *th* sound is somewhat difficult to make quickly in four consecutive words, so the phrase conveys a little of the difficulty of moving through tall grass. If the author is indeed trying to suggest this difficulty, then the alliteration is a success. Consider, however, the description of eyes as "glassy globes of glitter." This is definitely alliteration, since the initial *gl* sound is used three times. However, one might question whether this awkward sound is appropriate for a description of pretty eyes. The phrase is not especially pleasant to the ear, and therefore is probably not effective as alliteration. Related to alliteration are *assonance*, the repetition of vowel sounds, and *consonance*, the repetition of consonant sounds.

A **figure of speech**, sometimes termed a rhetorical figure or device, or elocution, is a word or phrase that departs from straightforward, literal language. Figures of speech are often used and crafted for emphasis, freshness of expression, or clarity. However, clarity may also suffer from their use.

> ➢ **Review Video:** <u>Figure of Speech</u>
> Visit **mometrix.com/academy** and enter *Code:* **111295**

Note that not all theories of meaning necessarily have a concept of "literal language" (see literal and figurative language). Under theories that do not, figure of speech is not an entirely coherent concept.

As an example of the figurative use of a word, consider the sentence, "I am going to crown you." It may mean:
I am going to place a literal crown on your head.
I am going to symbolically exalt you to the place of kingship.
I am going to punch you in the head with my clenched fist.
I am going to put a second checker on top of your checker to signify that it has become a king.

A **metaphor** is a type of figurative language in which the writer equates one thing with a different thing. For instance, in the sentence "The bird was an arrow arcing through the sky," the arrow is serving as a metaphor for the bird. The point of a metaphor is to encourage the reader to think about the thing being described in a different way. Using this example, we are being asked to envision the bird's flight as being similar to the arc of an arrow, so we will imagine it to be swift, bending, etc. Metaphors are a way for the author to describe without being direct and obvious. Metaphors are a more lyrical and suggestive way of providing information. Note that the thing to which a metaphor refers will not always be mentioned explicitly by the author. For instance, consider the following description of a forest in winter: "Swaying skeletons reached for the sky and groaned as the wind blew through them." The author is clearly using *skeletons* as a metaphor for leafless trees. This metaphor creates a spooky tone while inspiring the reader's imagination.

Metonymy is referring to one thing in terms of another, closely related thing. This is similar to metaphor, but there is less distance between the description and the thing being described. An example of metonymy is referring to the news media as the "press," when of course the press is only the device by which newspapers are printed. Metonymy is a way of referring to something without having to repeat its name constantly. **Synecdoche**, on the other hand, is referring to a whole by one of its parts. An example of synecdoche would be calling a police officer a "badge." Synecdoche, like metonymy, is a handy way of referring without having to overuse certain words. It also allows the writer to emphasize aspects of the thing being described. For instance, referring to businessmen as "suits" suggests professionalism, conformity, and drabness.

Hyperbole is overstatement for effect. The following sentence is an example of hyperbole: *He jumped ten feet in the air when he heard the good news*. Obviously, no person has the ability to jump ten feet in the air. The author hyperbolizes not because he believes the statement will be taken literally, but because the exaggeration conveys the extremity of emotion. Consider how much less colorful the sentence would be if the author simply said, "He jumped when he heard the good news." Hyperbole can be dangerous if the author does not exaggerate enough. For instance, if the author wrote, "He jumped two feet in the air when he heard the good news," the reader might not be sure whether this is actually true or just hyperbole. Of course, in many situations this distinction will not really matter. However, an author should avoid confusing or vague hyperbole when he needs to maintain credibility or authority with readers.

> ➤ **Review Video: Hyperbole and Understatement**
> Visit *mometrix.com/academy* and enter *Code:* **308470**

Understatement is the opposite of hyperbole: that is, it is describing something as less than it is, for effect. As an example, consider a person who climbs Mount Everest and then describes the journey as "a little stroll." This is an almost extreme example of understatement. Like other types of figurative language, understatement has a range of uses. It may convey self-deprecation or modesty, as in the above example. Of course, some people might interpret understatement as false modesty, a deliberate attempt to call attention to the magnitude of what is being discussed. For example, a woman is complimented on her enormous diamond engagement ring and says, "Oh, this little thing?" Her understatement might be viewed as snobby or insensitive. Understatement can have various effects, but it always calls attention to itself.

A **simile** is a figurative expression similar to a metaphor, though it requires the use of a distancing word like *like* or *as*. Some examples are "The sun was like an orange," "eager as a beaver," and "nimble as a mountain goat." Because a simile includes *like* or a*s*, it creates a little space between the description and the thing being described. If an author says that a house was "like a shoebox," the tone is slightly different than if the author said that the house *was* a shoebox. In a simile, the author indicates an awareness that the description is not the same thing as the thing being described. In a metaphor, there is no such distinction, even though one may safely assume that the author is aware of it. This is a subtle difference, but authors will alternately use metaphors and similes depending on their intended tone.

Another type of figurative language is **personification.** This is the description of the nonhuman as if it were human. Literally, the word means the process of making something into a person. There is a wide range of approaches to personification, from common expressions like "whispering wind" to full novels like *Animal Farm*, by George Orwell, in which the Bolshevik Revolution is reenacted by farmyard animals. The general intent of personification is to describe things in a manner that will be comprehensible to readers. When an author states that a tree "groans" in the wind, she of course does not mean that the tree is emitting a low, pained sound from its mouth. Instead, she means that the tree is making a noise similar to a human groan. Of course, this personification establishes a tone of sadness or suffering. A different tone would be established if the author said the tree was "swaying" or "dancing."

Irony is a statement that suggests its opposite. In other words, it is when an author or character says one thing but means another. For example, imagine a man walks in his front door, covered in mud and in tattered clothes. His wife asks him, "How was your day?" and he says "Great!" The man's comment is an example of irony. As in this example, irony often depends on information the reader obtains elsewhere. There is a fine distinction between irony and sarcasm. Irony is any statement in which the literal meaning is opposite from the intended meaning, while sarcasm is a statement of this type that is also insulting to the person at whom it is directed. A sarcastic statement suggests that the other person is stupid enough to believe an obviously false statement is true. Irony is a bit more subtle than sarcasm.

> **Review Video: Irony**
Visit **mometrix.com/academy** and enter **Code: 374204**

The more words a person is exposed to, the greater their vocabulary will become. By reading on a regular basis, a person can increase the number of ways they have seen a word in context. Based on experience, a person can recall how a word was used in the past and apply that knowledge to a new context. For example, a person may have seen the word *gull* used to mean a bird that is found near the seashore. However, a *gull* can also be a person who is easily tricked. If the word is used in context in reference to a character, the reader can recognize that the character is being called a bird that is not seen as extremely intelligent. Using what the reader knows about a word can be useful when making comparisons or figuring out the meaning of a new use of a word, as in figurative language, idioms, analogies, and multiple-meaning words.

Testing Tips

Nearly and Perfect Synonyms

You must determine which of five provided choices has the best similar definition as a certain word. Nearly similar may often be more correct, because the goal is to test your understanding of the nuances, or little differences, between words. A perfect match may not exist, so don't be concerned if your answer choice is not a complete synonym. Focus upon edging closer to the word. Eliminate the words that you know aren't correct first. Then narrow your search. Cross out the words that are the least similar to the main word until you are left with the one that is the most similar.

Prefixes

Take advantage of every clue that the word might include. Prefixes and suffixes can be a huge help. Usually they allow you to determine a basic meaning. Pre- means before, post- means after, pro – is positive, de- is negative. From these prefixes and suffixes, you can get an idea of the general meaning of the word and look for its opposite. Beware though of any traps. Just because con is the opposite of pro, doesn't necessarily mean congress is the opposite of progress! A list of the most common prefixes and suffixes is included in the appendix.

Positive vs. Negative

Many words can be easily determined to be a positive word or a negative word. Words such as despicable, gruesome, and bleak are all negative. Words such as ecstatic, praiseworthy, and magnificent are all positive. You will be surprised at how many words can be considered as either positive or negative. Once that is determined, you can quickly eliminate any other words with an opposite meaning and focus on those that have the other characteristic, whether positive or negative.

Word Strength

Part of the challenge is determining the most nearly similar word. This is particularly true when two words seem to be similar. When analyzing a word, determine how strong it is. For example, stupendous and good are both positive words. However, stupendous is a much stronger positive adjective than good. Also, towering or gigantic are stronger words than tall or large. Search for an answer choice that is similar and also has the same strength. If the main word is weak, look for similar words that are also weak. If the main word is strong, look for similar words that are also strong.

Type and Topic

Another key is what type of word is the main word. If the main word is an adjective describing height, then look for the answer to be an adjective describing height as well. Match both the type and topic of the main word. The type refers the parts of speech, whether the word is an adjective, adverb, or verb. The topic refers to what the definition of the word includes, such as sizes or fashion styles.

Form a Sentence

Many words seem more natural in a sentence. *Specious* reasoning, *irresistible* force, and *uncanny* resemblance are just a few of the word combinations that usually go together. When faced with an uncommon word that you barely understand (and on the ASVAB there will be many), try to put the word in a sentence that makes sense. It will help you to understand the word's meaning and make it easier to determine its opposite. Once you have a good descriptive sentence that utilizes the main word properly, plug in the answer choices and see if the sentence still has the same meaning with each answer choice. The answer choice that maintains the meaning of the sentence is correct!

Use Replacements

Using a sentence is a great help because it puts the word into a proper perspective. Since ASVAB actually gives you a sentence, sometimes you don't always have to create your own (though in many cases the sentence won't be helpful). Read the provided sentence with the underlined word. Then read the sentence again and again, each time replacing the underlined word with one of the answer choices. The correct answer should "sound" right and fit.

Example: The desert landscape was **desolate**.
    A. cheerful
    B. creepy
    C. excited
    D. forlorn

After reading the example sentence, begin replacing "desolate" with each of the answer choices. Does "the desert landscape was cheerful, creepy, excited, or forlorn" sound right? Deserts are typically hot, empty, and rugged environments, probably not cheerful, or excited. While creepy might sound right, that word would certainly be more appropriate for a haunted house. But "the desert landscape was forlorn" has a certain ring to it and would be correct.

Eliminate Similar Choices

If you don't know the word, don't worry. Look at the answer choices and just use them. Remember that three of the answer choices will always be wrong. If you can find a common relationship between any three answer choices, then you know they are wrong. Find the answer choice that does not have a common relationship to the other answer choices and it will be the correct answer.

Example: **Laconic** most nearly means
    A. wordy
    B. talkative
    C. expressive
    D. quiet

In this example the first three choices are all similar. Even if you don't know that laconic means the same as quiet, you know that "quiet" must be correct, because the other three choices were all virtually the same. They were all the same, so they must all be wrong. The one that is different must be correct. So, don't worry if you don't know a word. Focus on the answer choices that you do understand and see if you can identify similarities. Even identifying two words that are similar will allow you to eliminate those two answer choices, for they are both wrong, because they are either both right or both wrong (they're similar, remember), so since they can't both be right, they both must be wrong.

The Trap of Familiarity

Don't choose a word just because you recognize it. On difficult questions, you may only recognize one or two words. ASVAB doesn't put "make-believe" words on the test, so don't think that just because you only recognize one word means that word must be correct. If you don't recognize four words, then focus on the one that you do recognize. Is it correct? Try your best to determine if it fits the sentence. If it does, that is great, but if it doesn't, eliminate it. Each word you eliminate increases your chances of getting the question correct.

Read Carefully

Be sure to read all of the choices. You may find an answer choice that seems right at first, but continue reading and you may find a better choice.

Difficult words are usually synonyms or antonyms (opposites). Whenever you have extremely difficult words that you don't understand, look at the answer choices. Try and identify whether two or more of the answer choices are either synonyms or antonyms. Remember that if you can find two words that have the same relationship (for example, two answer choices are synonyms) then you can eliminate them both.

Work Quickly

Since you have 35 questions to answer in only 11 minutes, that means that you have between 18 and 19 seconds to spend per question. This section faces a greater time crunch that any other test you will take on the ASVAB. If you are stuck on one word, don't waste too much time. Eliminate the answers you could bet a quick $5 on and then pick the first one that remains. You can make a note in your book and if you have time you can always come back, but don't waste your time. You must work quickly!

Common Subject: Reading Comprehension

Comprehension Skills

One of the most important skills in reading comprehension is the identification of **topics** and **main ideas.** There is a subtle difference between these two features. The topic is the subject of a text, or what the text is about. The main idea, on the other hand, is the most important point being made by the author. The topic is usually expressed in a few words at the most, while the main idea often needs a full sentence to be completely defined. As an example, a short passage might have the topic of penguins and the main idea *Penguins are different from other birds in many ways.* In most nonfiction writing, the topic and the main idea will be stated directly, often in a sentence at the very beginning or end of the text. When being tested on an understanding of the author's topic, the reader can quickly *skim* the passage for the general idea, stopping to read only the first sentence of each paragraph. A paragraph's first sentence is often (but not always) the main topic sentence, and it gives you a summary of the content of the paragraph. However, there are cases in which the reader must figure out an unstated topic or main idea. In these instances, the student must read every sentence of the text, and try to come up with an overarching idea that is supported by each of those sentences.

> **Review Video: Topics and Main Ideas**
> Visit *mometrix.com/academy* and enter **Code: 407801**

While the main idea is the overall premise of a story, **supporting details** provide evidence and backing for the main point. In order to show that a main idea is correct, or valid, the author needs to add details that prove their point. All texts contain details, but they are only classified as supporting details when they serve to reinforce some larger point.

Supporting details are most commonly found in informative and persuasive texts. In some cases, they will be clearly indicated with words like *for example* or *for instance*, or they will be enumerated with words like *first*, *second*, and *last*. However, they may not be indicated with special words.

As a reader, it is important to consider whether the author's supporting details really back up his or her main point. Supporting details can be factual and correct but still not relevant to the author's point. Conversely, supporting details can seem pertinent but be ineffective because they are based on opinion or assertions that cannot be proven.

> **Review Video: Supporting Details**
> Visit *mometrix.com/academy* and enter **Code: 396297**

An example of a main idea is: "Giraffes live in the Serengeti of Africa." A supporting detail about giraffes could be: "A giraffe uses its long neck to reach twigs and leaves on trees." The main idea gives the general idea that the text is about giraffes. The supporting detail gives a specific fact about how the giraffes eat.

As opposed to a main idea, themes are seldom expressed directly in a text, so they can be difficult to identify. A **theme** is an issue, an idea, or a question raised by the text. For instance, a theme of William Shakespeare's *Hamlet* is indecision, as the title character explores his own psyche and the results of his failure to make bold choices. A great work of literature may have many themes, and the reader is justified in identifying any for which he or she can find support. One common characteristic of themes is that they raise more questions than they answer. In a good piece of fiction, the author is not always trying to convince the reader, but is instead trying to elevate the reader's perspective and encourage him to consider the themes more deeply. When reading, one can identify themes by constantly asking what general issues the text is addressing. A good way to evaluate an author's approach to a theme is to begin reading with a question in mind (for example, how does this text approach the theme of love?) and then look for evidence in the text that addresses that question.

> **Review Video: Theme**
> *Visit mometrix.com/academy and enter Code:* **732074**

Topic and summary sentences are a convenient way to encapsulate the main idea of a text. In some textbooks and academic articles, the author will place a topic or summary sentence at the beginning of each section as a means of preparing the reader for what is to come. Research suggests that the brain is more receptive to new information when it has been prepared by the presentation of the main idea or some key words. The phenomenon is somewhat akin to the primer coat of paint that allows subsequent coats of paint to absorb more easily. A good topic sentence will be clear and not contain any jargon. When topic or summary sentences are not provided, good readers can jot down their own so that they can find their place in a text and refresh their memory.

Purposes for Writing

In order to be an effective reader, one must pay attention to the author's **position** and purpose. Even those texts that seem objective and impartial, like textbooks, have some sort of position and bias. Readers need to take these positions into account when considering the author's message. When an author uses emotional language or clearly favors one side of an argument, his position is clear. However, the author's position may be evident not only in what he writes, but in what he doesn't write. For this reason, it is sometimes necessary to review some other texts on the same topic in order to develop a view of the author's position. If this is not possible, then it may be useful to acquire a little background personal information about the author. When the only source of information is the text, however, the reader should look for language and argumentation that seems to indicate a particular stance on the subject.

Identifying the **purpose** of an author is usually easier than identifying her position. In most cases, the author has no interest in hiding his or her purpose. A text that is meant to entertain, for instance, should be obviously written to please the reader. Most narratives, or stories, are written to entertain, though they may also inform or persuade. Informative texts are easy to identify as well. The most difficult purpose of a text to identify is persuasion, because the author has an interest in making this purpose hard to detect. When a person knows that the author is trying to convince him, he is automatically more wary and skeptical of the argument. For this reason persuasive texts often try to establish an entertaining tone, hoping to amuse the reader into agreement, or an informative tone, hoping to create an appearance of authority and objectivity.

An author's purpose is often evident in the organization of the text. For instance, if the text has headings and subheadings, if key terms are in bold, and if the author makes his main idea clear from the beginning, then the likely purpose of the text is to inform. If the author begins by making a claim and then makes various arguments to support that claim, the purpose is probably to persuade. If the author is telling a story, or is more interested in holding the attention of the reader than in making a particular point or delivering information, then his purpose is most likely to entertain. As a reader, it is best to judge an author on how well he accomplishes his purpose. In other words, it is not entirely fair to complain that a textbook is boring: if the text is clear and easy to understand, then the author has done his job. Similarly, a storyteller should not be judged too harshly for getting some facts wrong, so long as he is able to give pleasure to the reader.

> ➢ **Review Video: <u>Purpose</u>**
> *Visit **mometrix.com/academy** and enter **Code: 511819***

The author's purpose for writing will affect his writing style and the response of the reader. In a **persuasive essay**, the author is attempting to change the reader's mind or convince him of something he did not believe previously. There are several identifying characteristics of persuasive writing. One is opinion presented as fact. When an author attempts to persuade the reader, he often presents his or her opinions as if they were fact. A reader must be on guard for statements that sound factual but which cannot be subjected to research, observation, or experiment. Another characteristic of persuasive writing is emotional language. An author will often try to play on the reader's emotion by appealing to his sympathy or sense of morality. When an author uses colorful or evocative language with the intent of arousing the reader's passions, it is likely that he is attempting to persuade. Finally, in many cases a persuasive text will give an unfair explanation of opposing positions, if these positions are mentioned at all.

An **informative text** is written to educate and enlighten the reader. Informative texts are almost always nonfiction, and are rarely structured as a story. The intention of an informative text is to deliver information in the most comprehensible way possible, so the structure of the text is likely to be very clear. In an informative text, the thesis statement is often in the first sentence. The author may use some colorful language, but is likely to put more emphasis on clarity and precision. Informative essays do not typically appeal to the emotions. They often contain facts and figures, and rarely include the opinion of the author. Sometimes a persuasive essay can resemble an informative essay, especially if the author maintains an even tone and presents his or her views as if they were established fact.

The success or failure of an author's intent to **entertain** is determined by those who read the author's work. Entertaining texts may be either fiction or nonfiction, and they may describe real or imagined people, places, and events. Entertaining texts are often narratives, or stories. A text that is written to entertain is likely to contain colorful language that engages the imagination and the emotions. Such writing often features a great deal of figurative language, which typically enlivens its subject matter with images and analogies. Though an entertaining text is not usually written to persuade or inform, it may accomplish both of these tasks. An entertaining text may appeal to the reader's emotions and cause him or her to think differently about a particular subject. In any case, entertaining texts tend to showcase the personality of the author more so than do other types of writing.

When an author intends to **express feelings,** she may use colorful and evocative language. An author may write emotionally for any number of reasons. Sometimes, the author will do so because she is describing a personal situation of great pain or happiness. Sometimes an author is attempting to persuade the reader, and so will use emotion to stir up the passions. It can be easy to identify this kind of expression when the writer uses phrases like *I felt* and *I sense*. However, sometimes the author will simply describe feelings without introducing them. As a reader, it is important to recognize when an author is expressing emotion, and not to become overwhelmed by sympathy or passion. A reader should maintain some detachment so that he or she can still evaluate the strength of the author's argument or the quality of the writing.

In a sense, almost all writing is descriptive, insofar as it seeks to describe events, ideas, or people to the reader. Some texts, however, are primarily concerned with **description**. A descriptive text focuses on a particular subject, and attempts to depict it in a way that will be clear to the reader. Descriptive texts contain many adjectives and adverbs, words that give shades of meaning and create a more detailed mental picture for the reader. A descriptive text fails when it is unclear or vague to the reader. On the other hand, however, a descriptive text that compiles too much detail can be boring and overwhelming to the reader. A descriptive text will certainly be informative, and it may be persuasive and entertaining as well. Descriptive writing is a challenge for the author, but when it is done well, it can be fun to read.

Writing Devices

Authors will use different stylistic and writing devices to make their meaning more clearly understood. One of those devices is comparison and contrast. When an author describes the ways in which two things are alike, he or she is **comparing** them. When the author describes the ways in which two things are different, he or she is **contrasting** them. The "compare and contrast" essay is one of the most common forms in nonfiction. It is often signaled with certain words: a comparison may be indicated with such words as *both, same, like, too,* and *as well*; while a contrast may be indicated by words like *but, however, on the other hand, instead,* and *yet*. Of course, comparisons and contrasts may be implicit without using any such signaling language. A single sentence may both compare and contrast. Consider the sentence *Brian and Sheila love ice cream, but Brian prefers vanilla and Sheila prefers strawberry*. In one sentence, the author has described both a similarity (love of ice cream) and a difference (favorite flavor).

> ➢ **Review Video: Compare and Contrast**
> *Visit* **mometrix.com/academy** *and enter* **Code: 798319**

One of the most common text structures is **cause and effect**. A cause is an act or event that makes something happen, and an effect is the thing that happens as a result of that cause. A cause-and-effect relationship is not always explicit, but there are some words in English that signal causality, such as *since, because,* and *as a result*. As an example, consider the sentence *Because the sky was clear, Ron did not bring an umbrella*. The cause is the clear sky, and the effect is that Ron did not bring an umbrella. However, sometimes the cause-and-effect relationship will not be clearly noted. For instance, the sentence *He was late and missed the meeting* does not contain any signaling words, but it still contains a cause (he was late) and an effect (he missed the meeting). It is possible for a single cause to have multiple effects, or for a single effect to have multiple causes. Also, an effect can in turn be the cause of another effect, in what is known as a cause-and-effect chain.

Authors often use analogies to add meaning to the text. An **analogy** is a comparison of two things. The words in the analogy are connected by a certain, often undetermined relationship. Look at this analogy: moo is to cow as quack is to duck. This analogy compares the sound that a cow makes with the sound that a duck makes. Even if the word 'quack' was not given, one could figure out it is the correct word to complete the analogy based on the relationship between the words 'moo' and 'cow'. Some common relationships for analogies include synonyms, antonyms, part to whole, definition, and actor to action.

Another element that impacts a text is the author's point of view. The **point of view** of a text is the perspective from which it is told. The author will always have a point of view about a story before he draws up a plot line. The author will know what events they want to take place, how they want the characters to interact, and how the story will resolve. An author will also have an opinion on the topic, or series of events, which is presented in the story, based on their own prior experience and beliefs.

The two main points of view that authors use are first person and third person. If the narrator of the story is also the main character, or *protagonist*, the text is written in first-person point of view. In first person, the author writes with the word *I*. Third-person point of view is probably the most common point of view that authors use. Using third person, authors refer to each character using the words *he* or *she*. In third-person omniscient, the narrator is not a character in the story and tells the story of all of the characters at the same time.

> ➤ **Review Video: Point of View**
> *Visit **mometrix.com/academy** and enter **Code**: **383336***

A good writer will use **transitional words** and phrases to guide the reader through the text. You are no doubt familiar with the common transitions, though you may never have considered how they operate. Some transitional phrases (*after, before, during, in the middle of*) give information about time. Some indicate that an example is about to be given (*for example, in fact, for instance*). Writers use them to compare (*also, likewise*) and contrast (*however, but, yet*). Transitional words and phrases can suggest addition (*and, also, furthermore, moreover*) and logical relationships (*if, then, therefore, as a result, since*). Finally, transitional words and phrases can demarcate the steps in a process (*first, second, last*). You should incorporate transitional words and phrases where they will orient your reader and illuminate the structure of your composition.

> ➤ **Review Video: Transitional Words and Phrases**
> *Visit **mometrix.com/academy** and enter **Code**: **197796***

Types of Passages

A **narrative** passage is a story. Narratives can be fiction or nonfiction. However, there are a few elements that a text must have in order to be classified as a narrative. To begin with, the text must have a plot. That is, it must describe a series of events. If it is a good narrative, these events will be interesting and emotionally engaging to the reader. A narrative also has characters. These could be people, animals, or even inanimate objects, so long as they participate in the plot. A narrative passage often contains figurative language, which is meant to stimulate the imagination of the reader by making comparisons and observations. A metaphor, which is a description of one thing in terms of another, is a common piece of figurative language. *The moon was a frosty snowball* is an example of a metaphor: it is obviously untrue in the literal sense, but it suggests a certain mood for the reader. Narratives often proceed in a clear sequence, but they do not need to do so.

An **expository** passage aims to inform and enlighten the reader. It is nonfiction and usually centers around a simple, easily defined topic. Since the goal of exposition is to teach, such a passage should be as clear as possible. It is common for an expository passage to contain helpful organizing words, like *first, next, for example*, and *therefore*. These words keep the reader oriented in the text. Although expository passages do not need to feature colorful language and artful writing, they are often more effective when they do. For a reader, the challenge of expository passages is to maintain steady attention. Expository passages are not always about subjects in which a reader will naturally be interested, and the writer is often more concerned with clarity and comprehensibility than with engaging the reader. For this reason, many expository passages are dull. Making notes is a good way to maintain focus when reading an expository passage.

A **technical** passage is written to describe a complex object or process. Technical writing is common in medical and technological fields, in which complicated mathematical, scientific, and engineering ideas need to be explained simply and clearly. To ease comprehension, a technical passage usually proceeds in a very logical order. Technical passages often have clear headings and subheadings, which are used to keep the reader oriented in the text. It is also common for these passages to break sections up with numbers or letters. Many technical passages look more like an outline than a piece of prose. The amount of jargon or difficult vocabulary will vary in a technical passage depending on the intended audience. As much as possible, technical passages try to avoid language that the reader will have to research in order to understand the message. Of course, it is not always possible to avoid jargon.

A **persuasive** passage is meant to change the reader's mind or lead her into agreement with the author. The persuasive intent may be obvious, or it may be quite difficult to discern. In some cases, a persuasive passage will be indistinguishable from an informative passage: it will make an assertion and offer supporting details. However, a persuasive passage is more likely to make claims based on opinion and to appeal to the reader's emotions. Persuasive passages may not describe alternate positions and, when they do, they often display significant bias. It may be clear that a persuasive passage is giving the author's viewpoint, or the passage may adopt a seemingly objective tone. A persuasive passage is successful if it can make a convincing argument and win the trust of the reader.

A persuasive essay will likely focus on one central argument, but it may make many smaller claims along the way. These are subordinate arguments with which the reader must agree if he or she is going to agree with the central argument. The central argument will only be as strong as the subordinate claims. These claims should be rooted in fact and observation, rather than subjective judgment. The best persuasive essays provide enough supporting detail to justify claims without overwhelming the reader. Remember that a fact must be susceptible to independent verification: that is, it must be something the reader could confirm. Also, statistics are only effective when they take into account possible objections. For instance, a statistic on the number of foreclosed houses would only be useful if it was taken over a defined interval and in a defined area. Most readers are wary of statistics, because they are so often misleading. If possible, a persuasive essay should always include references so that the reader can obtain more information. Of course, this means that the writer's accuracy and fairness may be judged by the inquiring reader.

Opinions are formed by emotion as well as reason, and persuasive writers often appeal to the feelings of the reader. Although readers should always be skeptical of this technique, it is often used in a proper and ethical manner. For instance, there are many subjects that have an obvious emotional component, and therefore cannot be completely treated without an appeal to the emotions. Consider an article on drunk driving: it makes sense to include some specific examples that will alarm or sadden the reader. After all, drunk driving often has serious and tragic consequences. Emotional appeals are not appropriate, however, when they attempt to mislead the reader. For instance, in political advertisements it is common to emphasize the patriotism of the preferred candidate, because this will encourage the audience to link their own positive feelings about the country with their opinion of the candidate. However, these ads often imply that the other candidate is unpatriotic, which in most cases is far from the truth. Another common and improper emotional appeal is the use of loaded language, as for instance referring to an avidly religious person as a "fanatic" or a passionate environmentalist as a "tree hugger." These terms introduce an emotional component that detracts from the argument.

History and Culture

Historical context has a profound influence on literature: the events, knowledge base, and assumptions of an author's time color every aspect of his or her work. Sometimes, authors hold opinions and use language that would be considered inappropriate or immoral in a modern setting, but that was acceptable in the author's time. As a reader, one should consider how the historical context influenced a work and also how today's opinions and ideas shape the way modern readers read the works of the past. For instance, in most societies of the past, women were treated as second-class citizens. An author who wrote in 18th-century England might sound sexist to modern readers, even if that author was relatively feminist in his time. Readers should not have to excuse the faulty assumptions and prejudices of the past, but they should appreciate that a person's thoughts and words are, in part, a result of the time and culture in which they live or lived, and it is perhaps unfair to expect writers to avoid all of the errors of their times.

Even a brief study of world literature suggests that writers from vastly different cultures address similar themes. For instance, works like the *Odyssey* and *Hamlet* both tackle the individual's battle for self-control and independence. In every culture, authors address themes of personal growth and the struggle for maturity. Another universal theme is the conflict between the individual and society. In works as culturally disparate as *Native Son*, the *Aeneid*, and *1984*, authors dramatize how people struggle to maintain their personalities and dignity in large, sometimes oppressive groups. Finally, many cultures have versions of the hero's (or heroine's) journey, in which an adventurous person must overcome many obstacles in order to gain greater knowledge, power, and perspective. Some famous works that treat this theme are the *Epic of Gilgamesh*, Dante's *Divine Comedy*, and *Don Quixote*.

Authors from different genres (for instance poetry, drama, novel, short story) and cultures may address similar themes, but they often do so quite differently. For instance, poets are likely to address subject matter obliquely, through the use of images and allusions. In a play, on the other hand, the author is more likely to dramatize themes by using characters to express opposing viewpoints. This disparity is known as a dialectical approach. In a novel, the author does not need to express themes directly; rather, they can be illustrated through events and actions. In some regional literatures, like those of Greece or England, authors use more irony: their works have characters that express views and make decisions that are clearly disapproved of by the author. In Latin America, there is a great tradition of using supernatural events to illustrate themes about real life. In China and Japan, authors frequently use well-established regional forms (haiku, for instance) to organize their treatment of universal themes.

Responding to Literature

When reading good literature, the reader is moved to engage actively in the text. One part of being an active reader involves making predictions. A **prediction** is a guess about what will happen next. Readers are constantly making predictions based on what they have read and what they already know. Consider the following sentence: *Staring at the computer screen in shock, Kim blindly reached over for the brimming glass of water on the shelf to her side.* The sentence suggests that Kim is agitated and that she is not looking at the glass she is going to pick up, so a reader might predict that she is going to knock the glass over. Of course, not every prediction will be accurate: perhaps Kim will pick the glass up cleanly. Nevertheless, the author has certainly created the expectation that the water might be spilled. Predictions are always subject to revision as the reader acquires more information.

Readers are often required to understand text that claims and suggests ideas without stating them directly. An **inference** is a piece of information that is implied but not written outright by the author. For instance, consider the following sentence: *Mark made more money that week than he had in the previous year*. From this sentence, the reader can infer that Mark either has not made much money in the previous year or made a great deal of money that week. Often, a reader can use information he or she already knows to make inferences. Take as an example the sentence *When his coffee arrived, he looked around the table for the silver cup*. Many people know that cream is typically served in a silver cup, so using their own base of knowledge they can infer that the subject of this sentence takes his coffee with cream. Making inferences requires concentration, attention, and practice.

> **Review Video: Inference**
> Visit *mometrix.com/academy* and enter **Code: 379203**

Test-taking tip: While being tested on his ability to make correct inferences, the student must look for contextual clues. An answer can be *true* but not *correct*. The contextual clues will help you find the answer that is the best answer out of the given choices. Understand the context in which a phrase is stated. When asked for the implied meaning of a statement made in the passage, the student should immediately locate the statement and read the context in which it was made. Also, look for an answer choice that has a similar phrase to the statement in question.

A reader must be able to identify a text's **sequence**, or the order in which things happen. Often, and especially when the sequence is very important to the author, it is indicated with signal words like *first*, *then*, *next*, and *last*. However, sometimes a sequence is merely implied and must be noted by the reader. Consider the sentence *He walked in the front door and switched on the hall lamp*. Clearly, the man did not turn the lamp on before he walked in the door, so the implied sequence is that he first walked in the door and then turned on the lamp. Texts do not always proceed in an orderly sequence from first to last: sometimes, they begin at the end and then start over at the beginning. As a reader, it can be useful to make brief notes to clarify the sequence.

In addition to inferring and predicting things about the text, the reader must often **draw conclusions** about the information he has read. When asked for a *conclusion* that may be drawn, look for critical "hedge" phrases, such as *likely*, *may*, *can*, *will often*, among many others. When you are being tested on this knowledge, remember that question writers insert these hedge phrases to cover every possibility. Often an answer will be wrong simply because it leaves no room for exception. Extreme positive or negative answers (such as always, never, etc.) are usually not correct. The reader should not use any outside knowledge that is not gathered from the reading passage to answer the related questions. Correct answers can be derived straight from the reading passage.

Literary Genres

Literary genres refer to the basic generic types of poetry, drama, fiction, and nonfiction. Literary genre is a method of classifying and analyzing literature. There are numerous subdivisions within genre, including such categories as novels, novellas, and short stories in fiction. Drama may also be subdivided into comedy, tragedy, and many other categories. Poetry and nonfiction have their own distinct divisions.

Genres often overlap, and the distinctions among them are blurred, such as that between the nonfiction novel and docudrama, as well as many others. However, the use of genres is helpful to the reader as a set of understandings that guide our responses to a work. The generic norm sets expectations and forms the framework within which we read and evaluate a work. This framework will guide both our understanding and interpretation of the work. It is a useful tool for both literary criticism and analysis.

Fiction is a general term for any form of literary narrative that is invented or imagined rather than being factual. For those individuals who equate fact with truth, the imagined or invented character of fiction tends to render it relatively unimportant or trivial among the genres. Defenders of fiction are quick to point out that the fictional mode is an essential part of being. The ability to imagine or discuss what-if plots, characters, and events is clearly part of the human experience.

Prose is derived from the Latin and means "straightforward discourse." Prose fiction, although having many categories, may be divided into three main groups:
- **Short stories**: a fictional narrative, the length of which varies, usually under 20,000 words. Short stories usually have only a few characters and generally describe one major event or insight. The short story began in magazines in the late 1800s and has flourished ever since.
- **Novels**: a longer work of fiction, often containing a large cast of characters and extensive plotting. The emphasis may be on an event, action, social problems, or any experience. There is now a genre of nonfiction novels pioneered by Truman Capote's *In Cold Blood* in the 1960s. Novels may also be written in verse.
- **Novellas**: a work of narrative fiction longer than a short story but shorter than a novel. Novellas may also be called short novels or novelettes. They originated from the German tradition and have become common forms in all of the world's literature.

Many elements influence a work of prose fiction. Some important ones are:
- Speech and dialogue: Characters may speak for themselves or through the narrator. Dialogue may be realistic or fantastic, depending on the author's aim.
- Thoughts and mental processes: There may be internal dialogue used as a device for plot development or character understanding.
- Dramatic involvement: Some narrators encourage readers to become involved in the events of the story, whereas others attempt to distance readers through literary devices.
- Action: This is any information that advances the plot or involves new interactions between the characters.
- Duration: The time frame of the work may be long or short, and the relationship between described time and narrative time may vary.
- Setting and description: Is the setting critical to the plot or characters? How are the action scenes described?
- Themes: This is any point of view or topic given sustained attention.
- Symbolism: Authors often veil meanings through imagery and other literary constructions.

Fiction is much wider than simply prose fiction. Songs, ballads, epics, and narrative poems are examples of non-prose fiction. A full definition of fiction must include not only the work itself but also the framework in which it is read. Literary fiction can also be defined as not true rather than nonexistent, as many works of historical fiction refer to real people, places, and events that are treated imaginatively as if they were true. These imaginary elements enrich and broaden literary expression.

When analyzing fiction, it is important for the reader to look carefully at the work being studied. The plot or action of a narrative can become so entertaining that the language of the work is ignored. The language of fiction should not simply be a way to relate a plot—it should also yield many insights to the judicious reader. Some prose fiction is based on the reader's engagement with the language rather than the story. A studious reader will analyze the mode of expression as well as the narrative. Part of the reward of reading in this manner is to discover how the author uses different language to describe familiar objects, events, or emotions. Some works focus the reader on an author's unorthodox use of language, whereas others may emphasize characters or storylines. What happens in a story is not always the critical element in the work. This type of reading may be difficult at first but yields great rewards.

The **narrator** is a central part of any work of fiction, and can give insight about the purpose of the work and its main themes and ideas. The following are important questions to address to better understand the voice and role of the narrator and incorporate that voice into an overall understanding of the novel:
- Who is the narrator of the novel? What is the narrator's perspective, first person or third person? What is the role of the narrator in the plot? Are there changes in narrators or the perspective of narrators?
- Does the narrator explain things in the novel, or does meaning emerge from the plot and events? The personality of the narrator is important. She may have a vested interest in a character or event described. Some narratives follow the time sequence of the plot, whereas others do not. A narrator may express approval or disapproval about a character or events in the work.
- Tone is an important aspect of the narration. Who is actually being addressed by the narrator? Is the tone familiar or formal, intimate or impersonal? Does the vocabulary suggest clues about the narrator?

> **Review Video: The Narrator**
> Visit *mometrix.com/academy* and enter *Code:* **742528**

A **character** is a person intimately involved with the plot and development of the novel. Development of the novel's characters not only moves the story along but will also tell the reader a lot about the novel itself. There is usually a physical description of the character, but this is often omitted in modern and postmodern novels. These works may focus on the psychological state or motivation of the character. The choice of a character's name may give valuable clues to his role in the work.

Characters are said to be flat or round. Flat characters tend to be minor figures in the story, changing little or not at all. Round characters (those understood from a well-rounded view) are more central to the story and tend to change as the plot unfolds. Stock characters are similar to flat characters, filling out the story without influencing it.

Modern literature has been greatly affected by Freudian psychology, giving rise to such devices as the interior monologue and magical realism as methods of understanding characters in a work. These give the reader a more complex understanding of the inner lives of the characters and enrich the understanding of relationships between characters.

Another important genre is that of **drama**: a play written to be spoken aloud. The drama is in many ways inseparable from performance. Reading drama ideally involves using imagination to visualize and re-create the play with characters and settings. The reader stages the play in his imagination, watching characters interact and developments unfold. Sometimes this involves simulating a theatrical presentation; other times it involves imagining the events. In either case, the reader is imagining the unwritten to re-create the dramatic experience. Novels present some of the same problems, but a narrator will provide much more information about the setting, characters, inner dialogues, and many other supporting details. In drama, much of this is missing, and we are required to use our powers of projection and imagination to taste the full flavor of the dramatic work. There are many empty spaces in dramatic texts that must be filled by the reader to fully appreciate the work.

When reading drama in this way, there are some advantages over watching the play performed (though there is much criticism in this regard):
- Freedom of point of view and perspective: Text is free of interpretations of actors, directors, producers, and technical staging.
- Additional information: The text of a drama may be accompanied by notes or prefaces placing the work in a social or historical context. Stage directions may also provide relevant information about the author's purpose. None of this is typically available at live or filmed performances.
- Study and understanding: Difficult or obscure passages may be studied at leisure and supplemented by explanatory works. This is particularly true of older plays with unfamiliar language, which cannot be fully understood without an opportunity to study the material.

Critical elements of drama, especially when it is being read aloud or performed, include dialect, speech, and dialogue. Analysis of speech and dialogue is important in the critical study of drama. Some playwrights use speech to develop their characters. Speeches may be long or short, and written in as normal prose or blank verse. Some characters have a unique way of speaking which illuminates aspects of the drama. Emphasis and tone are both important, as well. Does the author make clear the tone in which lines are to be spoken, or is this open to interpretation? Sometimes there are various possibilities in tone with regard to delivering lines.

Dialect is any distinct variety of a language, especially one spoken in a region or part of a country. The criterion for distinguishing dialects from languages is that of mutual understanding. For example, people who speak Dutch cannot understand English unless they have learned it. But a speaker from Amsterdam can understand one from Antwerp; therefore, they speak different dialects of the same language. This is, however, a matter of degree; there are languages in which different dialects are unintelligible.

Dialect mixtures are the presence in one form of speech with elements from different neighboring dialects. The study of speech differences from one geographical area to another is called dialect geography. A dialect atlas is a map showing distribution of dialects in a given area. A dialect continuum shows a progressive shift in dialects across a territory, such that adjacent dialects are understandable, but those at the extremes are not.

Dramatic dialogue can be difficult to interpret and changes depending upon the tone used and which words are emphasized. Where the stresses, or meters, of dramatic dialogue fall can determine meaning. Variations in emphasis are only one factor in the manipulability of dramatic speech. Tone is of equal or greater importance and expresses a range of possible emotions and feelings that cannot be readily discerned from the script of a play. The reader must add tone to the words to understand the full meaning of a passage. Recognizing tone is a cumulative process as the reader begins to understand the characters and situations in the play. Other elements that influence the interpretation of dialogue include the setting, possible reactions of the characters to the speech, and possible gestures or facial expressions of the actor. There are no firm rules to guide the interpretation of dramatic speech. An open and flexible attitude is essential in interpreting dramatic dialogue.

Action is a crucial element in the production of a dramatic work. Many dramas contain little dialogue and much action. In these cases, it is essential for the reader to carefully study stage directions and visualize the action on the stage. Benefits of understanding stage directions include knowing which characters are on the stage at all times, who is speaking to whom, and following these patterns through changes of scene.

Stage directions also provide additional information, some of which is not available to a live audience. The nature of the physical space where the action occurs is vital, and stage directions help with this. The historical context of the period is important in understanding what the playwright was working with in terms of theaters and physical space. The type of staging possible for the author is a good guide to the spatial elements of a production.

Asides and soliloquies are devices that authors use in plot and character development. **Asides** indicate that not all characters are privy to the lines. This may be a method of advancing or explaining the plot in a subtle manner. **Soliloquies** are opportunities for character development, plot enhancement, and to give insight to characters' motives, feelings, and emotions. Careful study of these elements provides a reader with an abundance of clues to the major themes and plot of the work.

Art, music, and literature all interact in ways that contain many opportunities for the enrichment of all of the arts. Students could apply their knowledge of art and music by creating illustrations for a work or creating a musical score for a text. Students could discuss the meanings of texts and decide on their illustrations, or a score could amplify the meaning of the text.

Understanding the art and music of a period can make the experience of literature a richer, more rewarding experience. Students should be encouraged to use the knowledge of art and music to illuminate the text. Examining examples of dress, architecture, music, and dance of a period may be helpful in a fuller engagement of the text. Much of period literature lends itself to the analysis of the prevailing taste in art and music of an era, which helps place the literary work in a more meaningful context.

Critical Thinking Skills

Opinions, Facts, & Fallacies

Critical thinking skills are mastered through understanding various types of writing and the different purposes that authors have for writing the way they do. Every author writes for a purpose. Understanding that purpose, and how they accomplish their goal, will allow you to critique the writing and determine whether or not you agree with their conclusions.

Readers must always be conscious of the distinction between fact and opinion. A **fact** can be subjected to analysis and can be either proved or disproved. An **opinion**, on the other hand, is the author's personal feeling, which may not be alterable by research, evidence, or argument. If the author writes that the distance from New York to Boston is about two hundred miles, he is stating a fact. But if he writes that New York is too crowded, then he is giving an opinion, because there is no objective standard for overpopulation. An opinion may be indicated by words like *believe*, *think*, or *feel*. Also, an opinion may be supported by facts: for instance, the author might give the population density of New York as a reason for why it is overcrowded. An opinion supported by fact tends to be more convincing. When authors support their opinions with other opinions, the reader is unlikely to be moved.

> **Review Video: Fact or Opinion**
> Visit *mometrix.com/academy* and enter **Code: 870899**

Facts should be presented to the reader from reliable sources. An opinion is what the author thinks about a given topic. An opinion is not common knowledge or proven by expert sources, but it is information that the author believes and wants the reader to consider. To distinguish between fact and opinion, a reader needs to look at the type of source that is presenting information, what information backs-up a claim, and whether or not the author may be motivated to have a certain point of view on a given topic. For example, if a panel of scientists has conducted multiple studies on the effectiveness of taking a certain vitamin, the results are more likely to be factual than if a company selling a vitamin claims that taking the vitamin can produce positive effects. The company is motivated to sell its product, while the scientists are using the scientific method to prove a theory. If the author uses words such as "I think...", the statement is an opinion.

In their attempt to persuade, writers often make mistakes in their thinking patterns and writing choices. It's important to understand these so you can make an informed decision. Every author has a point of view, but when an author ignores reasonable counterarguments or distorts opposing viewpoints, she is demonstrating a **bias**. A bias is evident whenever the author is unfair or inaccurate in his or her presentation. Bias may be intentional or unintentional, but it should always alert the reader to be skeptical of the argument being made. It should be noted that a biased author may still be correct. However, the author will be correct in spite of her bias, not because of it. A **stereotype** is like a bias, except that it is specifically applied to a group or place. Stereotyping is considered to be particularly abhorrent because it promotes negative generalizations about people. Many people are familiar with some of the hateful stereotypes of certain ethnic, religious, and cultural groups. Readers should be very wary of authors who stereotype. These faulty assumptions typically reveal the author's ignorance and lack of curiosity.

Sometimes, authors will **appeal to the reader's emotions** in an attempt to persuade or to distract the reader from the weakness of the argument. For instance, the author may try to inspire the pity of the reader by delivering a heart-rending story. An author also might use the bandwagon approach, in which he suggests that his opinion is correct because it is held by the majority. Some authors resort to name-calling, in which insults and harsh words are delivered to the opponent in an attempt to distract. In advertising, a common appeal is the testimonial, in which a famous person endorses a product. Of course, the fact that a celebrity likes something should not really mean anything to the reader. These and other emotional appeals are usually evidence of poor reasoning and a weak argument.

> **Review Video: Appeal to Emotions**
> Visit *mometrix.com/academy* and enter **Code: 163442**

Certain *logical fallacies* are frequent in writing. A logical fallacy is a failure of reasoning. As a reader, it is important to recognize logical fallacies, because they diminish the value of the author's message. The four most common logical fallacies in writing are the false analogy, circular reasoning, false dichotomy, and overgeneralization. In a **false analogy**, the author suggests that two things are similar, when in fact they are different. This fallacy is often committed when the author is attempting to convince the reader that something unknown is like something relatively familiar. The author takes advantage of the reader's ignorance to make this false comparison. One example might be the following statement: *Failing to tip a waitress is like stealing money out of somebody's wallet.* Of course, failing to tip is very rude, especially when the service has been good, but people are not arrested for failing to tip as they would for stealing money from a wallet. To compare stingy diners with thieves is a false analogy.

> ➢ **Review Video: False Analogy**
> *Visit mometrix.com/academy and enter Code:* **865045**

Circular reasoning is one of the more difficult logical fallacies to identify, because it is typically hidden behind dense language and complicated sentences. Reasoning is described as circular when it offers no support for assertions other than restating them in different words. Put another way, a circular argument refers to itself as evidence of truth. A simple example of circular argument is when a person uses a word to define itself, such as saying *Niceness is the state of being nice*. If the reader does not know what *nice* means, then this definition will not be very useful. In a text, circular reasoning is usually more complex. For instance, an author might say *Poverty is a problem for society because it creates trouble for people throughout the community*. It is redundant to say that poverty is a problem because it creates trouble. When an author engages in circular reasoning, it is often because he or she has not fully thought out the argument, or cannot come up with any legitimate justifications.

> ➢ **Review Video: Circular Reasoning**
> *Visit mometrix.com/academy and enter Code:* **398925**

One of the most common logical fallacies is the **false dichotomy**, in which the author creates an artificial sense that there are only two possible alternatives in a situation. This fallacy is common when the author has an agenda and wants to give the impression that his view is the only sensible one. A false dichotomy has the effect of limiting the reader's options and imagination. An example of a false dichotomy is the statement *You need to go to the party with me, otherwise you'll just be bored at home.* The speaker suggests that the only other possibility besides being at the party is being bored at home. But this is not true, as it is perfectly possible to be entertained at home, or even to go somewhere other than the party. Readers should always be wary of the false dichotomy: when an author limits alternatives, it is always wise to ask whether he is being valid.

> ➢ **Review Video: False Dichotomy**
> *Visit mometrix.com/academy and enter Code:* **484397**

Overgeneralization is a logical fallacy in which the author makes a claim that is so broad it cannot be proved or disproved. In most cases, overgeneralization occurs when the author wants to create an illusion of authority, or when he is using sensational language to sway the opinion of the reader. For instance, in the sentence *Everybody knows that she is a terrible teacher*, the author makes an assumption that cannot really be believed. This kind of statement is made when the author wants to create the illusion of consensus when none actually exists: it may be that most people have a negative view of the teacher, but to say that *everybody* feels that way is an exaggeration. When a reader spots overgeneralization, she should become skeptical about the argument that is being made, because an author will often try to hide a weak or unsupported assertion behind authoritative language.

> ➢ **Review Video: Overgeneralization**
> *Visit mometrix.com/academy and enter Code:* **367357**

Two other types of logical fallacies are **slippery slope** arguments and **hasty generalizations**. In a slippery slope argument, the author says that if something happens, it automatically means that something else will happen as a result, even though this may not be true. (i.e., just because you study hard does not mean you are going to ace the test). "Hasty generalization" is drawing a conclusion too early, without finishing analyzing the details of the argument. Writers of persuasive texts often use these techniques because they are very effective. In order to **identify logical fallacies**, readers need to read carefully and ask questions as they read. Thinking critically means not taking everything at face value. Readers need to critically evaluate an author's argument to make sure that the logic used is sound.

Organization of the Text

The way a text is organized can help the reader to understand more clearly the author's intent and his conclusions. There are various ways to organize a text, and each one has its own purposes and uses.

Some nonfiction texts are organized to **present a problem** followed by a solution. In this type of text, it is common for the problem to be explained before the solution is offered. In some cases, as when the problem is well known, the solution may be briefly introduced at the beginning. The entire passage may focus on the solution, and the problem will be referenced only occasionally. Some texts will outline multiple solutions to a problem, leaving the reader to choose among them. If the author has an interest or an allegiance to one solution, he may fail to mention or may describe inaccurately some of the other solutions. Readers should be careful of the author's agenda when reading a problem-solution text. Only by understanding the author's point of view and interests can one develop a proper judgment of the proposed solution.

Authors need to organize information logically so the reader can follow it and locate information within the text. Two common organizational structures are cause and effect and chronological order. When using **chronological order**, the author presents information in the order that it happened. For example, biographies are written in chronological order; the subject's birth and childhood are presented first, followed by their adult life, and lastly by the events leading up to the person's death.

In **cause and effect**, an author presents one thing that makes something else happen. For example, if one were to go to bed very late, they would be tired. The cause is going to bed late, with the effect of being tired the next day.

It can be tricky to identify the cause-and-effect relationships in a text, but there are a few ways to approach this task. To begin with, these relationships are often signaled with certain terms. When an author uses words like *because*, *since*, *in order*, and *so*, she is likely describing a cause-and-effect relationship. Consider the sentence, "He called her because he needed the homework." This is a simple causal relationship, in which the cause was his need for the homework and the effect was his phone call. Not all cause-and-effect relationships are marked in this way, however. Consider the sentences, "He called her. He needed the homework." When the cause-and-effect relationship is not indicated with a keyword, it can be discovered by asking why something happened. He called her: why? The answer is in the next sentence: He needed the homework.

Persuasive essays, in which an author tries to make a convincing argument and change the reader's mind, usually include cause-and-effect relationships. However, these relationships should not always be taken at face value. An author frequently will assume a cause or take an effect for granted. To read a persuasive essay effectively, one needs to judge the cause-and-effect relationships the author is presenting. For instance, imagine an author wrote the following: "The parking deck has been unprofitable because people would prefer to ride their bikes." The relationship is clear: the cause is that people prefer to ride their bikes, and the effect is that the parking deck has been unprofitable. However, a reader should consider whether this argument is conclusive. Perhaps there are other reasons for the failure of the parking deck: a down economy, excessive fees, etc. Too often, authors present causal relationships as if they are fact rather than opinion. Readers should be on the alert for these dubious claims.

Thinking critically about ideas and conclusions can seem like a daunting task. One way to make it easier is to understand the basic elements of ideas and writing techniques. Looking at the way different ideas relate to each other can be a good way for the reader to begin his analysis. For instance, sometimes writers will write about two different ideas that are in opposition to each other. The analysis of these opposing ideas is known as **contrast**. Contrast is often marred by the author's obvious partiality to one of the ideas. A discerning reader will be put off by an author who does not engage in a fair fight. In an analysis of opposing ideas, both ideas should be presented in their clearest and most reasonable terms. If the author does prefer a side, he should avoid indicating this preference with pejorative language. An analysis of opposing ideas should proceed through the major differences point by point, with a full explanation of each side's view. For instance, in an analysis of capitalism and communism, it would be important to outline each side's view on labor, markets, prices, personal responsibility, etc. It would be less effective to describe the theory of communism and then explain how capitalism has thrived in the West. An analysis of opposing views should present each side in the same manner.

Many texts follow the **compare-and-contrast** model, in which the similarities and differences between two ideas or things are explored. Analysis of the similarities between ideas is called comparison. In order for a comparison to work, the author must place the ideas or things in an equivalent structure. That is, the author must present the ideas in the same way. Imagine an author wanted to show the similarities between cricket and baseball. The correct way to do so would be to summarize the equipment and rules for each game. It would be incorrect to summarize the equipment of cricket and then lay out the history of baseball, since this would make it impossible for the reader to see the similarities. It is perhaps too obvious to say that an analysis of similar ideas should emphasize the similarities. Of course, the author should take care to include any differences that must be mentioned. Often, these small differences will only reinforce the more general similarity.

Drawing Conclusions

Authors should have a clear purpose in mind while writing. Especially when reading informational texts, it is important to understand the logical conclusion of the author's ideas. **Identifying this logical conclusion** can help the reader understand whether he agrees with the writer or not. Identifying a logical conclusion is much like making an inference: it requires the reader to combine the information given by the text with what he already knows to make a supportable assertion. If a passage is written well, then the conclusion should be obvious even when it is unstated. If the author intends the reader to draw a certain conclusion, then all of his argumentation and detail should be leading toward it. One way to approach the task of drawing conclusions is to make brief notes of all the points made by the author. When these are arranged on paper, they may clarify the logical conclusion. Another way to approach conclusions is to consider whether the reasoning of the author raises any pertinent questions. Sometimes it will be possible to draw several conclusions from a passage, and on occasion these will be conclusions that were never imagined by the author. It is essential, however, that these conclusions be supported directly by the text.

The term **text evidence** refers to information that supports a main point or points in a story, and can help lead the reader to a conclusion. Information used as *text evidence* is precise, descriptive, and factual. A main point is often followed by supporting details that provide evidence to back-up a claim. For example, a story may include the claim that winter occurs during opposite months in the Northern and Southern hemispheres. *Text evidence* based on this claim may include countries where winter occurs in opposite months, along with reasons that winter occurs at different times of the year in separate hemispheres (due to the tilt of the Earth as it rotates around the sun).

Readers interpret text and respond to it in a number of ways. Using textual support helps defend your response or interpretation because it roots your thinking in the text. You are interpreting based on information in the text and not simply your own ideas. When crafting a response, look for important quotes and details from the text to help bolster your argument. If you are writing about a character's personality trait, for example, use details from the text to show that the character acted in such a way. You can also include statistics and facts from a nonfiction text to strengthen your response. For example, instead of writing, "A lot of people use cell phones," use statistics to provide the exact number. This strengthens your argument because it is more precise.

The text used to support an argument can be the argument's downfall if it is not credible. A text is **credible**, or believable, when the author is knowledgeable and objective, or unbiased. The author's motivations for writing the text play a critical role in determining the credibility of the text and must be evaluated when assessing that credibility. The author's motives should be for the dissemination of information. The purpose of the text should be to inform or describe, not to persuade. When an author writes a persuasive text, he has the motivation that the reader will do what they want. The extent of the author's knowledge of the topic and their motivation must be evaluated when assessing the credibility of a text. Reports written about the Ozone layer by an environmental scientist and a hairdresser will have a different level of credibility.

After determining your own opinion and evaluating the credibility of your supporting text, it is sometimes necessary to communicate your ideas and findings to others. When **writing a response to a text**, it is important to use elements of the text to support your assertion or defend your position. Using supporting evidence from the text strengthens the argument because the reader can see how in depth the writer read the original piece and based their response on the details and facts within that text. Elements of text that can be used in a response include: facts, details, statistics, and direct quotations from the text. When writing a response, one must make sure they indicate which information comes from the original text and then base their discussion, argument, or defense around this information.

A reader should always be drawing conclusions from the text. Sometimes conclusions are implied from written information, and other times the information is **stated directly** within the passage. It is always more comfortable to draw conclusions from information stated within a passage, rather than to draw them from mere implications. At times an author may provide some information and then describe a counterargument. The reader should be alert for direct statements that are subsequently rejected or weakened by the author. The reader should always read the entire passage before drawing conclusions. Many readers are trained to expect the author's conclusions at either the beginning or the end of the passage, but many texts do not adhere to this format.

Drawing conclusions from information implied within a passage requires confidence on the part of the reader. **Implications** are things the author does not state directly, but which can be assumed based on what the author does say. For instance, consider the following simple passage: "I stepped outside and opened my umbrella. By the time I got to work, the cuffs of my pants were soaked." The author never states that it is raining, but this fact is clearly implied. Conclusions based on implication must be well supported by the text. In order to draw a solid conclusion, a reader should have multiple pieces of evidence, or, if he only has one, must be assured that there is no other possible explanation than his conclusion. A good reader will be able to draw many conclusions from information implied by the text, which enriches the reading experience considerably.

As an aid to drawing conclusions, the reader should be adept at **outlining** the information contained in the passage; an effective outline will reveal the structure of the passage, and will lead to solid conclusions. An effective outline will have a title that refers to the basic subject of the text, though it need not recapitulate the main idea. In most outlines, the main idea will be the first major section. It will have each major idea of the passage established as the head of a category. For instance, the most common outline format calls for the main ideas of the passage to be indicated with Roman numerals. In an effective outline of this kind, each of the main ideas will be represented by a Roman numeral and none of the Roman numerals will designate minor details or secondary ideas. Moreover, all supporting ideas and details should be placed in the appropriate place on the outline. An outline does not need to include every detail listed in the text, but it should feature all of those that are central to the argument or message. Each of these details should be listed under the appropriate main idea.

It is also helpful to **summarize** the information you have read in a paragraph or passage format. This process is similar to creating an effective outline. To begin with, a summary should accurately define the main idea of the passage, though it does not need to explain this main idea in exhaustive detail. It should continue by laying out the most important supporting details or arguments from the passage. All of the significant supporting details should be included, and none of the details included should be irrelevant or insignificant. Also, the summary should accurately report all of these details. Too often, the desire for brevity in a summary leads to the sacrifice of clarity or veracity. Summaries are often difficult to read, because they omit all of graceful language, digressions, and asides that distinguish great writing. However, if the summary is effective, it should contain much the same message as the original text.

Paraphrasing is another method the reader can use to aid in comprehension. When paraphrasing, one puts what they have read into their own words, rephrasing what the author has written to make it their own, to "translate" all of what the author says to their own words, including as many details as they can.

Testing Tips

Skimming

Your first task when you begin reading is to answer the question "What is the topic of the selection?" This can best be answered by quickly skimming the passage for the general idea, stopping to read only the first sentence of each paragraph. A paragraph's first sentence is usually the main topic sentence, and it gives you a summary of the content of the paragraph.

Once you've skimmed the passage, stopping to read only the first sentences, you will have a general idea about what it is about, as well as what is the expected topic in each paragraph.

Each question will contain clues as to where to find the answer in the passage. Do not just randomly search through the passage for the correct answer to each question. Search scientifically. Find key word(s) or ideas in the question that are going to either contain or be near the correct answer. These are typically nouns, verbs, numbers, or phrases in the question that will probably be duplicated in the passage. Once you have identified those key word(s) or idea, skim the passage quickly to find where those key word(s) or idea appears. The correct answer choice will be nearby.

Example: What caused Martin to suddenly return to Paris?
The key word is Paris. Skim the passage quickly to find where this word appears. The answer will be close by that word. However, sometimes key words in the question are not repeated in the passage. In those cases, search for the general idea of the question.

Example: Which of the following was the psychological impact of the author's childhood upon the remainder of his life?
Key words are "childhood" or "psychology". While searching for those words, be alert for other words or phrases that have similar meaning, such as "emotional effect" or "mentally" which could be used in the passage, rather than the exact word "psychology". Numbers or years can be particularly good key words to skim for, as they stand out from the rest of the text.

Example: Which of the following best describes the influence of Monet's work in the 20th century?
20th contains numbers and will easily stand out from the rest of the text. Use 20th as the key word to skim for in the passage. Other good key word(s) may be in quotation marks. These identify a word or phrase that is copied directly from the passage. In those cases, the word(s) in quotation marks are exactly duplicated in the passage.

Example: In her college years, what was meant by Margaret's "drive for excellence"?
"Drive for excellence" is a direct quote from the passage and should be easy to find.

Once you've quickly found the correct section of the passage to find the answer, focus upon the answer choices. Sometimes a choice will repeat word for word a portion of the passage near the answer. However, beware of such duplication – it may be a trap! More than likely, the correct choice will paraphrase or summarize the related portion of the passage, rather than being exactly the same wording.

For the answers that you think are correct, read them carefully and make sure that they answer the question. An answer can be factually correct, but it MUST answer the question asked. Additionally, two answers can both be seemingly correct, so be sure to read all of the answer choices, and make sure that you get the one that BEST answers the question. Some questions will not have a key word.
Example: Which of the following would the author of this passage likely agree with?

In these cases, look for key words in the answer choices. Then skim the passage to find where the answer choice occurs. By skimming to find where to look, you can minimize the time required.

Sometimes it may be difficult to identify a good key word in the question to skim for in the passage. In those cases, look for a key word in one of the answer choices to skim for. Often the answer choices can all be found in the same paragraph, which can quickly narrow your search.

Paragraph Focus

Focus upon the first sentence of each paragraph, which is the most important. The main topic of the paragraph is usually there.

Once you've read the first sentence in the paragraph, you have a general idea about what each paragraph will be about. As you read the questions, try to determine which paragraph will have the answer. Paragraphs have a concise topic. The answer should either obviously be there or obviously not. It will save time if you can jump straight to the paragraph, so try to remember what you learned from the first sentences.
Example: The first paragraph is about poets; the second is about poetry. If a question asks about poetry, where will the answer be? *The second paragraph.*

The main idea of a passage is typically spread across all or most of its paragraphs. Whereas the main idea of a paragraph may be completely different than the main idea of the very next paragraph, a main idea for a passage affects all of the paragraphs in one form or another.
Example: What is the main idea of the passage?

For each answer choice, try to see how many paragraphs are related. It can help to count how many sentences are affected by each choice, but it is best to see how many paragraphs are affected by the choice. Typically the answer choices will include incorrect choices that are main ideas of individual paragraphs, but not the entire passage. That is why it is crucial to choose ideas that are supported by the most paragraphs possible.

Eliminate Choices

Some choices can quickly be eliminated. "Andy Warhol lived there." Is Andy Warhol even mentioned in the article? If not, quickly eliminate it.

When trying to answer a question such as "the passage indicates all of the following EXCEPT" quickly skim the paragraph searching for references to each choice. If the reference exists, scratch it off as a choice. Similar choices may be crossed off simultaneously if they are close enough.

In choices that ask you to choose "which answer choice does NOT describe?" or "all of the following answer choices are identifiable characteristics, EXCEPT which?" look for answers that are similarly worded. Since only one answer can be correct, if there are two answers that appear to mean the same thing, they must BOTH be incorrect, and can be eliminated.

Example:
  A. changing values and attitudes
  B. large population of mobile or uprooted people

These answer choices are similar; they both describe a fluid culture. Because of their similarity, they can be linked together. Since the answer can have only one choice, they can also be eliminated together.

Contextual Clues

Look for contextual clues. An answer can be right but not correct. The contextual clues will help you find the answer that is most right and is correct. Understand the context in which a phrase is stated.

When asked for the implied meaning of a statement made in the passage, immediately go find the statement and read the context it was made in. Also, look for an answer choice that has a similar phrase to the statement in question.
Example: In the passage, what is implied by the phrase "Churches have become more or less part of the furniture"?

Find an answer choice that is similar or describes the phrase "part of the furniture" as that is the key phrase in the question. "Part of the furniture" is a saying that means something is fixed, immovable, or set in their ways. Those are all similar ways of saying "part of the furniture." As such, the correct answer choice will probably include a similar rewording of the expression.
Example: Why was John described as "morally desperate"?
　The answer will probably have some sort of definition of morals in it. "Morals" refers to a code of right and wrong behavior, so the correct answer choice will likely have words that mean something like that.

Fact/Opinion

When asked about which statement is a fact or opinion, remember that answer choices that are facts will typically have no ambiguous words. For example, how long is a long time? What defines an ordinary person? These ambiguous words of "long" and "ordinary" should not be in a factual statement. However, if all of the choices have ambiguous words, go to the context of the passage. Often a factual statement may be set out as a research finding.
Example: "The scientist found that the eye reacts quickly to change in light."

Opinions may be set out in the context of words like thought, believed, understood, or wished.
Example: "He thought the Yankees should win the World Series."

Opposites

Answer choices that are direct opposites are usually correct. The paragraph will often contain established relationships (when this goes up, that goes down). The question may ask you to draw conclusions for this and will give two similar answer choices that are opposites.
Example:
　A. a decrease in housing starts
　B. an increase in housing starts

Time Management

In technical passages, do not get lost on the technical terms. Skip them and move on. You want a general understanding of what is going on, not a mastery of the passage. When you encounter material in the selection that seems difficult to understand, it often may not be necessary and can be skipped. Only spend time trying to understand it if it is going to be relevant for a question. Understand difficult phrases only as a last resort.

Answer general questions before detail questions. A reader with a good understanding of the whole passage can often answer general questions without rereading a word. Get the easier questions out of the way before tackling the more time consuming ones.

Identify each question by type. Usually the wording of a question will tell you whether you can find the answer by referring directly to the passage or by using your reasoning powers. You alone know which question types you customarily handle with ease and which give you trouble and will require more time. Save the difficult questions for last.

Hedge Phrases Revisited

Once again, watch out for critical "hedge" phrases, such as likely, may, can, will often, sometimes, etc, often, almost, mostly, usually, generally, rarely, sometimes. Question writers insert these hedge phrases, to cover every possibility. Often an answer will be wrong simply because it leaves no room for exception.

Example: Animals live longer in cold places than animals in warm places.

This answer choice is wrong, because there are exceptions in which certain warm climate animals live longer. This answer choice leaves no possibility of exception. It states that every animal species in cold places live longer than animal species in warm places. Correct answer choices will typically have a key hedge word to leave room for exceptions.

Example: In severe cold, a polar bear cub is likely to survive longer than an adult polar bear.

This answer choice is correct, because not only does the paragraph imply that younger animals survive better in the cold, it also allows for exceptions to exist. The use of the word "likely" leaves room for cases in which a polar bear cub might not survive longer than the adult polar bear.

Word Usage Questions

When asked how a word is used in the paragraph, don't use your existing knowledge of the word. The question is being asked precisely because there is some strange or unusual usage of the word in the paragraph. Go to the paragraph and use contextual clues to determine the answer. Don't simply use the popular definition you already know.

Switchback Words

Stay alert for "switchbacks". These are the words and phrases frequently used to alert you to shifts in thought. The most common switchback word is "but". Others include although, however, nevertheless, on the other hand, even though, while, in spite of, despite, regardless of.

Avoid "Fact Traps"

Once you know which paragraph the answer will be in, focus on that paragraph. However, don't get distracted by a choice that is factually true about the paragraph. Your search is for the answer that answers the question, which may be about a tiny aspect in the paragraph. Stay focused and don't fall for an answer that describes the larger picture of the paragraph. Always go back to the question and make sure you're choosing an answer that actually answers the question and is not just a true statement.

Milk the Paragraph

Some of the paragraphs may throw you completely off. They might deal with a subject you have not been exposed to, or one that you haven't reviewed in years. While your lack of knowledge about the subject will be a hindrance, the paragraph itself can give you many clues that will help you find the correct answer. Read the paragraph carefully, and look for clues. Watch particularly for adjectives and nouns describing difficult terms or words that you don't recognize. Regardless of if you understand a word or not, replacing it with the synonyms used for it in the paragraph may help you to understand what the questions are asking.

Example: A bacteriophage is a virus that infects bacteria.

While you may not know much about the characteristics of a bacteriophage, the fifth word into the paragraph told you that a bacteriophage is a virus. Wherever you see the word "bacteriophage," you can mentally replace it with the word "virus". Your more general knowledge of viruses may enable you to answer the question.

Look carefully for these descriptive synonyms (nouns) and adjectives and use them to help you understand the difficult terms. Rather than wracking your mind about specific detail information concerning a difficult term in the paragraph, use the more general description or synonym provided to make it easier for you.

Make Predictions

One convenience of questions with short paragraphs full of information is that you can easily remember the few facts presented, compared to a much longer passage full of much more information. As you read and understand the paragraph and then the question, try to guess what the answer will be. Remember that three of the four answer choices are wrong, and once you begin reading them, your mind will immediately become cluttered with answer choices designed to throw you off. Your mind is typically the most focused immediately after you have read the paragraph and question and digested its contents. If you can, try to predict what the correct answer will be. You may be surprised at what you can predict.

Quickly scan the choices and see if your prediction is in the listed answer choices. If it is, then you can be quite confident that you have the right answer. It still won't hurt to check the other answer choices, but most of the time, you've got it!

Answer the Question

It may seem obvious to only pick answer choices that answer the question, but ASVAB can create some excellent answer choices that are wrong. Don't pick an answer just because it sounds right, or you believe it to be true. It MUST answer the question. Once you've made your selection, always go back and check it against the question and make sure that you didn't misread the question, and the answer choice does answer the question posed.

Benchmark

After you read the first answer choice, decide if you think it sounds correct or not. If it doesn't, move on to the next answer choice. If it does, tentatively mark in your answer book beside that choice. This doesn't mean that you've definitely selected it as your answer choice; it just means that it's the best you've seen thus far. Go ahead and read the next choice. If the next choice is worse than the one you've already selected, keep going to the next answer choice. If the next choice is better than the choice you've already selected, mark the new answer choice as your best guess.

The first answer choice that you select becomes your standard. Every other answer choice must be benchmarked against that standard. That choice is correct until proven otherwise by another answer choice beating it out. Once you've decided that no other answer choice seems as good, do one final check to ensure that it answers the question posed.

New Information

Correct answers will usually contain the information listed in the paragraph and question. Rarely will completely new information be inserted into a correct answer choice. Occasionally the new information may be related in a manner that ASVAB is asking for you to interpret, but seldom.
Example:
The argument above is dependent upon which of the following assumptions?
A. Charles's Law was used to interpret the relationship.

If Charles's Law is not mentioned at all in the referenced paragraph and argument, then it is unlikely that this choice is correct. All of the information needed to answer the question is provided for you, and so you should not have to make guesses that are unsupported or choose answer choices that have unknown information that cannot be reasoned.

Valid Information

Don't discount any of the information provided in short paragraphs. They are short to begin with and every piece of information may be necessary to determine the correct answer. None of the information in the paragraph is there to throw you off (while the answer choices will certainly have information to throw you off). If two seemingly unrelated topics are discussed, don't ignore either. You can be confident there is a relationship, or it wouldn't be included in the paragraph, and you are probably going to have to determine what is that relationship for the answer.

Don't Fall for the Obvious

When in doubt of the answer, it is easy to go with what you are familiar with. If you are familiar with one of the answer choices and know it is correct, then you may be inclined to guess at that term. Be careful though, and don't go with familiar answers simply because they are familiar.
Example: What happened when the temperature changed to 212° F?
   A. The solution began to boil.
   B. The reaction would become stabilized.
   C. The solution would become saturated.
   D. The reaction would be more easily controlled.

You know that 212° F is the boiling point of pure water. Therefore choice A is familiar, because there is a link between the temperature 212° F and the word "boiling". If you are unsure of the correct answer, you may decide upon choice A simply because of its familiarity. Don't be deceived though. Think through the other answer choices before making your final selection. Just because you have a mental link between the question and an answer choice, doesn't make that answer choice correct.

Random Tips

- For questions that you're not clear on the answer, use the process of elimination. Weed out the answer choices that you know are wrong before choosing an answer.
- Don't fall for "bizarre" choices, mentioning things that are not relevant to the paragraph. Also avoid answers that sound "smart." Again, if you're willing to bet $5, ignore the tips and go with your bet.

Common Subject: Mathematics

Math Basics

Numbers and Their Classifications

Numbers are the basic building blocks of mathematics. Specific features of numbers are identified by the following terms:

Integers – The set of whole positive and negative numbers, including zero. Integers do not include fractions ($\frac{1}{3}$), decimals (0.56), or mixed numbers ($7\frac{3}{4}$).

Prime number – A whole number greater than 1 that has only two factors, itself and 1; that is, a number that can be divided evenly only by 1 and itself.

Composite number – A whole number greater than 1 that has more than two different factors; in other words, any whole number that is not a prime number. For example: The composite number 8 has the factors of 1, 2, 4, and 8.

Even number – Any integer that can be divided by 2 without leaving a remainder. For example: 2, 4, 6, 8, and so on.

Odd number – Any integer that cannot be divided evenly by 2. For example: 3, 5, 7, 9, and so on.

Decimal number – a number that uses a decimal point to show the part of the number that is less than one. Example: 1.234.

Decimal point – a symbol used to separate the ones place from the tenths place in decimals or dollars from cents in currency.

Decimal place – the position of a number to the right of the decimal point. In the decimal 0.123, the 1 is in the first place to the right of the decimal point, indicating tenths; the 2 is in the second place, indicating hundredths; and the 3 is in the third place, indicating thousandths.

> ➢ **Review Video: Numbers and Their Classifications**
> Visit **mometrix.com/academy** and enter **Code: 461071**

The decimal, or base 10, system is a number system that uses ten different digits (0, 1, 2, 3, 4, 5, 6, 7, 8, 9). An example of a number system that uses something other than ten digits is the binary, or base 2, number system, used by computers, which uses only the numbers 0 and 1. It is thought that the decimal system originated because people had only their 10 fingers for counting.

Rational, irrational, and real numbers can be described as follows:

Rational numbers include all integers, decimals, and fractions. Any terminating or repeating decimal number is a rational number.

Irrational numbers cannot be written as fractions or decimals because the number of decimal places is infinite and there is no recurring pattern of digits within the number. For example, pi (π) begins with 3.141592 and continues without terminating or repeating, so pi is an irrational number.

Real numbers are the set of all rational and irrational numbers.

Operations

There are four basic mathematical operations:
Addition increases the value of one quantity by the value of another quantity. Example: $2 + 4 = 6; 8 + 9 = 17$. The result is called the sum. With addition, the order does not matter. $4 + 2 = 2 + 4$.
Subtraction is the opposite operation to addition; it decreases the value of one quantity by the value of another quantity. Example: $6 - 4 = 2; 17 - 8 = 9$. The result is called the difference. Note that with subtraction, the order does matter. $6 - 4 \neq 4 - 6$.

Multiplication can be thought of as repeated addition. One number tells how many times to add the other number to itself. Example: 3×2 (three times two) $= 2 + 2 + 2 = 6$. With multiplication, the order does not matter. $2 \times 3 = 3 \times 2$ or $3 + 3 = 2 + 2 + 2$.
Division is the opposite operation to multiplication; one number tells us how many parts to divide the other number into. Example: $20 \div 4 = 5$; if 20 is split into 4 equal parts, each part is 5. With division, the order of the numbers does matter. $20 \div 4 \neq 4 \div 20$.

An exponent is a superscript number placed next to another number at the top right. It indicates how many times the base number is to be multiplied by itself. Exponents provide a shorthand way to write what would be a longer mathematical expression. Example: $a^2 = a \times a$; $2^4 = 2 \times 2 \times 2 \times 2$. A number with an exponent of 2 is said to be "squared," while a number with an exponent of 3 is said to be "cubed." The value of a number raised to an exponent is called its power. So, 8^4 is read as "8 to the 4th power," or "8 raised to the power of 4." A negative exponent is the same as the reciprocal of a positive exponent. Example: $a^{-2} = \frac{1}{a^2}$.

> **Review Video: Exponents**
> *Visit mometrix.com/academy and enter Code*: **600998**

Parentheses are used to designate which operations should be done first when there are multiple operations. Example: 4 – (2 + 1) = 1; the parentheses tell us that we must add 2 and 1, and then subtract the sum from 4, rather than subtracting 2 from 4 and then adding 1 (this would give us an answer of 3).

Order of Operations is a set of rules that dictates the order in which we must perform each operation in an expression so that we will evaluate at accurately. If we have an expression that includes multiple different operations, Order of Operations tells us which operations to do first. The most common mnemonic for Order of Operations is PEMDAS, or "Please Excuse My Dear Aunt Sally." PEMDAS stands for Parentheses, Exponents, Multiplication, Division, Addition, Subtraction. It is important to understand that multiplication and division have equal precedence, as do addition and subtraction, so those pairs of operations are simply worked from left to right in order.

> **Review Video: Order of Operations**
> *Visit mometrix.com/academy and enter Code*: **259675**

Example: Evaluate the expression $5 + 20 \div 4 \times (2 + 3)^2 - 6$ using the correct order of operations.
P: Perform the operations inside the parentheses, $(2 + 3) = 5$.
E: Simplify the exponents, $(5)^2 = 25$.
The equation now looks like this: $5 + 20 \div 4 \times 25 - 6$.
MD: Perform multiplication and division from left to right, $20 \div 4 = 5$; then $5 \times 25 = 125$.
The equation now looks like this: $5 + 125 - 6$.
AS: Perform addition and subtraction from left to right, $5 + 125 = 130$; then $130 - 6 = 124$.

The laws of exponents are as follows:
1) Any number to the power of 1 is equal to itself: $a^1 = a$.
2) The number 1 raised to any power is equal to 1: $1^n = 1$.
3) Any number raised to the power of 0 is equal to 1: $a^0 = 1$.
4) Add exponents to multiply powers of the same base number: $a^n \times a^m = a^{n+m}$.
5) Subtract exponents to divide powers of the same number; that is $a^n \div a^m = a^{n-m}$.
6) Multiply exponents to raise a power to a power: $(a^n)^m = a^{n \times m}$.
7) If multiplied or divided numbers inside parentheses are collectively raised to a power, this is the same as each individual term being raised to that power: $(a \times b)^n = a^n \times b^n$; $(a \div b)^n = a^n \div b^n$.
Note: Exponents do not have to be integers. Fractional or decimal exponents follow all the rules above as well. Example: $5^{\frac{1}{4}} \times 5^{\frac{3}{4}} = 5^{\frac{1}{4}+\frac{3}{4}} = 5^1 = 5$.

A root, such as a square root, is another way of writing a fractional exponent. Instead of using a superscript, roots use the radical symbol ($\sqrt{}$) to indicate the operation. A radical will have a number underneath the bar, and may sometimes have a number in the upper left: $\sqrt[n]{a}$, read as "the nth root of a." The relationship between radical notation and exponent notation can be described by this equation: $\sqrt[n]{a} = a^{\frac{1}{n}}$. The two special cases of $n = 2$ and $n = 3$ are called square roots and cube roots. If there is no number to the upper left, it is understood to be a square root ($n = 2$). Nearly all of the roots you encounter will be square roots. A square root is the same as a number raised to the one-half power. When we say that a is the square root of b ($a = \sqrt{b}$), we mean that a multiplied by itself equals b: ($a \times a = b$).

A perfect square is a number that has an integer for its square root. There are 10 perfect squares from 1 to 100: 1, 4, 9, 16, 25, 36, 49, 64, 81, 100 (the squares of integers 1 through 10).

> **Review Video: Square Root and Perfect Square**
> *Visit mometrix.com/academy and enter Code: 648063*

Scientific notation is a way of writing large numbers in a shorter form. The form $a \times 10^n$ is used in scientific notation, where a is greater than or equal to 1, but less than 10, and n is the number of places the decimal must move to get from the original number to a.

Example: The number 230,400,000 is cumbersome to write. To write the value in scientific notation, place a decimal point between the first and second numbers, and include all digits through the last non-zero digit ($a = 2.304$). To find the appropriate power of 10, count the number of places the decimal point had to move ($n = 8$). The number is positive if the decimal moved to the left, and negative if it moved to the right. We can then write 230,400,000 as 2.304×10^8. If we look instead at the number 0.00002304, we have the same value for a, but this time the decimal moved 5 places to the right ($n = -5$). Thus, 0.00002304 can be written as 2.304×10^{-5}. Using this notation makes it simple to compare very large or very small numbers. By comparing exponents, it is easy to see that 3.28×10^4 is smaller than 1.51×10^5, because 4 is less than 5.

> ➤ **Review Video: Scientific Notation**
> *Visit* **mometrix.com/academy** *and enter* **Code: 976454**

Factors and Multiples

Factors are numbers that are multiplied together to obtain a product. For example, in the equation $2 \times 3 = 6$, the numbers 2 and 3 are factors. A prime number has only two factors (1 and itself), but other numbers can have many factors.

A common factor is a number that divides exactly into two or more other numbers. For example, the factors of 12 are 1, 2, 3, 4, 6, and 12, while the factors of 15 are 1, 3, 5, and 15. The common factors of 12 and 15 are 1 and 3.

A prime factor is also a prime number. Therefore, the prime factors of 12 are 2 and 3. For 15, the prime factors are 3 and 5.

The greatest common factor (GCF) is the largest number that is a factor of two or more numbers. For example, the factors of 15 are 1, 3, 5, and 15; the factors of 35 are 1, 5, 7, and 35. Therefore, the greatest common factor of 15 and 35 is 5.

The least common multiple (LCM) is the smallest number that is a multiple of two or more numbers. For example, the multiples of 3 include 3, 6, 9, 12, 15, etc.; the multiples of 5 include 5, 10, 15, 20, etc. Therefore, the least common multiple of 3 and 5 is 15.

> ➤ **Review Video: Greatest Common Factor (GCF)**
> *Visit* **mometrix.com/academy** *and enter* **Code: 838699**

Fractions, Percentages, and Related Concepts

A fraction is a number that is expressed as one integer written above another integer, with a dividing line between them $\left(\frac{x}{y}\right)$. It represents the quotient of the two numbers "x divided by y." It can also be thought of as *x* out of *y* equal parts.

The top number of a fraction is called the numerator, and it represents the number of parts under consideration. The 1 in $\frac{1}{4}$ means that 1 part out of the whole is being considered in the calculation. The bottom number of a fraction is called the denominator, and it represents the total number of equal parts. The 4 in $\frac{1}{4}$ means that the whole consists of 4 equal parts. A fraction cannot have a denominator of zero; this is referred to as "undefined."

Fractions can be manipulated, without changing the value of the fraction, by multiplying or dividing (but not adding or subtracting) both the numerator and denominator by the same number. If you divide both numbers by a common factor, you are reducing or simplifying the fraction. Two fractions that have the same value, but are expressed differently are known as equivalent fractions.

For example, $\frac{2}{10}, \frac{3}{15}, \frac{4}{20}$, and $\frac{5}{25}$ are all equivalent fractions. They can also all be reduced or simplified to $\frac{1}{5}$.

> ➤ **Review Video: Fractions**
> Visit **mometrix.com/academy** and enter **Code: 262335**

When two fractions are manipulated so that they have the same denominator, this is known as finding a common denominator. The number chosen to be that common denominator should be the least common multiple of the two original denominators. Example: $\frac{3}{4}$ and $\frac{5}{6}$; the least common multiple of 4 and 6 is 12. Manipulating to achieve the common denominator: $\frac{3}{4} = \frac{9}{12}; \frac{5}{6} = \frac{10}{12}$.

If two fractions have a common denominator, they can be added or subtracted simply by adding or subtracting the two numerators and retaining the same denominator. Example: $\frac{1}{2} + \frac{1}{4} = \frac{2}{4} + \frac{1}{4} = \frac{3}{4}$. If the two fractions do not already have the same denominator, one or both of them must be manipulated to achieve a common denominator before they can be added or subtracted.

Two fractions can be multiplied by multiplying the two numerators to find the new numerator and the two denominators to find the new denominator. Example: $\frac{1}{3} \times \frac{2}{3} = \frac{1 \times 2}{3 \times 3} = \frac{2}{9}$. Two fractions can be divided flipping the numerator and denominator of the second fraction and then proceeding as though it were a multiplication. Example: $\frac{2}{3} \div \frac{3}{4} = \frac{2}{3} \times \frac{4}{3} = \frac{8}{9}$.

A fraction whose denominator is greater than its numerator is known as a proper fraction, while a fraction whose numerator is greater than its denominator is known as an improper fraction. Proper fractions have values less than one and improper fractions have values greater than one.

A mixed number is a number that contains both an integer and a fraction. Any improper fraction can be rewritten as a mixed number. Example: $\frac{8}{3} = \frac{6}{3} + \frac{2}{3} = 2 + \frac{2}{3} = 2\frac{2}{3}$. Similarly, any mixed number can be rewritten as an improper fraction. Example: $1\frac{3}{5} = 1 + \frac{3}{5} = \frac{5}{5} + \frac{3}{5} = \frac{8}{5}$.

Percentages can be thought of as fractions that are based on a whole of 100; that is, one whole is equal to 100%. The word percent means "per hundred." Fractions can be expressed as percents by finding equivalent fractions with a denominator of 100. Example: $\frac{7}{10} = \frac{70}{100} = 70\%$; $\frac{1}{4} = \frac{25}{100} = 25\%$.

To express a percentage as a fraction, divide the percentage number by 100 and reduce the fraction to its simplest possible terms. Example: $60\% = \frac{60}{100} = \frac{3}{5}$; $96\% = \frac{96}{100} = \frac{24}{25}$.

Converting decimals to percentages and percentages to decimals is as simple as moving the decimal point. To convert from a decimal to a percent, move the decimal point two places to the right. To convert from a percent to a decimal, move it two places to the left. Example: 0.23 = 23%; 5.34 = 534%; 0.007 = 0.7%; 700% = 7.00; 86% = 0.86; 0.15% = 0.0015.

It may be helpful to remember that the percentage number will always be larger than the equivalent decimal number.

A percentage problem can be presented three main ways: (1) Find what percentage of some number another number is. Example: What percentage of 40 is 8? (2) Find what number is some percentage of a given number. Example: What number is 20% of 40? (3) Find what number another number is a given percentage of. Example: What number is 8 20% of? The three components in all of these cases are the same: a whole (W), a part (P), and a percentage (%). These are related by the equation: $P = W \times \%$. This is the form of the equation you would use to solve problems of type (2). To solve types (1) and (3), you would use these two forms: $\% = \frac{P}{W}$ and $W = \frac{P}{\%}$.

> ➢ **Review Video: <u>Percentages</u>**
> *Visit **mometrix.com/academy** and enter **Code**:* **141911**

The thing that frequently makes percentage problems difficult is that they are most often also word problems, so a large part of solving them is figuring out which quantities are what. Example: In a school cafeteria, 7 students choose pizza, 9 choose hamburgers, and 4 choose tacos. Find the percentage that chooses tacos. To find the whole, you must first add all of the parts: 7 + 9 + 4 = 20. The percentage can then be found by dividing the part by the whole ($\% = \frac{P}{W}$): $\frac{4}{20} = \frac{20}{100} = 20\%$.

A ratio is a comparison of two quantities in a particular order. Example: If there are 14 computers in a lab, and the class has 20 students, there is a student to computer ratio of 20 to 14, commonly written as 20:14. Ratios are normally reduced to their smallest whole number representation, so 20:14 would be reduced to 10:7 by dividing both sides by 2.

A proportion is a relationship between two quantities that dictates how one changes when the other changes. A direct proportion describes a relationship in which a quantity increases by a set amount for every increase in the other quantity, or decreases by that same amount for every decrease in the other quantity. Example: Assuming a constant driving speed, the time required for a car trip increases as the distance of the trip increases. The distance to be traveled and the time required to travel are directly proportional.

Inverse proportion is a relationship in which an increase in one quantity is accompanied by a decrease in the other, or vice versa. Example: the time required for a car trip decreases as the speed increases, and increases as the speed decreases, so the time required is inversely proportional to the speed of the car.

Systems of Equations

Systems of Equations are a set of simultaneous equations that all use the same variables. A solution to a system of equations must be true for each equation in the system. *Consistent Systems* are those with at least one solution. *Inconsistent Systems* are systems of equations that have no solution.

> ➢ **Review Video: <u>Systems of Equations</u>**
> *Visit **mometrix.com/academy** and enter **Code**:* **658153**

To solve a system of linear equations by *substitution*, start with the easier equation and solve for one of the variables. Express this variable in terms of the other variable. Substitute this expression in the other equation, and solve for the other variable. The solution should be expressed in the form (x, y). Substitute the values into both of the original equations to check your answer. Consider the following problem.

Solve the system using substitution:
$$x + 6y = 15$$
$$3x - 12y = 18$$

Solve the first equation for x:
$$x = 15 - 6y$$

Substitute this value in place of x in the second equation, and solve for y:
$$3(15 - 6y) - 12y = 18$$
$$45 - 18y - 12y = 18$$
$$30y = 27$$
$$y = \frac{27}{30} = \frac{9}{10} = 0.9$$

Plug this value for y back into the first equation to solve for x:
$$x = 15 - 6(0.9) = 15 - 5.4 = 9.6$$

Check both equations if you have time:
$$9.6 + 6(0.9) = 9.6 + 5.4 = 15$$
$$3(9.6) - 12(0.9) = 28.8 - 10.8 = 18$$

Therefore, the solution is (9.6, 0.9).

To solve a system of equations using *elimination*, begin by rewriting both equations in standard form $Ax + By = C$. Check to see if the coefficients of one pair of like variables add to zero. If not, multiply one or both of the equations by a non-zero number to make one set of like variables add to zero. Add the two equations to solve for one of the variables. Substitute this value into one of the original equations to solve for the other variable. Check your work by substituting into the other equation. Next we will solve the same problem as above, but using the addition method.

Solve the system using elimination:
$$x + 6y = 15$$
$$3x - 12y = 18$$

If we multiply the first equation by 2, we can eliminate the y terms:
$$2x + 12y = 30$$
$$3x - 12y = 18$$

Add the equations together and solve for x:
$$5x = 48$$
$$x = \frac{48}{5} = 9.6$$

Plug the value for x back into either of the original equations and solve for y:
$$9.6 + 6y = 15$$
$$y = \frac{15 - 9.6}{6} = 0.9$$

Check both equations if you have time:
$$9.6 + 6(0.9) = 9.6 + 5.4 = 15$$
$$3(9.6) - 12(0.9) = 28.8 - 10.8 = 18$$
Therefore, the solution is (9.6, 0.9).

Polynomial Algebra

To multiply two binomials, follow the *FOIL* method. FOIL stands for:
- First: Multiply the first term of each binomial
- Outer: Multiply the outer terms of each binomial
- Inner: Multiply the inner terms of each binomial
- Last: Multiply the last term of each binomial

Using FOIL, $(Ax + By)(Cx + Dy) = ACx^2 + ADxy + BCxy + BDy^2$.

> ➢ **Review Video: Multiplying Terms Using the FOIL Method**
> Visit *mometrix.com/academy* and enter **Code: 854792**

To divide polynomials, begin by arranging the terms of each polynomial in order of one variable. You may arrange in ascending or descending order, but be consistent with both polynomials. To get the first term of the quotient, divide the first term of the dividend by the first term of the divisor. Multiply the first term of the quotient by the entire divisor and subtract that product from the dividend. Repeat for the second and successive terms until you either get a remainder of zero or a remainder whose degree is less than the degree of the divisor. If the quotient has a remainder, write the answer as a mixed expression in the form: quotient + $\frac{\text{remainder}}{\text{divisor}}$.

Rational Expressions are fractions with polynomials in both the numerator and the denominator; the value of the polynomial in the denominator cannot be equal to zero. To add or subtract rational expressions, first find the common denominator, then rewrite each fraction as an equivalent fraction with the common denominator. Finally, add or subtract the numerators to get the numerator of the answer, and keep the common denominator as the denominator of the answer. When multiplying rational expressions, factor each polynomial and cancel like factors (a factor which appears in both the numerator and the denominator). Then, multiply all remaining factors in the numerator to get the numerator of the product, and multiply the remaining factors in the denominator to get the denominator of the product. Remember – cancel entire factors, not individual terms. To divide rational expressions, take the reciprocal of the divisor (the rational expression you are dividing by) and multiply by the dividend.

Below are patterns of some special products to remember: *perfect trinomial squares*, the *difference between two squares*, the *sum and difference of two cubes*, and *perfect cubes*.

- Perfect Trinomial Squares: $x^2 + 2xy + y^2 = (x + y)^2$ or $x^2 - 2xy + y^2 = (x - y)^2$
- Difference between Two Squares: $x^2 - y^2 = (x + y)(x - y)$

- Sum of Two Cubes: $x^3 + y^3 = (x+y)(x^2 - xy + y^2)$

    Note: the second factor is NOT the same as a perfect trinomial square, so do not try to factor it further.

- Difference between Two Cubes: $x^3 - y^3 = (x-y)(x^2 + xy + y^2)$

    Again, the second factor is NOT the same as a perfect trinomial square.

- Perfect Cubes: $x^3 + 3x^2y + 3xy^2 + y^3 = (x+y)^3$ and $x^3 - 3x^2y + 3xy^2 - y^3 = (x-y)^3$

In order to *factor* a polynomial, first check for a common monomial factor. When the greatest common monomial factor has been factored out, look for patterns of special products: differences of two squares, the sum or difference of two cubes for binomial factors, or perfect trinomial squares for trinomial factors. If the factor is a trinomial but not a perfect trinomial square, look for a factorable form, such as $x^2 + (a+b)x + ab = (x+a)(x+b)$ or $(ac)x^2 + (ad+bc)x + bd = (ax+b)(cx+d)$. For factors with four terms, look for groups to factor. Once you have found the factors, write the original polynomial as the product of all the factors. Make sure all of the polynomial factors are prime. Monomial factors may be prime or composite. Check your work by multiplying the factors to make sure you get the original polynomial.

Solving Quadratic Equations

The *Quadratic Formula* is used to solve quadratic equations when other methods are more difficult. To use the quadratic formula to solve a quadratic equation, begin by rewriting the equation in standard form $ax^2 + bx + c = 0$, where a, b, and c are coefficients. Once you have identified the values of the coefficients, substitute those values into the quadratic formula $= \frac{-b \pm \sqrt{b^2 - 4ac}}{2a}$. Evaluate the equation and simplify the expression. Again, check each root by substituting into the original equation. In the quadratic formula, the portion of the formula under the radical $(b^2 - 4ac)$ is called the *Discriminant*. If the discriminant is zero, there is only one root: zero. If the discriminant is positive, there are two different real roots. If the discriminant is negative, there are no real roots.

To solve a quadratic equation by *Factoring*, begin by rewriting the equation in standard form, if necessary. Factor the side with the variable then set each of the factors equal to zero and solve the resulting linear equations. Check your answers by substituting the roots you found into the original equation. If, when writing the equation in standard form, you have an equation in the form $x^2 + c = 0$ or $x^2 - c = 0$, set $x^2 = -c$ or $x^2 = c$ and take the square root of c. If $c = 0$, the only real root is zero. If c is positive, there are two real roots—the positive and negative square root values. If c is negative, there are no real roots because you cannot take the square root of a negative number.

> **Review Video: Factoring Quadratic Equations**
> Visit *mometrix.com/academy* and enter **Code: 336566**

To solve a quadratic equation by *Completing the Square*, rewrite the equation so that all terms containing the variable are on the left side of the equal sign, and all the constants are on the right side of the equal sign. Make sure the coefficient of the squared term is 1. If there is a coefficient with the squared term, divide each term on both sides of the equal side by that number. Next, work with the coefficient of the single-variable term. Square half of this coefficient, and add that value to both sides. Now you can factor the left side (the side containing the variable) as the square of a binomial. $x^2 + 2ax + a^2 = C \Rightarrow (x+a)^2 = C$, where x is the variable, and a and C are constants. Take the square root of both sides and solve for the variable. Substitute the value of the variable in the original problem to check your work.

Geometry Basics

Below are some terms that are commonly used in geometric studies. Most of these concepts are foundational to geometry, so understanding them is a necessary first step to studying geometry.

A point is a fixed location in space; has no size or dimensions; commonly represented by a dot.

A line is a set of points that extends infinitely in two opposite directions. It has length, but no width or depth. A line can be defined by any two distinct points that it contains. A line segment is a portion of a line that has definite endpoints. A ray is a portion of a line that extends from a single point on that line in one direction along the line. It has a definite beginning, but no ending.

A plane is a two-dimensional flat surface defined by three non-collinear points. A plane extends an infinite distance in all directions in those two dimensions. It contains an infinite number of points, parallel lines and segments, intersecting lines and segments, as well as parallel or intersecting rays. A plane will never contain a three-dimensional figure or skew lines. Two given planes will either be parallel or they will intersect to form a line. A plane may intersect a circular conic surface, such as a cone, to form conic sections, such as the parabola, hyperbola, circle or ellipse.

Perpendicular lines are lines that intersect at right angles. They are represented by the symbol ⊥. The shortest distance from a line to a point not on the line is a perpendicular segment from the point to the line.

Parallel lines are lines in the same plane that have no points in common and never meet. It is possible for lines to be in different planes, have no points in common, and never meet, but they are not parallel because they are in different planes.

A bisector is a line or line segment that divides another line segment into two equal lengths. A perpendicular bisector of a line segment is composed of points that are equidistant from the endpoints of the segment it is dividing.

Intersecting lines are lines that have exactly one point in common. Concurrent lines are multiple lines that intersect at a single point.

A transversal is a line that intersects at least two other lines, which may or may not be parallel to one another. A transversal that intersects parallel lines is a common occurrence in geometry.

Angles

An angle is formed when two lines or line segments meet at a common point. It may be a common starting point for a pair of segments or rays, or it may be the intersection of lines. Angles are represented by the symbol ∠.

The vertex is the point at which two segments or rays meet to form an angle. If the angle is formed by intersecting rays, lines, and/or line segments, the vertex is the point at which four angles are formed. The pairs of angles opposite one another are called vertical angles, and their measures are equal. In the figure below, angles ABC and DBE are congruent, as are angles ABD and CBE.

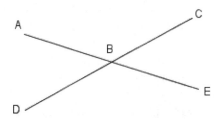

An acute angle is an angle with a degree measure less than 90°.
A right angle is an angle with a degree measure of exactly 90°.
An obtuse angle is an angle with a degree measure greater than 90° but less than 180°.
A straight angle is an angle with a degree measure of exactly 180°. This is also a semicircle.
A reflex angle is an angle with a degree measure greater than 180° but less than 360°.
A full angle is an angle with a degree measure of exactly 360°.

Two angles whose sum is exactly 90° are said to be complementary. The two angles may or may not be adjacent. In a right triangle, the two acute angles are complementary.

Two angles whose sum is exactly 180° are said to be supplementary. The two angles may or may not be adjacent. Two intersecting lines always form two pairs of supplementary angles. Adjacent supplementary angles will always form a straight line.

Two angles that have the same vertex and share a side are said to be adjacent. Vertical angles are not adjacent because they share a vertex but no common side.

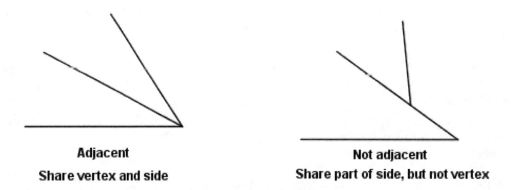

When two parallel lines are cut by a transversal, the angles that are between the two parallel lines are interior angles. In the diagram below, angles 3, 4, 5, and 6 are interior angles.

When two parallel lines are cut by a transversal, the angles that are outside the parallel lines are exterior angles. In the diagram below, angles 1, 2, 7, and 8 are exterior angles.

When two parallel lines are cut by a transversal, the angles that are in the same position relative to the transversal and a parallel line are corresponding angles. The diagram below has four pairs of corresponding angles: angles 1 and 5; angles 2 and 6; angles 3 and 7; and angles 4 and 8. Corresponding angles formed by parallel lines are congruent.

When two parallel lines are cut by a transversal, the two interior angles that are on opposite sides of the transversal are called alternate interior angles. In the diagram below, there are two pairs of alternate interior angles: angles 3 and 6, and angles 4 and 5. Alternate interior angles formed by parallel lines are congruent.

When two parallel lines are cut by a transversal, the two exterior angles that are on opposite sides of the transversal are called alternate exterior angles. In the diagram below, there are two pairs of alternate exterior angles: angles 1 and 8, and angles 2 and 7. Alternate exterior angles formed by parallel lines are congruent.

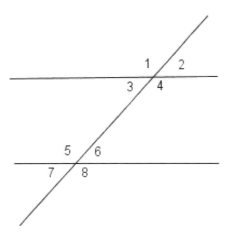

Circles

The center is the single point inside the circle that is equidistant from every point on the circle. (Point O in the diagram below.)

> ➤ **Review Video: Points of a Circle**
> Visit **mometrix.com/academy** and enter **Code: 420746**

The radius is a line segment that joins the center of the circle and any one point on the circle. All radii of a circle are equal. (Segments OX, OY, and OZ in the diagram below.)

The diameter is a line segment that passes through the center of the circle and has both endpoints on the circle. The length of the diameter is exactly twice the length of the radius. (Segment XZ in the diagram below.)

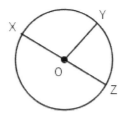

A circle is inscribed in a polygon if each of the sides of the polygon is tangent to the circle. A polygon is inscribed in a circle if each of the vertices of the polygon lies on the circle.

A circle is circumscribed about a polygon if each of the vertices of the polygon lies on the circle. A polygon is circumscribed about the circle if each of the sides of the polygon is tangent to the circle.

If one figure is inscribed in another, then the other figure is circumscribed about the first figure.

Circle circumscribed about a pentagon
Pentagon inscribed in a circle

Polygons

A polygon is a planar shape formed from line segments called sides that are joined together at points called vertices (singular: vertex). Specific polygons are named by the number of angles or sides they have. Regular polygons are polygons whose sides are all equal and whose angles are all congruent.

An interior angle is any of the angles inside a polygon where two sides meet at a vertex. The sum of the interior angles of a polygon is dependent only on the number of sides. For example, all 5-sided polygons have interior angles that sum to 540°, regardless of the particular shape.

A diagonal is a line that joins two nonconsecutive vertices of a polygon. The number of diagonals that can be drawn on an n-sided polygon is $d = \frac{n(n-3)}{2}$.

The following list presents several different types of polygons:
Triangle – 3 sides
Quadrilateral – 4 sides
Pentagon – 5 sides
Hexagon – 6 sides
Heptagon – 7 sides
Octagon – 8 sides
Nonagon – 9 sides
Decagon – 10 sides
Dodecagon – 12 sides

More generally, an n-gon is a polygon that has n angles and n sides.

The sum of the interior angles of an n-sided polygon is $(n - 2)180°$. For example, in a triangle n = 3, so the sum of the interior angles is $(3 - 2)180° = 180°$. In a quadrilateral, n = 4, and the sum of the angles is $(4 - 2)180° = 360°$. The sum of the interior angles of a polygon is equal to the sum of the interior angles of any other polygon with the same number of sides.

Below are descriptions for several common quadrilaterals. Recall that a quadrilateral is a four-sided polygon.

Trapezoid – quadrilateral with exactly one pair of parallel sides (opposite one another); in an isosceles trapezoid, the two non-parallel sides have equal length and both pairs of non-opposite angles are congruent
Parallelogram – quadrilateral with two pairs of parallel sides (opposite one another), and two pairs of congruent angles (opposite one another)
Rhombus – parallelogram with four equal sides
Rectangle – parallelogram with four congruent angles (right angles)
Square – parallelogram with four equal sides and four congruent angles (right angles)

Triangles

A triangle is a polygon with three sides and three angles. Triangles can be classified according to the length of their sides or magnitude of their angles.

An acute triangle is a triangle whose three angles are all less than 90°. If two of the angles are equal, the acute triangle is also an isosceles triangle. If the three angles are all equal, the acute triangle is also an equilateral triangle.

A right triangle is a triangle with exactly one angle equal to 90°. All right triangles follow the Pythagorean Theorem. A right triangle can never be acute or obtuse.

An obtuse triangle is a triangle with exactly one angle greater than 90°. The other two angles may or may not be equal. If the two remaining angles are equal, the obtuse triangle is also an isosceles triangle.

An equilateral triangle is a triangle with three congruent sides. An equilateral triangle will also have three congruent angles, each 60°. All equilateral triangles are also acute triangles.

An isosceles triangle is a triangle with two congruent sides. An isosceles triangle will also have two congruent angles opposite the two congruent sides.

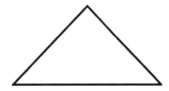

A scalene triangle is a triangle with no congruent sides. A scalene triangle will also have three angles of different measures. The angle with the largest measure is opposite the longest side, and the angle with the smallest measure is opposite the shortest side.

The Triangle Inequality Theorem states that the sum of the measures of any two sides of a triangle is always greater than the measure of the third side. If the sum of the measures of two sides were equal to the third side, a triangle would be impossible because the two sides would lie flat across the third side and there would be no vertex. If the sum of the measures of two of the sides was less than the third side, a closed figure would be impossible because the two shortest sides would never meet.

Similar triangles are triangles whose corresponding angles are congruent to one another. Their corresponding sides may or may not be equal, but they are proportional to one another. Since the angles in a triangle always sum to 180°, it is only necessary to determine that two pairs of corresponding angles are congruent, since the third will be also in that case.

> ➤ **Review Video: Similar Triangles**
> *Visit mometrix.com/academy and enter Code:* **398538**

Congruent triangles are similar triangles whose corresponding sides are all equal. Congruent triangles can be made to fit on top of one another by rotation, reflection, and/or translation. When trying to determine whether two triangles are congruent, there are several criteria that can be used.

Side-side-side (SSS): if all three sides of one triangle are equal to all three sides of another triangle, they are congruent by SSS.
Side-angle-side (SAS): if two sides and the adjoining angle in one triangle are equal to two sides and the adjoining angle of another triangle, they are congruent by SAS.
Additionally, if two triangles can be shown to be similar, then there need only be one pair of corresponding equal sides to show congruence.

One of the most important theorems in geometry is the Pythagorean Theorem. Named after the sixth-century Greek mathematician Pythagoras, this theorem states that, for a right triangle, the square of the hypotenuse (the longest side of the triangle, always opposite the right angle) is equal to the sum of the squares of the other two sides. Written symbolically, the Pythagorean Theorem can be expressed as $a^2 + b^2 = c^2$, where c is the hypotenuse and a and b are the remaining two sides.

The theorem is most commonly used to find the length of an unknown side of a right triangle, given the lengths of the other two sides. For example, given that the hypotenuse of a right triangle is 5 and one side is 3, the other side can be found using the formula: $a^2 + b^2 = c^2$, $3^2 + b^2 = 5^2$, $9 + b^2 = 25$, $b^2 = 25 - 9 = 16$, $b = \sqrt{16} = 4$.

The theorem can also be used "in reverse" to show that when the square of one side of a triangle is equal to the sum of the squares of the other two sides, the triangle must be a right triangle.

The Law of Sines states that $\frac{\sin A}{a} = \frac{\sin B}{b} = \frac{\sin C}{c}$, where A, B, and C are the angles of a triangle, and a, b, and c are the sides opposite their respective angles. This formula will work with all triangles, not just right triangles.

The Law of Cosines is given by the formula $c^2 = a^2 + b^2 - 2ab(\cos C)$, where a, b, and c are the sides of a triangle, and C is the angle opposite side c. This formula is similar to the Pythagorean Theorem, but unlike the Pythagorean Theorem, it can be used on any triangle.

Symmetry

Symmetry is a property of a shape in which the shape can be transformed by either reflection or rotation without losing its original shape and orientation. A shape that has reflection symmetry can be reflected across a line with the result being the same shape as before the reflection. A line of symmetry divides a shape into two parts, with each part being a mirror image of the other. A shape can have more than one line of symmetry. A circle, for instance, has an infinite number of lines of symmetry. When reflection symmetry is extended to three-dimensional space, it is taken to describe a solid that can be divided into mirror image parts by a plane of symmetry. Rotational symmetry describes a shape that can be rotated about a point and achieve its original shape and orientation with less than a 360° rotation. When rotational symmetry is extended to three-dimensional space, it describes a solid that can be rotated about a line with the same conditions. Many shapes have both reflection and rotational symmetry.

Area formulas

Rectangle: $A = wl$, where w is the width and l is the length

Square: $A = s^2$, where s is the length of a side.

Triangle: $A = \frac{1}{2}bh$, where b is the length of one side (base) and h is the distance from that side to the opposite vertex measured perpendicularly (height).

Parallelogram: $A = bh$, where b is the length of one side (base) and h is the perpendicular distance between that side and its parallel side (height).

Trapezoid: $A = \frac{1}{2}(b_1 + b_2)h$, where b_1 and b_2 are the lengths of the two parallel sides (bases), and h is the perpendicular distance between them (height).

Circle: $A = \pi r^2$, where π is the mathematical constant approximately equal to 3.14 and r is the distance from the center of the circle to any point on the circle (radius).

> ➢ **Review Video: <u>Finding Areas in Geometry</u>**
> *Visit **mometrix.com/academy** and enter Code: **663492***

Volume Formulas

For some of these shapes, it is necessary to find the area of the base polygon before the volume of the solid can be found. This base area is represented in the volume equations as B.

Pyramid – consists of a polygon base, and triangles connecting each side of that polygon to a vertex. The volume can be calculated as $V = \frac{1}{3}Bh$, where h is the distance between the vertex and the base polygon, measured perpendicularly.

Prism – consists of two identical polygon bases, attached to one another on corresponding sides by parallelograms. The volume can be calculated as $V = Bh$, where h is the perpendicular distance between the two bases.

Cube – a special type of prism in which the two bases are the same shape as the side faces. All faces are squares. The volume can be calculated as $V = s^3$, where s is the length of any side.

Sphere – a round solid consisting of one continuous, uniformly-curved surface. The volume can be calculated as $V = \frac{4}{3}\pi r^3$, where r is the distance from the center of the sphere to any point on the surface (radius).

> ➢ **Review Video: Finding Volume in Geometry**
> Visit **mometrix.com/academy** and enter **Code: 754774**

Probability Basics

Probability is a branch of statistics that deals with the likelihood of something taking place. One classic example is a coin toss. There are only two possible results: heads or tails. The likelihood, or probability, that the coin will land as heads is 1 out of 2 (1/2, 0.5, 50%). Tails has the same probability. Another common example is a 6-sided die roll. There are six possible results from rolling a single die, each with an equal chance of happening, so the probability of any given number coming up is 1 out of 6.

Terms frequently used in probability:
Event – a situation that produces results of some sort (a coin toss)
Compound event – event that involves two or more independent events (rolling a pair of dice; taking the sum)
Outcome – a possible result in an experiment or event (heads, tails)
Desired outcome (or success) – an outcome that meets a particular set of criteria (a roll of 1 or 2 if we are looking for numbers less than 3)
Independent events – two or more events whose outcomes do not affect one another (two coins tossed at the same time)
Dependent events – two or more events whose outcomes affect one another (two cards drawn consecutively from the same deck)
Certain outcome – probability of outcome is 100% or 1
Impossible outcome – probability of outcome is 0% or 0
Mutually exclusive outcomes – two or more outcomes whose criteria cannot all be satisfied in a single event (a coin coming up heads and tails on the same toss)

Probability is the likelihood of a certain outcome occurring for a given event. The **theoretical probability** can usually be determined without actually performing the event. The likelihood of a outcome occurring, or the probability of an outcome occurring, is given by the formula

$$P(A) = \frac{\text{Number of acceptable outcomes}}{\text{Number of possible outcomes}}$$

where $P(A)$ is the probability of an outcome A occurring, and each outcome is just as likely to occur as any other outcome. If each outcome has the same probability of occurring as every other possible outcome, the outcomes are said to be equally likely to occur. The total number of acceptable outcomes must be less than or equal to the total number of possible outcomes. If the two are equal, then the outcome is certain to occur and the probability is 1. If the number of acceptable outcomes is zero, then the outcome is impossible and the probability is 0.
Example:
There are 20 marbles in a bag and 5 are red. The theoretical probability of randomly selecting a red marble is 5 out of 20, (5/20 = 1/4, 0.25, or 25%).

> ➢ **Review Video: Simple Probability**
> *Visit **mometrix.com/academy** and enter **Code: 212374***

When trying to calculate the probability of an event using the $\frac{desired\ outcomes}{total\ outcomes}$ formula, you may frequently find that there are too many outcomes to individually count them. Permutation and combination formulas offer a shortcut to counting outcomes. A permutation is an arrangement of a specific number of a set of objects in a specific order. The number of **permutations** of r items given a set of n items can be calculated as $_nP_r = \frac{n!}{(n-r)!}$. Combinations are similar to permutations, except there are no restrictions regarding the order of the elements. While ABC is considered a different permutation than BCA, ABC and BCA are considered the same combination. The number of **combinations** of r items given a set of n items can be calculated as $_nC_r = \frac{n!}{r!(n-r)!}$ or $_nC_r = \frac{_nP_r}{r!}$.
Example: Suppose you want to calculate how many different 5-card hands can be drawn from a deck of 52 cards. This is a combination since the order of the cards in a hand does not matter. There are 52 cards available, and 5 to be selected. Thus, the number of different hands is $_{52}C_5 = \frac{52!}{5! \times 47!} = 2{,}598{,}960$.

Sometimes it may be easier to calculate the possibility of something not happening, or the **complement of an event**. Represented by the symbol $\bar{A}$, the complement of A is the probability that event A does not happen. When you know the probability of event A occurring, you can use the formula $P(\bar{A}) = 1 - P(A)$, where $P(\bar{A})$ is the probability of event A not occurring, and $P(A)$ is the probability of event A occurring.

The **addition rule** for probability is used for finding the probability of a compound event. Use the formula $P(A \text{ or } B) = P(A) + P(B) - P(A \text{ and } B)$, where $P(A \text{ and } B)$ is the probability of both events occurring to find the probability of a compound event. The probability of both events occurring at the same time must be subtracted to eliminate any overlap in the first two probabilities.

Conditional probability is the probability of an event occurring once another event has already occurred. Given event A and dependent event B, the probability of event B occurring when event A

has already occurred is represented by the notation $P(A|B)$. To find the probability of event B occurring, take into account the fact that event A has already occurred and adjust the total number of possible outcomes. For example, suppose you have ten balls numbered 1–10 and you want ball number 7 to be pulled in two pulls. On the first pull, the probability of getting the 7 is $\frac{1}{10}$ because there is one ball with a 7 on it and 10 balls to choose from. Assuming the first pull did not yield a 7, the probability of pulling a 7 on the second pull is now $\frac{1}{9}$ because there are only 9 balls remaining for the second pull.

The **multiplication rule** can be used to find the probability of two independent events occurring using the formula $P(A \text{ and } B) = P(A) \times P(B)$, where $P(A \text{ and } B)$ is the probability of two independent events occurring, $P(A)$ is the probability of the first event occurring, and $P(B)$ is the probability of the second event occurring.

The multiplication rule can also be used to find the probability of two dependent events occurring using the formula $P(A \text{ and } B) = P(A) \times P(B|A)$, where $P(A \text{ and } B)$ is the probability of two dependent events occurring and $P(B|A)$ is the probability of the second event occurring after the first event has already occurred.
Before using the multiplication rule, you MUST first determine whether the two events are dependent or independent.

Use a combination of the multiplication rule and the rule of complements to find the probability that at least one outcome of the element will occur. This given by the general formula $P(\text{at least one event occurring}) = 1 - P(\text{no outcomes occurring})$. For example, to find the probability that at least one even number will show when a pair of dice is rolled, find the probability that two odd numbers will be rolled (no even numbers) and subtract from one. You can always use a tree diagram or make a chart to list the possible outcomes when the sample space is small, such as in the dice-rolling example, but in most cases it will be much faster to use the multiplication and complement formulas.

Expected value is a method of determining expected outcome in a random situation. It is really a sum of the weighted probabilities of the possible outcomes. Multiply the probability of an event occurring by the weight assigned to that probability (such as the amount of money won or lost). A practical application of the expected value is to determine whether a game of chance is really fair. If the sum of the weighted probabilities is equal to zero, the game is generally considered fair because the player has a fair chance to at least to break even. If the expected value is less than zero, then players lose more than they win. For example, a lottery drawing might allow the player to choose any three-digit number, 000–999. The probability of choosing the winning number is 1:1000. If it costs \$1 to play, and a winning number receives \$500, the expected value is $\left(-\$1 \cdot \frac{999}{1,000}\right) + \left(\$500 \cdot \frac{1}{1,000}\right) = -0.499$ or $-\$0.50$. You can expect to lose on average 50 cents for every dollar you spend.

Most of the time, when we talk about probability, we mean theoretical probability. **Empirical probability**, or experimental probability or relative frequency, is the number of times an outcome occurs in a particular experiment or a certain number of observed events. While theoretical probability is based on what *should* happen, experimental probability is based on what *has* happened. Experimental probability is calculated in the same way as theoretical, except that actual outcomes are used instead of possible outcomes.

Theoretical and experimental probability do not always line up with one another. Theoretical probability says that out of 20 coin tosses, 10 should be heads. However, if we were actually to toss 20 coins, we might record just 5 heads. This doesn't mean that our theoretical probability is incorrect; it just means that this particular experiment had results that were different from what was predicted. A practical application of empirical probability is the insurance industry. There are no set functions that define life span, health, or safety. Insurance companies look at factors from hundreds of thousands of individuals to find patterns that they then use to set the formulas for insurance premiums.

Statistics Basics

Statistics is the branch of mathematics that deals with collecting, recording, interpreting, illustrating, and analyzing large amounts of data. The following terms are often used in the discussion of data and statistics:
Data – the collective name for pieces of information (singular is datum).
Quantitative data – measurements (such as length, mass, and speed) that provide information about quantities in numbers
Qualitative data – information (such as colors, scents, tastes, and shapes) that cannot be measured using numbers
Discrete data – information that can be expressed only by a specific value, such as whole or half numbers; For example, since people can be counted only in whole numbers, a population count would be discrete data.
Continuous data – information (such as time and temperature) that can be expressed by any value within a given range
Primary data – information that has been collected directly from a survey, investigation, or experiment, such as a questionnaire or the recording of daily temperatures; Primary data that has not yet been organized or analyzed is called raw data.
Secondary data – information that has been collected, sorted, and processed by the researcher
Ordinal data – information that can be placed in numerical order, such as age or weight
Nominal data – information that cannot be placed in numerical order, such as names or places

Measures of Central Tendency

The quantities of mean, median, and mode are all referred to as measures of central tendency. They can each give a picture of what the whole set of data looks like with just a single number. Knowing what each of these values represents is vital to making use of the information they provide.

The mean, also known as the arithmetic mean or average, of a data set is calculated by summing all of the values in the set and dividing that sum by the number of values. For example, if a data set has 6 numbers and the sum of those 6 numbers is 30, the mean is calculated as $30/6 = 5$.

The median is the middle value of a data set. The median can be found by putting the data set in numerical order, and locating the middle value. In the data set (1, 2, 3, 4, 5), the median is 3. If there is an even number of values in the set, the median is calculated by taking the average of the two middle values. In the data set, (1, 2, 3, 4, 5, 6), the median would be $(3 + 4)/2 = 3.5$.

The mode is the value that appears most frequently in the data set. In the data set (1, 2, 3, 4, 5, 5, 5), the mode would be 5 since the value 5 appears three times. If multiple values appear the same number of times, there are multiple values for the mode. If the data set were (1, 2, 2, 3, 4, 4, 5, 5), the modes would be 2, 4, and 5. If no value appears more than any other value in the data set, then there is no mode.

Measures of Dispersion

The standard deviation expresses how spread out the values of a distribution are from the mean. Standard deviation is given in the same units as the original data and is represented by a lower case sigma (σ).
A high standard deviation means that the values are very spread out. A low standard deviation means that the values are close together.
If every value in a distribution is increased or decreased by the same amount, the mean, median, and mode are increased or decreased by that amount, but the standard deviation stays the same.
If every value in a distribution is multiplied or divided by the same number, the mean, median, mode, and standard deviation will all be multiplied or divided by that number.

The range of a distribution is the difference between the highest and lowest values in the distribution. For example, in the data set (1, 3, 5, 7, 9, 11), the highest and lowest values are 11 and 1, respectively. The range then would be calculated as 11 – 1 = 10.
The three quartiles are the three values that divide a data set into four equal parts. Quartiles are generally only calculated for data sets with a large number of values. As a simple example, for the data set consisting of the numbers 1 through 99, the first quartile (Q1) would be 25, the second quartile (Q2), always equal to the median, would be 50, and the third quartile (Q3) would be 75. The difference between Q1 and Q3 is known as the interquartile range.

Displaying Data

A bar graph is a graph that uses bars to compare data, as if each bar were a ruler being used to measure the data. The graph includes a scale that identifies the units being measured.

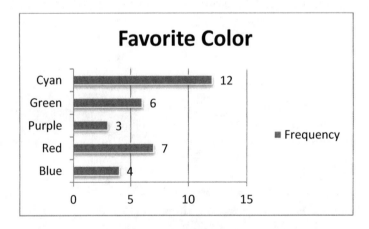

> **Review Video: Bar Graphs**
> *Visit* **mometrix.com/academy** *and enter* **Code: 226729**

A line graph is a graph that connects points to show how data increases or decreases over time. The time line is the horizontal axis. The connecting lines between data points on the graph are a way to more clearly show how the data changes.

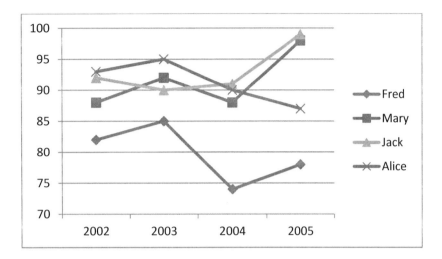

A pictograph is a graph that uses pictures or symbols to show data. The pictograph will have a key to identify what each symbol represents. Generally, each symbol stands for one or more objects.

A pie chart or circle graph is a diagram used to compare parts of a whole. The full pie represents the whole, and it is divided into sectors that each represent something that is a part of the whole. Each sector or slice of the pie is either labeled to indicate what it represents, or explained on a key associated with the chart. The size of each slice is determined by the percentage of the whole that the associated quantity represents. Numerically, the angle measurement of each sector can be computed by solving the proportion: x/360 = part/whole.

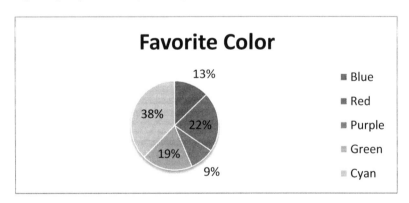

> **Review Video: Pie Charts**
> *Visit **mometrix.com/academy** and enter **Code: 895285***

A histogram is a special type of bar graph where the data are grouped in intervals (for example 20-29, 30-39, 40-49, etc.). The frequency, or number of times a value occurs in each interval, is indicated by the height of the bar. The intervals do not have to be the same amount but usually are (all data in ranges of 10 or all in ranges of 5, for example). The smaller the intervals, the more detailed the information.

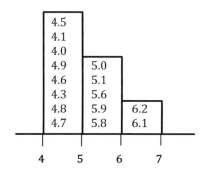

A stem-and-leaf plot is a way to organize data visually so that the information is easy to understand. A stem-and-leaf plot is simple to construct because a simple line separates the stem (the part of the plot listing the tens digit, if displaying two-digit data) from the leaf (the part that shows the ones digit). Thus, the number 45 would appear as 4 | 5. The stem-and-leaf plot for test scores of a group of 11 students might look like the following:

9 | 5
8 | 1, 3, 8
7 | 0, 2, 4, 6, 7
6 | 2, 8

A stem-and-leaf plot is similar to a histogram or other frequency plot, but with a stem-and-leaf plot, all the original data is preserved. In this example, it can be seen at a glance that nearly half the students scored in the 70's, yet all the data has been maintained. These plots can be used for larger numbers as well, but they tend to work better for small sets of data as they can become unwieldy with larger sets.

Final Notes

Word problems describing shapes should always be drawn out. Remember the old adage that a picture is worth a thousand words. If geometric shapes are described (line segments, circles, squares, etc) draw them out rather than trying to visualize how they should look.

On problems with variables, "plug and chug" by picking a number such as 1 or 2 and seeing if that would solve the problem. A 1 or 2 are good numbers to start with because they are easy to solve for with multiplication or division. If the 1 or 2 doesn't answer the problem, you can try either a larger or a smaller number, until you finally reach the result.

Approach problems systematically. Take time to understand what is being asked for. In many cases there is a drawing or graph that you can write on. Draw lines, jot notes, do whatever is necessary to create a visual picture and to allow you to understand what is being asked.

Even if you have always done well in math, you're not guaranteed to succeed on the ASVAB math sections. While math tests in high school and college test specific competencies in specific subjects, the ASVAB may test your ability to apply math concepts from different math subjects in one problem. Solid preparation and practice is the key to acing the ASVAB.

Common Subject: Science

Earth and Space Science

Geology

Minerals are naturally occurring, inorganic solids with a definite chemical composition and an orderly internal crystal structure. A polymorph is two minerals with the same chemical composition, but a different crystal structure. Rocks are aggregates of one or more minerals, and may also contain mineraloids (minerals lacking a crystalline structure) and organic remains. The three types of rocks are sedimentary, igneous, and metamorphic. Rocks are classified based on their formation and the minerals they contain. Minerals are classified by their chemical composition. Geology is the study of the planet Earth as it pertains to the composition, structure, and origin of its rocks. Petrology is the study of rocks, including their composition, texture, structure, occurrence, mode of formation, and history. Mineralogy is the study of minerals.

Sedimentary rocks are formed by the process of lithification, which involves compaction, the expulsion of liquids from pores, and the cementation of the pre-existing rock. It is pressure and temperature that are responsible for this process. Sedimentary rocks are often formed in layers in the presence of water, and may contain organic remains, such as fossils. Sedimentary rocks are organized into three groups: detrital, biogenic, and chemical. Texture refers to the size, shape, and grains of sedimentary rock. Texture can be used to determine how a particular sedimentary rock was created. Composition refers to the types of minerals present in the rock. The origin of sedimentary rock refers to the type of water that was involved in its creation. Marine deposits, for example, likely involved ocean environments, while continental deposits likely involved dry land and lakes.

Igneous rock is formed from magma, which is molten material originating from beneath the Earth's surface. Depending upon where magma cools, the resulting igneous rock can be classified as intrusive, plutonic, hypabyssal, extrusive, or volcanic. Magma that solidifies at a depth is intrusive, cools slowly, and has a coarse grain as a result. An example is granite. Magma that solidifies at or near the surface is extrusive, cools quickly, and usually has a fine grain. An example is basalt. Magma that actually flows out of the Earth's surface is called lava. Some extrusive rock cools so quickly that crystals do not have time to form. These rocks have a glassy appearance. An example is obsidian. Hypabyssal rock is igneous rock that is formed at medium depths.

Metamorphic rock is that which has been changed by great heat and pressure. This results in a variety of outcomes, including deformation, compaction, destruction of the characteristics of the original rock, bending, folding, and formation of new minerals because of chemical reactions, and changes in the size and shape of the mineral grain. For example, the igneous rock ferromagnesian can be changed into schist and gneiss. The sedimentary rock carbonaceous can be changed into marble. The texture of metamorphic rocks can be classified as foliated and unfoliated. Foliation, or layering, occurs when rock is compressed along one axis during recrystallization. This can be seen in schist and shale. Unfoliated rock does not include this banding. Rocks that are compressed equally from all sides or lack specific minerals will be unfoliated. An example is marble.

Fossils are preservations of plants, animals, their remains, or their traces that date back to about 10,000 years ago. Fossils and where they are found in rock strata makes up the fossil record. Fossils are formed under a very specific set of conditions. The fossil must not be damaged by predators and scavengers after death, and the fossil must not decompose. Usually, this happens when the organism is quickly covered with sediment. This sediment builds up and molecules in the organism's body are replaced by minerals. Fossils come in an array of sizes, from single-celled organisms to large dinosaurs.

Plate Tectonics

The Earth is ellipsoid, not perfectly spherical. This means the diameter is different through the poles and at the equator. Through the poles, the Earth is about 12,715 km in diameter. The approximate center of the Earth is at a depth of 6,378 km. The Earth is divided into a crust, mantle, and core. The core consists of a solid inner portion. Moving outward, the molten outer core occupies the space from about a depth of 5,150 km to a depth of 2,890 km. The mantle consists of a lower and upper layer. The lower layer includes the D' (D prime) and D" (D double-prime) layers. The solid portion of the upper mantle and crust together form the lithosphere, or rocky sphere. Below this, but still within the mantle, is the asthenosphere, or weak sphere. These layers are distinguishable because the lithosphere is relatively rigid, while the asthenosphere resembles a thick liquid.

The theory of plate tectonics states that the lithosphere, the solid portion of the mantle and Earth's crust, consists of major and minor plates. These plates are on top of and move with the viscous upper mantle, which is heated because of the convection cycle that occurs in the interior of the Earth. There are different estimates as to the exact number of major and minor plates. The number of major plates is believed to be between 9 and 15, and it is thought that there may be as many as 40 minor plates. The United States is atop the North American plate. The Pacific Ocean is atop the Pacific plate. The point at which these two plates slide horizontally along the San Andreas fault is an example of a transform plate boundary. The other two types of boundaries are divergent (plates that are spreading apart and forming new crust) and convergent (the process of subduction causes one plate to go under another). The movement of plates is what causes other features of the Earth's crust, such as mountains, volcanoes, and earthquakes.

Volcanoes can occur along any type of tectonic plate boundary. At a divergent boundary, as plates move apart, magma rises to the surface, cools, and forms a ridge. An example of this is the mid-Atlantic ridge. Convergent boundaries, where one plate slides under another, are often areas with a lot of volcanic activity. The subduction process creates magma. When it rises to the surface, volcanoes can be created. Volcanoes can also be created in the middle of a plate over hot spots. Hot spots are locations where narrow plumes of magma rise through the mantle in a fixed place over a long period of time. The Hawaiian Islands and Midway are examples. The plate shifts and the island moves. Magma continues to rise through the mantle, however, which produces another island. Volcanoes can be active, dormant, or extinct. Active volcanoes are those that are erupting or about to erupt. Dormant volcanoes are those that might erupt in the future and still have internal volcanic activity. Extinct volcanoes are those that will not erupt.

Geography

For the purposes of tracking time and location, the Earth is divided into sections with imaginary lines. Lines that run vertically around the globe through the poles are lines of longitude, sometimes called meridians. The Prime Meridian is the longitudinal reference point of 0. Longitude is measured in 15-degree increments toward the east or west. Degrees are further divided into 60 minutes, and each minute is divided into 60 seconds. Lines of latitude run horizontally around the Earth parallel to the equator, which is the 0 reference point and the widest point of the Earth. Latitude is the distance north or south from the equator, and is also measured in degrees, minutes, and seconds.

Tropic of Cancer: This is located at 23.5 degrees north. The Sun is directly overhead at noon on June 21st in the Tropic of Cancer, which marks the beginning of summer in the Northern Hemisphere.
Tropic of Capricorn: This is located at 23.5 degrees south. The Sun is directly overhead at noon on December 21st in the Tropic of Capricorn, which marks the beginning of winter in the Northern Hemisphere.
Arctic Circle: This is located at 66.5 degrees north, and marks the start of when the Sun is not visible above the horizon. This occurs on December 21st, the same day the Sun is directly over the Tropic of Capricorn.
Antarctic Circle: This is located at 66.5 degrees south, and marks the start of when the Sun is not visible above the horizon. This occurs on June 21st, which marks the beginning of winter in the Southern Hemisphere and is when the Sun is directly over the Tropic of Cancer.

Latitude is a measurement of the distance from the equator. The distance from the equator indicates how much solar radiation a particular area receives. The equator receives more sunlight, while polar areas receive less. The Earth tilts slightly on its rotational axis. This tilt determines the seasons and affects weather. There are eight biomes or ecosystems with particular climates that are associated with latitude. Those in the high latitudes, which get the least sunlight, are tundra and taiga. Those in the mid latitudes are grassland, temperate forest, and chaparral. Those in latitudes closest to the equator are the warmest. The biomes are desert and tropical rain forest. The eighth biome is the ocean, which is unique because it consists of water and spans the entire globe. Insolation refers to incoming solar radiation. Diurnal variations refer to the daily changes in insolation. The greatest insolation occurs at noon.

The tilt of the Earth on its axis is 23.5°. This tilt causes the seasons and affects the temperature because it affects the amount of Sun the area receives. When the Northern or Southern Hemispheres are tilted toward the Sun, the hemisphere tilted toward the sun experiences summer and the other hemisphere experiences winter. This reverses as the Earth revolves around the Sun. Fall and spring occur between the two extremes.

The equator gets the same amount of sunlight every day of the year, about 12 hours, and doesn't experience seasons. Both poles have days during the winter when they are tilted away from the Sun and receive no daylight. The opposite effect occurs during the summer. There are 24 hours of daylight and no night. The summer solstice, the day with the most amount of sunlight, occurs on June 21st in the Northern Hemisphere and on December 21st in the Southern Hemisphere. The winter solstice, the day with the least amount of sunlight, occurs on December 21st in the Northern Hemisphere and on June 21st in the Southern Hemisphere.

Weather, Atmosphere, Water Cycle

Meteorology is the study of the atmosphere, particularly as it pertains to forecasting the weather and understanding its processes. Weather is the condition of the atmosphere at any given moment. Most weather occurs in the troposphere. Weather includes changing events such as clouds, storms, and temperature, as well as more extreme events such as tornadoes, hurricanes, and blizzards.

Climate refers to the average weather for a particular area over time, typically at least 30 years. Latitude is an indicator of climate. Changes in climate occur over long time periods.

The hydrologic, or water, cycle refers to water movement on, above, and in the Earth. Water can be in any one of its three states during different phases of the cycle. The three states of water are liquid water, frozen ice, and water vapor. Processes involved in the hydrologic cycle include precipitation, canopy interception, snow melt, runoff, infiltration, subsurface flow, evaporation, sublimation, advection, condensation, and transpiration. Precipitation is when condensed water vapor falls to Earth.

Examples include rain, fog drip, and various forms of snow, hail, and sleet. Canopy interception is when precipitation lands on plant foliage instead of falling to the ground and evaporating. Snow melt is runoff produced by melting snow. Infiltration occurs when water flows from the surface into the ground. Subsurface flow refers to water that flows underground. Evaporation is when water in a liquid state changes to a gas. Sublimation is when water in a solid state (such as snow or ice) changes to water vapor without going through a liquid phase. Advection is the movement of water through the atmosphere. Condensation is when water vapor changes to liquid water. Transpiration is when water vapor is released from plants into the air.

The ocean is the salty body of water that encompasses the Earth. It has a mass of 1.4×10^{24} grams. Geographically, the ocean is divided into three large oceans: the Pacific Ocean, the Atlantic Ocean, and the Indian Ocean. There are also other divisions, such as gulfs, bays, and various types of seas, including Mediterranean and marginal seas. Ocean distances can be measured by latitude, longitude, degrees, meters, miles, and nautical miles. The ocean accounts for 70.8% of the surface of the Earth, amounting to 361,254,000 km². The ocean's depth is greatest at Challenger Deep in the Mariana Trench. The ocean floor here is 10,924 meters below sea level. The depths of the ocean are mapped by echo sounders and satellite altimeter systems. Echo sounders emit a sound pulse from the surface and record the time it takes to return. Satellite altimeters provide better maps of the ocean floor.

The atmosphere consists of 78% nitrogen, 21% oxygen, and 1% argon. It also includes traces of water vapor, carbon dioxide and other gases, dust particles, and chemicals from Earth. The atmosphere becomes thinner the farther it is from the Earth's surface. It becomes difficult to breathe at about 3 km above sea level. The atmosphere gradually fades into space. The lowest layer of the atmosphere is called the troposphere. Its thickness varies at the poles and the equator, varying from about 7 to 17 km. This is where most weather occurs. The stratosphere is next, and continues to an elevation of about 51 km. The mesosphere extends from the stratosphere to an elevation of about 81 km. It is the coldest layer and is where meteors tend to ablate. The next layer is the thermosphere. It is where the International Space Station orbits. The exosphere is the outermost layer, extends to 10,000 km, and mainly consists of hydrogen and helium.

Earth's atmosphere has five main layers. From lowest to highest, these are the troposphere, the stratosphere, the mesosphere, the thermosphere, and the exosphere. Between each pair of layers is a transition layer called a pause. The troposphere includes the tropopause, which is the transitional layer of the stratosphere. Energy from Earth's surface is transferred to the troposphere. Temperature decreases with altitude in this layer. In the stratosphere, the temperature is inverted, meaning that it increases with altitude. The stratosphere includes the ozone layer, which helps block ultraviolet light from the Sun. The stratopause is the transitional layer to the mesosphere. The temperature of the mesosphere decreases with height. It is considered the coldest place on Earth, and has an average temperature of -85 degrees Celsius. Temperature increases with altitude in the thermosphere, which includes the thermopause. Just past the thermosphere is the exobase, the base layer of the exosphere. Beyond the five main layers are the ionosphere, homosphere, heterosphere, and magnetosphere.

Most clouds can be classified according to the altitude of their base above Earth's surface. High clouds occur at altitudes between 5,000 and 13,000 meters. Middle clouds occur at altitudes between 2,000 and 7,000 meters. Low clouds occur from the Earth's surface to altitudes of 2,000 meters. Types of high clouds include cirrus (Ci), thin wispy mare's tails that consist of ice; cirrocumulus (Cc), small, pillow-like puffs that often appear in rows; and cirrostratus (Cs), thin, sheetlike clouds that often cover the entire sky. Types of middle clouds include altocumulus (Ac), gray-white clouds that consist of liquid water; and altostratus (As), grayish or blue-gray clouds that span the sky. Types of low clouds include stratus (St), gray and fog-like clouds consisting of water droplets that take up the whole sky; stratocumulus (Sc), low-lying, lumpy gray clouds; and nimbostratus (Ns), dark gray clouds with uneven bases that indicate rain or snow. Two types of clouds, cumulus (Cu) and cumulonimbus (Cb), are capable of great vertical growth. They can start at a wide range of altitudes, from the Earth's surface to altitudes of 13,000 meters.

Astronomy

Astronomy is the scientific study of celestial objects and their positions, movements, and structures. Celestial does not refer to the Earth in particular, but does include its motions as it moves through space. Other objects include the Sun, the Moon, planets, satellites, asteroids, meteors, comets, stars, galaxies, the universe, and other space phenomena. The term astronomy has its roots in the Greek words "astro" and "nomos," which means "laws of the stars."

What can be seen of the universe is believed to be at least 93 billion light years across. To put this into perspective, the Milky Way galaxy is about 100,000 light years across. Our view of matter in the universe is that it forms into clumps. Matter is organized into stars, galaxies, clusters of galaxies, superclusters, and the Great Wall of galaxies. Galaxies consist of stars, some with planetary systems. Some estimates state that the universe is about 13 billion years old. It is not considered dense, and is believed to consist of 73 percent dark energy, 23 percent cold dark matter, and 4 percent regular matter. Cosmology is the study of the universe. Interstellar medium (ISM) is the gas and dust in the interstellar space between a galaxy's stars.

The solar system is a planetary system of objects that exist in an ecliptic plane. Objects orbit around and are bound by gravity to a star called the Sun. Objects that orbit around the Sun include: planets, dwarf planets, moons, asteroids, meteoroids, cosmic dust, and comets. The definition of planets has changed. At one time, there were nine planets in the solar system. There are now eight. Planetary objects in the solar system include four inner, terrestrial planets: Mercury, Venus, Earth, and Mars. They are relatively small, dense, rocky, lack rings, and have few or no moons. The four outer, or Jovian, planets are Jupiter, Saturn, Uranus, and Neptune, which are large and have low densities, rings, and moons. They are also known as gas giants. Between the inner and outer planets is the asteroid belt. Beyond Neptune is the Kuiper belt. Within these belts are five dwarf planets: Ceres, Pluto, Haumea, Makemake, and Eris.

The Sun is at the center of the solar system. It is composed of 70% hydrogen (H) and 28% helium (He). The remaining 2% is made up of metals. The Sun is one of 100 billion stars in the Milky Way galaxy. Its diameter is 1,390,000 km, its mass is 1.989×10^{30} kg, its surface temperature is 5,800 K, and its core temperature is 15,600,000 K. The Sun represents more than 99.8% of the total mass of the solar system. At the core, the temperature is 15.6 million K, the pressure is 250 billion atmospheres, and the density is more than 150 times that of water. The surface is called the photosphere. The chromosphere lies above this, and the corona, which extends millions of kilometers into space, is next. Sunspots are relatively cool regions on the surface with a temperature of 3,800 K. Temperatures in the corona are over 1,000,000 K. Its magnetosphere, or heliosphere, extends far beyond Pluto.

Mercury: Mercury is the closest to the Sun and is also the smallest planet. It orbits the Sun every 88 days, has no satellites or atmosphere, has a Moon-like surface with craters, appears bright, and is dense and rocky with a large iron core.

Venus: Venus is the second planet from the Sun. It orbits the Sun every 225 days, is very bright, and is similar to Earth in size, gravity, and bulk composition. It has a dense atmosphere composed of carbon dioxide and some sulfur. It is covered with reflective clouds made of sulfuric acid and exhibits signs of volcanism. Lightning and thunder have been recorded on Venus's surface.

Earth: Earth is the third planet from the Sun. It orbits the Sun every 365 days. Approximately 71% of its surface is salt-water oceans. The Earth is rocky, has an atmosphere composed mainly of oxygen and nitrogen, has one moon, and supports millions of species. It contains the only known life in the solar system.

Mars: Mars it the fourth planet from the Sun. It appears reddish due to iron oxide on the surface, has a thin atmosphere, has a rotational period similar to Earth's, and has seasonal cycles. Surface features of Mars include volcanoes, valleys, deserts, and polar ice caps. Mars has impact craters and the tallest mountain, largest canyon, and perhaps the largest impact crater yet discovered.

Jupiter: Jupiter is the fifth planet from the Sun and the largest planet in the solar system. It consists mainly of hydrogen, and 25% of its mass is made up of helium. It has a fast rotation and has clouds in the tropopause composed of ammonia crystals that are arranged into bands sub-divided into lighter-hued zones and darker belts causing storms and turbulence. Jupiter has wind speeds of 100 m/s, a planetary ring, 63 moons, and a Great Red Spot, which is an anticyclonic storm.

Saturn: Saturn is the sixth planet from the Sun and the second largest planet in the solar system. It is composed of hydrogen, some helium, and trace elements. Saturn has a small core of rock and ice, a thick layer of metallic hydrogen, a gaseous outer layer, wind speeds of up to 1,800 km/h, a system of rings, and 61 moons.

Uranus: Uranus is the seventh planet from the Sun. Its atmosphere is composed mainly of hydrogen and helium, and also contains water, ammonia, methane, and traces of hydrocarbons. With a minimum temperature of 49 K, Uranus has the coldest atmosphere. Uranus has a ring system, a magnetosphere, and 13 moons.

Neptune: Neptune is the eighth planet from the Sun and is the planet with the third largest mass. It has 12 moons, an atmosphere similar to Uranus, a Great Dark Spot, and the strongest sustained winds of any planet (wind speeds can be as high as 2,100 km/h). Neptune is cold (about 55 K) and has a fragmented ring system.

The Earth is about 12,765 km (7,934 miles) in diameter. The Moon is about 3,476 km (2,160 mi) in diameter. The distance between the Earth and the Moon is about 384,401 km (238,910 mi). The diameter of the Sun is approximately 1,390,000 km (866,000 mi). The distance from the Earth to the Sun is 149,598,000 km, also known as 1 Astronomical Unit (AU). The star that is nearest to the solar system is Proxima Centauri. It is about 270,000 AU away. Some distant galaxies are so far away that their light takes several billion years to reach the Earth. In other words, people on Earth see them as they looked billions of years ago.

It takes about one month for the Moon to go through all its phases. Waxing refers to the two weeks during which the Moon goes from a new moon to a full moon. About two weeks is spent waning, going from a full moon to a new moon. The lit part of the Moon always faces the Sun. The phases of waxing are: new moon, during which the Moon is not illuminated and rises and sets with the Sun; crescent moon, during which a tiny sliver is lit; first quarter, during which half the Moon is lit and the phase of the Moon is due south on the meridian; gibbous, during which more than half of the Moon is lit and has a shape similar to a football; right side, during which the Moon is lit; and full moon, during which the Moon is fully illuminated, rises at sunset, and sets at sunrise. After a full moon, the Moon is waning. The phases of waning are: gibbous, during which the left side is lit and the Moon rises after sunset and sets after sunrise; third quarter, during which the Moon is half lit and rises at midnight and sets at noon; crescent, during which a tiny sliver is lit; and new moon, during which the Moon is not illuminated and rises and sets with the Sun.

Biology

Cells

The main difference between eukaryotic and prokaryotic cells is that eukaryotic cells have a nucleus and prokaryotic cells do not. Eukaryotic cells are considered more complex, while prokaryotic cells are smaller and simpler. Eukaryotic cells have membrane-bound organelles that perform various functions and contribute to the complexity of these types of cells. Prokaryotic cells do not contain membrane-bound organelles. In prokaryotic cells, the genetic material (DNA) is not contained within a membrane-bound nucleus. Instead, it aggregates in the cytoplasm in a nucleoid. In eukaryotic cells, DNA is mostly contained in chromosomes in the nucleus, although there is some DNA in mitochondria and chloroplasts. Prokaryotic cells usually divide by binary fission and are haploid. Eukaryotic cells divide by mitosis and are diploid. Prokaryotic structures include plasmids, ribosomes, cytoplasm, a cytoskeleton, granules of nutritional substances, a plasma membrane, flagella, and a few others. They are single-celled organisms. Bacteria are prokaryotic cells.

The functions of plant and animal cells vary greatly, and the functions of different cells within a single organism can also be vastly different. Animal and plant cells are similar in structure in that they are eukaryotic, which means they contain a nucleus. The nucleus is a round structure that controls the activities of the cell and contains chromosomes. Both types of cells have cell membranes, cytoplasm, vacuoles, and other structures.

The main difference between the two is that plant cells have a cell wall made of cellulose that can handle high levels of pressure within the cell, which can occur when liquid enters a plant cell. Plant cells have chloroplasts that are used during the process of photosynthesis, which is the conversion of sunlight into food. Plant cells usually have one large vacuole, whereas animal cells can have many smaller ones. Plant cells have a regular shape, while the shapes of animal cell can vary.

Plant cells can be much larger than animal cells, ranging from 10 to 100 micrometers. Animal cells are 10 to 30 micrometers in size. Plant cells can have much larger vacuoles that occupy a large portion of the cell. They also have cell walls, which are thick barriers consisting of protein and sugars. Animal cells lack cell walls. Chloroplasts in plants that perform photosynthesis absorb sunlight and convert it into energy. Mitochondria produce energy from food in animal cells.

Plant and animal cells are both eukaryotic, meaning they contain a nucleus. Both plant and animal cells duplicate genetic material, separate it, and then divide in half to reproduce. Plant cells build a cell plate between the two new cells, while animal cells make a cleavage furrow and pinch in half. Microtubules are components of the cytoskeleton in both plant and animal cells. Microtubule organizing centers (MTOCs) make microtubules in plant cells, while centrioles make microtubules in animal cells.

Photosynthesis is the conversion of sunlight into energy in plant cells, and also occurs in some types of bacteria and protists. Carbon dioxide and water are converted into glucose during photosynthesis, and light is required during this process. Cyanobacteria are thought to be the descendants of the first organisms to use photosynthesis about 3.5 billion years ago. Photosynthesis is a form of cellular respiration. It occurs in chloroplasts that use thylakoids, which are structures in the membrane that contain light reaction chemicals. Chlorophyll is a pigment that absorbs light. During the process, water is used and oxygen is released. The equation for the chemical reaction that occurs during photosynthesis is $6H_2O + 6CO_2 \rightarrow C_6H_{12}O_6 + 6O_2$. During photosynthesis, six molecules of water and six molecules of carbon dioxide react to form one molecule of sugar and six molecules of oxygen.

The term cell cycle refers to the process by which a cell reproduces, which involves cell growth, the duplication of genetic material, and cell division. Complex organisms with many cells use the cell cycle to replace cells as they lose their functionality and wear out. The entire cell cycle in animal cells can take 24 hours. The time required varies among different cell types. Human skin cells, for example, are constantly reproducing. Some other cells only divide infrequently. Once neurons are mature, they do not grow or divide. The two ways that cells can reproduce are through meiosis and mitosis. When cells replicate through mitosis, the "daughter cell" is an exact replica of the parent cell. When cells divide through meiosis, the daughter cells have different genetic coding than the parent cell. Meiosis only happens in specialized reproductive cells called gametes.

Mitosis is the process of cell reproduction in which a eukaryotic cell splits into two separate, but completely identical, cells. This process is divided into a number of different phases.

Interphase: The cell prepares for division by replicating its genetic and cytoplasmic material. Interphase can be further divided into G1, S, and G2.
Prophase: The chromatin thickens into chromosomes and the nuclear membrane begins to disintegrate. Pairs of centrioles move to opposite sides of the cell and spindle fibers begin to form. The mitotic spindle, formed from cytoskeleton parts, moves chromosomes around within the cell.
Metaphase: The spindle moves to the center of the cell and chromosome pairs align along the center of the spindle structure.
Anaphase: The pairs of chromosomes, called sisters, begin to pull apart, and may bend. When they are separated, they are called daughter chromosomes. Grooves appear in the cell membrane.
Telophase: The spindle disintegrates, the nuclear membranes reform, and the chromosomes revert to chromatin. In animal cells, the membrane is pinched. In plant cells, a new cell wall begins to form.
Cytokinesis: This is the physical splitting of the cell (including the cytoplasm) into two cells. Some believe this occurs following telophase. Others say it occurs from anaphase, as the cell begins to furrow, through telophase, when the cell actually splits into two.

Meiosis is another process by which eukaryotic cells reproduce. However, meiosis is used by more complex life forms such as plants and animals and results in four unique cells rather than two identical cells as in mitosis. Meiosis has the same phases as mitosis, but they happen twice. In addition, different events occur during some phases of meiosis than mitosis. The events that occur during the first phase of meiosis are interphase (I), prophase (I), metaphase (I), anaphase (I), telophase (I), and cytokinesis (I). During this first phase of meiosis, chromosomes cross over, genetic material is exchanged, and tetrads of four chromatids are formed. The nuclear membrane dissolves. Homologous pairs of chromatids are separated and travel to different poles. At this point, there has been one cell division resulting in two cells. Each cell goes through a second cell division, which consists of prophase (II), metaphase (II), anaphase (II), telophase (II), and cytokinesis (II). The result is four daughter cells with different sets of chromosomes. The daughter cells are haploid, which means they contain half the genetic material of the parent cell. The second phase of meiosis is similar to the process of mitosis. Meiosis encourages genetic diversity.

Genetics

Chromosomes consist of genes, which are single units of genetic information. Genes are made up of deoxyribonucleic acid (DNA). DNA is a nucleic acid located in the cell nucleus. There is also DNA in the mitochondria. DNA replicates to pass on genetic information. The DNA in almost all cells is the same. It is also involved in the biosynthesis of proteins. The model or structure of DNA is described as a double helix. A helix is a curve, and a double helix is two congruent curves connected by horizontal members. The model can be likened to a spiral staircase. It is right-handed. The British scientist Rosalind Elsie Franklin is credited with taking the x-ray diffraction image in 1952 that was used by Francis Crick and James Watson to formulate the double-helix model of DNA and speculate about its important role in carrying and transferring genetic information.

DNA has a double helix shape, resembles a twisted ladder, and is compact. It consists of nucleotides. Nucleotides consist of a five-carbon sugar (pentose), a phosphate group, and a nitrogenous base. Two bases pair up to form the rungs of the ladder. The "side rails" or backbone consists of the covalently bonded sugar and phosphate. The bases are attached to each other with hydrogen bonds, which are easily dismantled so replication can occur. Each base is attached to a phosphate and to a sugar. There are four types of nitrogenous bases: adenine (A), guanine (G), cytosine (C), and thymine (T). There are about 3 billion bases in human DNA. The bases are mostly the same in everybody, but their order is different. It is the order of these bases that creates diversity in people. Adenine (A) pairs with thymine (T), and cytosine (C) pairs with guanine (G).

A gene is a portion of DNA that identifies how traits are expressed and passed on in an organism. A gene is part of the genetic code. Collectively, all genes form the genotype of an individual. The genotype includes genes that may not be expressed, such as recessive genes. The phenotype is the physical, visual manifestation of genes. It is determined by the basic genetic information and how genes have been affected by their environment. An allele is a variation of a gene. Also known as a trait, it determines the manifestation of a gene. This manifestation results in a specific physical appearance of some facet of an organism, such as eye color or height. For example the genetic information for eye color is a gene. The gene variations responsible for blue, green, brown, or black eyes are called alleles. Locus (pl. loci) refers to the location of a gene or alleles.

Mendel's laws are the law of segregation (the first law) and the law of independent assortment (the second law). The law of segregation states that there are two alleles and that half of the total number of alleles are contributed by each parent organism. The law of independent assortment states that traits are passed on randomly and are not influenced by other traits. The exception to this is linked traits. A Punnett square can illustrate how alleles combine from the contributing genes to form various phenotypes. One set of a parent's genes are put in columns, while the genes from the other parent are placed in rows. The allele combinations are shown in each cell. When two different alleles are present in a pair, the dominant one is expressed. A Punnett square can be used to predict the outcome of crosses.

Gene traits are represented in pairs with an upper case letter for the dominant trait (A) and a lower case letter for the recessive trait (a). Genes occur in pairs (AA, Aa, or aa). There is one gene on each chromosome half supplied by each parent organism. Since half the genetic material is from each parent, the offspring's traits are represented as a combination of these. A dominant trait only requires one gene of a gene pair for it to be expressed in a phenotype, whereas a recessive requires both genes in order to be manifested. For example, if the mother's genotype is Dd and the father's is dd, the possible combinations are Dd and dd. The dominant trait will be manifested if the genotype is DD or Dd. The recessive trait will be manifested if the genotype is dd. Both DD and dd are homozygous pairs. Dd is heterozygous.

Evolution

Scientific evidence supporting the theory of evolution can be found in biogeography, comparative anatomy and embryology, the fossil record, and molecular evidence. Biogeography studies the geographical distribution of animals and plants. Evidence of evolution related to the area of biogeography includes species that are well suited for extreme environments. The fossil record shows that species lived only for a short time period before becoming extinct. The fossil record can also show the succession of plants and animals. Living fossils are existing species that have not changed much morphologically and are very similar to ancient examples in the fossil record. Examples include the horseshoe crab and ginko. Comparative embryology studies how species are similar in the embryonic stage, but become increasingly specialized and diverse as they age. Vestigial organs are those that still exist, but become nonfunctional. Examples include the hind limbs of whales and the wings of birds that can no longer fly, such as ostriches.

The rate of evolution is affected by the variability of a population. Variability increases the likelihood of evolution. Variability in a population can be increased by mutations, immigration, sexual reproduction (as opposed to asexual reproduction), and size. Natural selection, emigration, and smaller populations can lead to decreased variability. Sexual selection affects evolution. If fewer genes are available, it will limit the number of genes passed on to subsequent generations. Some animal mating behaviors are not as successful as others. A male that does not attract a female because of a weak mating call or dull feathers, for example, will not pass on its genes. Mechanical isolation, which refers to sex organs that do not fit together very well, can also decrease successful mating.

Natural selection: This theory developed by Darwin states that traits that help give a species a survival advantage are passed on to subsequent generations. Members of a species that do not have the advantageous trait die before they reproduce. Darwin's four principles are: from generation to generation, there are various individuals within a species; genes determine variations; more individuals are born than survive to maturation; and specific genes enable an organism to better survive.
Gradualism: This can be contrasted with punctuationism. It is an idea that evolution proceeds at a steady pace and does not include sudden developments of new species or features from one generation to the next.
Punctuated Equilibrium: This can be contrasted with gradualism. It is the idea in evolutionary biology that states that evolution involves long time periods of no change (stasis) accompanied by relatively brief periods (hundreds of thousands of years) of rapid change.

Three types of evolution are divergent, convergent, and parallel. Divergent evolution refers to two species that become different over time. This can be caused by one of the species adapting to a different environment. Convergent evolution refers to two species that start out fairly different, but evolve to share many similar traits. Parallel evolution refers to species that are not similar and do not become more or less similar over time. Mechanisms of evolution include descent (the passing on of genetic information), mutation, migration, natural selection, and genetic variation and drift. The biological definition of species refers to a group of individuals that can mate and reproduce. Speciation refers to the evolution of a new biological species. The biological species concept (BSC) basically states that a species is a community of individuals that can reproduce and have a niche in nature.

One theory of how life originated on Earth is that life developed from nonliving materials. The first stage of this transformation happened when abiotic (nonliving) synthesis took place, which is the formation of monomers like amino acids and nucleotides. Next, monomers joined together to create polymers such as proteins and nucleic acids. These polymers are then believed to have formed into protobionts. The last stage was the development of the process of heredity. Supporters of this theory believe that RNA was the first genetic material. Another theory postulates that hereditary systems came about before the origination of nucleic acids. Another theory is that life, or the precursors for it, were transported to Earth from a meteorite or other object from space. There is no real evidence to support this theory.

A number of scientists have made significant contributions to the theory of evolution:
Cuvier (1744-1829): Cuvier was a French naturalist who used the fossil record (paleontology) to compare the anatomies of extinct species and existing species to make conclusions about extinction. He believed in the catastrophism theory more strongly than the theory of evolution.
Lamarck (1769-1832): Lamarck was a French naturalist who believed in the idea of evolution and thought it was a natural occurrence influenced by the environment. He studied medicine and botany. Lamarck put forth a theory of evolution by inheritance of acquired characteristics. He theorized that organisms became more complex by moving up a ladder of progress.
Lyell (1797-1875): Lyell was a British geologist who believed in geographical uniformitarianism, which can be contrasted with catastrophism.
Charles Robert Darwin (1809-1882): Darwin was an English naturalist known for his belief that evolution occurred by natural selection. He believed that species descend from common ancestors.
Alfred Russell Wallace (1823-1913): He was a British naturalist who independently developed a theory of evolution by natural selection. He believed in the transmutation of species (that one species develops into another).

Organism Classification

The groupings in the five kingdom classification system are kingdom, phylum/division, class, order, family, genus, and species. A memory aid for this is: King Phillip Came Over For Good Soup. The five kingdoms are Monera, Protista, Fungi, Plantae, and Animalia. The kingdom is the top level classification in this system. Below that are the following groupings: phylum, class, order, family, genus, and species. The Monera kingdom includes about 10,000 known species of prokaryotes, such as bacteria and cyanobacteria. Members of this kingdom can be unicellular organisms or colonies. The next four kingdoms consist of eukaryotes. The Protista kingdom includes about 250,000 species of unicellular protozoans and unicellular and multicellular algae. The Fungi kingdom includes about 100,000 species. A recently introduced system of classification includes a three domain grouping above kingdom. The domain groupings are Archaea, Bacteria (which both consist of prokaryotes), and Eukarya, which include eukaryotes. According to the five kingdom classification system, humans are: kingdom Animalia, phylum Chordata, subphylum Vertebrata, class Mammalia, order Primate, family Hominidae, genus Homo, and species Sapiens.

An organism is a living thing. A unicellular organism is an organism that has only one cell. Examples of unicellular organisms are bacteria and paramecium. A multicellular organism is one that consists of many cells. Humans are a good example. By some estimates, the human body is made up of billions of cells. Others think the human body has more than 75 trillion cells. The term microbe refers to small organisms that are only visible through a microscope. Examples include viruses, bacteria, fungi, and protozoa. Microbes are also referred to as microorganisms, and it is these that are studied by microbiologists. Bacteria can be rod shaped, round (cocci), or spiral (spirilla). These shapes are used to differentiate among types of bacteria. Bacteria can be identified by staining them. This particular type of stain is called a gram stain. If bacteria are gram-positive, they absorb the stain and become purple. If bacteria are gram-negative, they do not absorb the stain and become a pinkish color.

Organisms in the Protista kingdom are classified according to their methods of locomotion, their methods of reproduction, and how they get their nutrients. Protists can move by the use of a flagellum, cilia, or pseudopod. Flagellates have flagellum, which are long tails or whip-like structures that are rotated to help the protist move. Ciliates use cilia, which are smaller hair-like structures on the exterior of a cell that wiggle to help move the surrounding matter. Amoeboids use pseudopodia to move. Bacteria reproduce either sexually or asexually. Binary fission is a form of asexual reproduction whereby bacteria divide in half to produce two new organisms that are clones of the parent. In sexual reproduction, genetic material is exchanged. When kingdom members are categorized according to how they obtain nutrients, the three types of protists are photosynthetic, consumers, and saprophytes. Photosynthetic protists convert sunlight into energy. Organisms that use photosynthesis are considered producers. Consumers, also known as heterotrophs, eat or consume other organisms. Saprophytes consume dead or decaying substances.

Mycology is the study of fungi. The Fungi kingdom includes about 100,000 species. They are further delineated as mushrooms, yeasts, molds, rusts, mildews, stinkhorns, puffballs, and truffles. Fungi are characterized by cell walls that have chitin, a long chain polymer carbohydrate. Fungi are different from species in the Plant kingdom, which have cell walls consisting of cellulose. Fungi are thought to have evolved from a single ancestor. Although they are often thought of as a type of plant, they are more similar to animals than plants. Fungi are typically small and numerous, and have a diverse morphology among species. They can have bright red cups and be orange jellylike masses, and their shapes can resemble golf balls, bird nests with eggs, starfish, parasols, and male genitalia. Some members of the stinkhorn family emit odors similar to dog scat to attract flies that help transport spores that are involved in reproduction. Fungi of this family are also consumed by humans.

Chlorophyta are green algae. Bryophyta are nonvascular mosses and liverworts. They have root-like parts called rhizoids. Since they do not have the vascular structures to transport water, they live in moist environments. Lycophyta are club mosses. They are vascular plants. They use spores and need water to reproduce. Equisetopsida (sphenophyta) are horsetails. Like lycophyta, they need water to reproduce with spores. They have rhizoids and needle-like leaves. The pteridophytes (filicopsida) are ferns. They have stems (rhizomes). Spermatopsida are the seed plants. Gymnosperms are a conifer, which means they have cones with seeds that are used in reproduction. Plants with seeds require less water. Cycadophyta are cone-bearing and look like palms. Gnetophyta are plants that live in the desert. Coniferophyta are pine trees, and have both cones and needles. Ginkgophyta are gingkos. Anthophyta is the division with the largest number of plant species, and includes flowering plants with true seeds.

Only plants in the division bryophyta (mosses and liverworts) are nonvascular, which means they do not have xylem to transport water. All of the plants in the remaining divisions are vascular, meaning they have true roots, stems, leaves, and xylem. Pteridophytes are plants that use spores and not seeds to reproduce. They include the following divisions: Psilophyta (whisk fern), Lycophyta (club mosses), Sphenophyta (horsetails), and Pterophyta (ferns). Spermatophytes are plants that use seeds to reproduce. Included in this category are gymnosperms, which are flowerless plants that use naked seeds, and angiosperms, which are flowering plants that contain seeds in or on a fruit. Gymnosperms include the following divisions: cycadophyta (cycads), ginkgophyta (maidenhair tree), gnetophyta (ephedra and welwitschia), and coniferophyta (which includes pinophyta conifers). Angiosperms comprise the division anthophyta (flowering plants).

Plants are autotrophs, which mean they make their own food. In a sense, they are self sufficient. Three major processes used by plants are photosynthesis, transpiration, and respiration. Photosynthesis involves using sunlight to make food for plants. Transpiration evaporates water out of plants. Respiration is the utilization of food that was produced during photosynthesis.

Two major systems in plants are the shoot and the root system. The shoot system includes leaves, buds, and stems. It also includes the flowers and fruits in flowering plants. The shoot system is located above the ground. The root system is the component of the plant that is underground, and includes roots, tubers, and rhizomes. Meristems form plant cells by mitosis. Cells then differentiate into cell types to form the three types of plant tissues, which are dermal, ground, and vascular. Dermal refers to tissues that form the covering or outer layer of a plant. Ground tissues consist of parenchyma, collenchyma, and/or sclerenchyma cells.

There are at least 230,000 species of flowering plants. They represent about 90 percent of all plants. Angiosperms have a sexual reproduction phase that includes flowering. When growing plants, one may think they develop in the following order: seeds, growth, flowers, and fruit. The reproductive cycle has the following order: flowers, fruit, and seeds. In other words, seeds are the products of successful reproduction. The colors and scents of flowers serve to attract pollinators. Flowers and other plants can also be pollinated by wind. When a pollen grain meets the ovule and is successfully fertilized, the ovule develops into a seed. A seed consists of three parts: the embryo, the endosperm, and a seed coat. The embryo is a small plant that has started to develop, but this development is paused. Germination is when the embryo starts to grow again. The endosperm consists of proteins, carbohydrates, or fats. It typically serves as a food source for the embryo. The seed coat provides protection from disease, insects, and water.

The animal kingdom is comprised of more than one million species in about 30 divisions (the plant kingdom uses the term phyla). There about 800,000 species of insects alone, representing half of all animal species. The characteristics that distinguish members of the animal kingdom from members of other kingdoms are that they are multicellular, are heterotrophic, reproduce sexually (there are some exceptions), have cells that do not contain cell walls or photosynthetic pigments, can move at some stage of life, and can rapidly respond to the environment as a result of specialized tissues like nerve and muscle. Heterotrophic refers to the method of getting energy by eating food that has energy releasing substances. Plants, on the other hand, are autotrophs, which mean they make their own energy. During reproduction, animals have a diploid embryo in the blastula stage. This structure is unique to animals. The blastula resembles a fluid-filled ball.

The animal kingdom includes about one million species. Metazoans are multicellular animals. Food is ingested and enters a mesoderm-lined coelom (body cavity). Phylum porifera and coelenterate are exceptions. The taxonomy of animals involves grouping them into phyla according to body symmetry and plan, as well as the presence of or lack of segmentation. The more complex phyla that have a coelom and a digestive system are further classified as protostomes or deuterostomes according to blastula development. In protostomes, the blastula's blastopore (opening) forms a mouth. In deuterostomes, the blastopore forms an anus. Taxonomy schemes vary, but there are about 36 phyla of animals. The corresponding term for plants at this level is division. The most notable phyla include chordata, mollusca, porifera, cnidaria, platyhelminthes, nematoda, annelida, arthropoda, and echinodermata, which account for about 96 percent of all animal species.

These four animal phyla lack a coelom or have a pseudocoelom.
Porifera: These are sponges. They lack a coelom and get food as water flows through them. They are usually found in marine and sometimes in freshwater environments. They are perforated and diploblastic, meaning there are two layers of cells.
Cnidaria: Members of this phylum are hydrozoa, jellyfish, and obelia. They have radial symmetry, sac-like bodies, and a polyp or medusa (jellyfish) body plan. They are diploblastic, possessing both an ectoderm and an endoderm. Food can get in through a cavity, but members of this phylum do not have an anus.
Platyhelminthes: These are also known as flatworms. Classes include turbellaria (planarian) and trematoda (which include lung, liver, and blood fluke parasites). They have organs and bilateral symmetry. They have three layers of tissue: an ectoderm, a mesoderm, and an endoderm.
Nematoda: These are roundworms. Hookworms and many other parasites are members of this phylum. They have a pseudocoelom, which means the coelom is not completely enclosed within the mesoderm. They also have a digestive tract that runs directly from the mouth to the anus. They are nonsegmented.

Members of the protostomic phyla have mouths that are formed from blastopores.
Mollusca: Classes include bivalvia (organisms with two shells, such as clams, mussels, and oysters), gastropoda (snails and slugs), cephalopoda (octopus, squid, and chambered nautilus), scaphopoda, amphineura (chitons), and monoplacophora.
Annelida: This phylum includes the classes oligochaeta (earthworms), polychaeta (clam worms), and hirudinea (leeches). They have true coeloms enclosed within the mesoderm. They are segmented, have repeating units, and have a nerve trunk.
Arthropoda: The phylum is diverse and populous. Members can be found in all types of environments. They have external skeletons, jointed appendages, bilateral symmetry, and nerve cords. They also have open circulatory systems and sense organs. Subphyla include crustacea (lobster, barnacles, pill bugs, and daphnia), hexapoda (all insects, which have three body segments, six legs, and usual wings), myriapoda (centipedes and millipedes), and chelicerata (the horseshoe crab and arachnids). Pill bugs have gills. Bees, ants, and wasps belong to the order hymenoptera. Like several other insect orders, they undergo complete metamorphosis.

Members of the deuterostomic phyla have anuses that are formed from blastopores.
Echinodermata: Members of this phylum have radial symmetry, are marine organisms, and have a water vascular system. Classes include echinoidea (sea urchins and sand dollars), crinoidea (sea lilies), asteroidea (starfish), ophiuroidea (brittle stars), and holothuroidea (sea cucumbers).
Chordata: This phylum includes humans and all other vertebrates, as well as a few invertebrates (urochordata and cephalochordata). Members of this phylum include agnatha (lampreys and hagfish), gnathostomata, chondrichthyes (cartilaginous fish-like sharks, skates, and rays), osteichthyes (bony fishes, including ray-finned fish that humans eat), amphibians (frogs, salamander, and newts), reptiles (lizards, snakes, crocodiles, and dinosaurs), birds, and mammals.

Anatomy

Extrinsic refers to homeostatic systems that are controlled from outside the body. In higher animals, the nervous system and endocrine system help regulate body functions by responding to stimuli. Hormones in animals regulate many processes, including growth, metabolism, reproduction, and fluid balance. The names of hormones tend to end in "-one." Endocrine hormones are proteins or steroids. Steroid hormones (anabolic steroids) help control the manufacture of protein in muscles and bones.

Invertebrates do not have a backbone, whereas vertebrates do. The great majority of animal species (an estimated 98 percent) are invertebrates, including worms, jellyfish, mollusks, slugs, insects, and spiders. They comprise 30 phyla in all. Vertebrates belong to the phylum chordata. The vertebrate body has two cavities. The thoracic cavity holds the heart and lungs and the abdominal cavity holds the digestive organs. Animals with exoskeletons have skeletons on the outside. Examples are crabs and turtles. Animals with endoskeletons have skeletons on the inside. Examples are humans, tigers, birds, and reptiles.

The 11 major organ systems are: skeletal, muscular, nervous, digestive, respiratory, circulatory, skin, excretory, immune, endocrine, and reproductive.

Skeletal: This consists of the bones and joints. The skeletal system provides support for the body through its rigid structure, provides protection for internal organs, and works to make organisms motile. Growth hormone affects the rate of reproduction and the size of body cells, and also helps amino acids move through membranes.

Muscular: This includes the muscles. The muscular system allows the body to move and respond to its environment.

Nervous: This includes the brain, spinal cord, and nerves. The nervous system is a signaling system for intrabody communications among systems, responses to stimuli, and interaction within an environment. Signals are electrochemical. Conscious thoughts and memories and sense interpretation occur in the nervous system. It also controls involuntary muscles and functions, such as breathing and the beating of the heart.

Digestive: This includes the mouth, pharynx, esophagus, stomach, intestines, rectum, anal canal, teeth, salivary glands, tongue, liver, gallbladder, pancreas, and appendix. The system helps change food into a form that the body can process and use for energy and nutrients. Food is eventually eliminated as solid waste. Digestive processes can be mechanical, such as chewing food and churning it in the stomach, and chemical, such as secreting hydrochloric acid to kill bacteria and converting protein to amino acids. The overall system converts large food particles into molecules so the body can use them. The small intestine transports the molecules to the circulatory system. The large intestine absorbs nutrients and prepares the unused portions of food for elimination.

Carbohydrates are the primary source of energy as they can be easily converted to glucose. Fats (oils or lipids) are usually not very water soluble, and vitamins A, D, E, and K are fat soluble. Fats are needed to help process these vitamins and can also store energy. Fats have the highest calorie value per gram (9,000 calories). Dietary fiber, or roughage, helps the excretory system. In humans, fiber can help regulate blood sugar levels, reduce heart disease, help food pass through the digestive system, and add bulk. Dietary minerals are chemical elements that are involved with biochemical functions in the body. Proteins consist of amino acids. Proteins are broken down in the body into amino acids that are used for protein biosynthesis or fuel. Vitamins are compounds that are not made by the body, but obtained through the diet. Water is necessary to prevent dehydration since water is lost through the excretory system and perspiration.

Respiratory: This includes the nose, pharynx, larynx, trachea, bronchi, and lungs. It is involved in gas exchange, which occurs in the alveoli. Fish have gills instead of lungs.

Circulatory: This includes the heart, blood, and blood vessels, such as veins, arteries, and capillaries. Blood transports oxygen and nutrients to cells and carbon dioxide to the lungs.

Skin (integumentary): This includes skin, hair, nails, sense receptors, sweat glands, and oil glands. The skin is a sense organ, provides an exterior barrier against disease, regulates body temperature through perspiration, manufactures chemicals and hormones, and provides a place for nerves from the nervous system and parts of the circulation system to travel through. Skin has three layers: epidermis, dermis, and subcutaneous. The epidermis is the thin, outermost, waterproof layer. Basal cells are located in the epidermis. The dermis contains the sweat glands, oil glands, and hair follicles. The subcutaneous layer has connective tissue, and also contains adipose (fat) tissue, nerves, arteries, and veins.

Excretory: This includes the kidneys, ureters, bladder, and urethra. The excretory system helps maintain the amount of fluids in the body. Wastes from the blood system and excess water are removed in urine. The system also helps remove solid waste.

Immune: This includes the lymphatic system, lymph nodes, lymph vessels, thymus, and spleen. Lymph fluid is moved throughout the body by lymph vessels that provide protection against disease. This system protects the body from external intrusions, such as microscopic organisms and foreign substances. It can also protect against some cancerous cells.

Endocrine: This includes the pituitary gland, pineal gland, hypothalamus, thyroid gland, parathyroids, thymus, adrenals, pancreas, ovaries, and testes. It controls systems and processes by secreting hormones into the blood system. Exocrine glands are those that secrete fluid into ducts. Endocrine glands secrete hormones directly into the blood stream without the use of ducts. Prostaglandin (tissue hormones) diffuses only a short distance from the tissue that created it, and influences nearby cells only. Adrenal glands are located above each kidney. The cortex secretes some sex hormones, as well as mineralocorticoids and glucocorticoids involved in immune suppression and stress response. The medulla secretes epinephrine and norepinephrine. Both elevate blood sugar, increase blood pressure, and accelerate heart rate. Epinephrine also stimulates heart muscle. The islets of Langerhans are clumped within the pancreas and secrete glucagon and insulin, thereby regulating blood sugar levels. The four parathyroid glands at the rear of the thyroid secrete parathyroid hormone.

Reproductive: In the male, this system includes the testes, vas deferens, urethra, prostate, penis, and scrotum. In the female, this system includes the ovaries, fallopian tubes (oviduct and uterine tubes), cervix, uterus, vagina, vulva, and mammary glands. Sexual reproduction helps provide genetic diversity as gametes from each parent contribute half the DNA to the zygote offspring. The system provides a method of transporting the male gametes to the female. It also allows for the growth and development of the embryo. Hormones involved are testosterone, interstitial cell stimulating hormone (ICSH), luteinizing hormone (LH), follicle stimulating hormone (FSH), and estrogen. Estrogens secreted from the ovaries include estradiol, estrone, and estriol. They encourage growth, among other things. Progesterone helps prepare the endometrium for pregnancy.

Based on whether or not and when an organism uses meiosis or mitosis, the three possible cycles of reproduction are haplontic, diplontic, and haplodiplontic. Fungi, green algae, and protozoa are haplontic. Animals and some brown algae and fungi are diplontic. Plants and some fungi are haplodiplontic. Diplontic organisms, like multicelled animals, have a dominant diploid life cycle. The haploid generation is simply the egg and sperm. Monoecious species are bisexual (hermaphroditic). In this case, the individual has both male and female organs: sperm-bearing testicles and egg-bearing ovaries. Hermaphroditic species can self fertilize. Some worms are hermaphroditic. Cross fertilization is when individuals exchange genetic information. Most animal species are dioecious, meaning individuals are distinctly male or female.

Biological Relationships

As heterotrophs, animals can be further classified as carnivores, herbivores, omnivores, and parasites. Predation refers to a predator that feeds on another organism, which results in its death. Detritivory refers to heterotrophs that consume organic dead matter. Carnivores are animals that are meat eaters. Herbivores are plant eaters, and omnivores eat both meat and plants. A parasite's food source is its host. A parasite lives off of a host, which does not benefit from the interaction. Nutrients can be classified as carbohydrates, fats, fiber, minerals, proteins, vitamins, and water. Each supply a specific substance required for various species to survive, grow, and reproduce. A calorie is a measurement of heat energy. It can be used to represent both how much energy a food can provide and how much energy an organism needs to live.

Biochemical cycles are how chemical elements required by living organisms cycle between living and nonliving organisms. Elements that are frequently required are phosphorus, sulfur, oxygen, carbon, gaseous nitrogen, and water. Elements can go through gas cycles, sedimentary cycles, or both. Elements circulate through the air in a gas cycle and from land to water in a sedimentary one.

A food chain is a linking of organisms in a community that is based on how they use each other as food sources. Each link in the chain consumes the link above it and is consumed by the link below it. The exceptions are the organism at the top of the food chain and the organism at the bottom. Biomagnification (bioamplification): This refers to an increase in concentration of a substance within a food chain. Examples are pesticides or mercury. Mercury is emitted from coal-fired power plants and gets into the water supply, where it is eaten by a fish. A larger fish eats smaller fish, and humans eat fish. The concentration of mercury in humans has now risen. Biomagnification is affected by the persistence of a chemical, whether it can be broken down and negated, food chain energetics, and whether organisms can reduce or negate the substance.

A food web consists of interconnected food chains in a community. The organisms can be linked to show the direction of energy flow. Energy flow in this sense is used to refer to the actual caloric flow through a system from trophic level to trophic level. Trophic level refers to a link in a food chain or a level of nutrition. The 10% rule is that from trophic level to level, about 90% of the energy is lost (in the form of heat, for example). The lowest trophic level consists of primary producers (usually plants), then primary consumers, then secondary consumers, and finally tertiary consumers (large carnivores). The final link is decomposers, which break down the consumers at the top. Food chains usually do not contain more than six links. These links may also be referred to as ecological pyramids.

Ecosystem stability is a concept that states that a stable ecosystem is perfectly efficient. Seasonal changes or expected climate fluctuations are balanced by homeostasis. It also states that interspecies interactions are part of the balance of the system. Four principles of ecosystem stability are that waste disposal and nutrient replenishment by recycling is complete, the system uses sunlight as an energy source, biodiversity remains, and populations are stable in that they do not over consume resources. Ecologic succession is the concept that states that there is an orderly progression of change within a community. An example of primary succession is that over hundreds of years bare rock decomposes to sand, which eventually leads to soil formation, which eventually leads to the growth of grasses and trees. Secondary succession occurs after a disturbance or major event that greatly affects a community, such as a wild fire or construction of a dam.

Population is a measure of how many individuals exist in a specific area. It can be used to measure the size of human, plant, or animal groups. Population growth depends on many factors. Factors that can limit the number of individuals in a population include lack of resources such as food and water, space, habitat destruction, competition, disease, and predators. Exponential growth refers to an unlimited rising growth rate. This kind of growth can be plotted on a chart in the shape of a J. Carrying capacity is the population size that can be sustained. The world's population is about 6.8 billion and growing. The human population has not yet reached its carrying capacity. Population dynamics refers to how a population changes over time and the factors that cause changes. An S-shaped curve shows that population growth has leveled off. Biotic potential refers to the maximum reproductive capacity of a population given ideal environmental conditions.

Biological concepts:
Territoriality: This refers to members of a species protecting areas from other members of their species and from other species. Species members claim specific areas as their own.
Dominance: This refers to the species in a community that is the most populous.
Altruism: This is when a species or individual in a community exhibits behaviors that benefit another individual at a cost to itself. In biology, altruism does not have to be a conscious sacrifice.
Threat display: This refers to behavior by an organism that is intended to intimidate or frighten away members of its own or another species.

The principle of **competitive exclusion** (Gause's Law) states that if there are limited or insufficient resources and species are competing for them, these species will not be able to co-exist. The result is that one of the species will become extinct or be forced to undergo a behavioral or evolutionary change. Another way to say this is that "complete competitors cannot coexist."

A **community** is any number of species interacting within a given area. A **niche** is the role of a species within a community. **Species diversity** refers to the number of species within a community and their populations. A **biome** refers to an area in which species are associated because of climate. The six major biomes in North America are desert, tropical rain forest, grassland, coniferous forest, deciduous forest, and tundra.

Biotic: Biotic factors are the living factors, such as other organisms, that affect a community or population. Abiotic factors are nonliving factors that affect a community or population, such as facets of the environment.
Ecology: Ecology is the study of plants, animals, their environments, and how they interact.
Ecosystem: An ecosystem is a community of species and all of the environment factors that affect them.
Biomass: In ecology, biomass refers to the mass of one or all of the species (species biomass) in an ecosystem or area.

Predation, parasitism, commensalism, and mutualism are all types of species interactions that affect species populations. **Intraspecific relationships** are relationships among members of a species. **Interspecific relationships** are relationships between members of different species.

Predation: This is a relationship in which one individual feeds on another (the prey), causing the prey to die. **Mimicry** is an adaptation developed as a response to predation. It refers to an organism that has a similar appearance to another species, which is meant to fool the predator into thinking the organism is more dangerous than it really is. Two examples are the drone fly and the io moth. The fly looks like a bee, but cannot sting. The io moth has markings on its wings that make it look like an owl. The moth can startle predators and gain time to escape. Predators can also use mimicry to lure their prey.

Commensalism: This refers to interspecific relationships in which one of the organisms benefits. Mutualism, competition, and parasitism are all types of commensalism.

Mutualism: This is a relationship in which both organisms benefit from an interaction.
Competition: This is a relationship in which both organisms are harmed.
Parasitism: This is a relationship in which one organism benefits and the other is harmed.

Chemistry

Atoms

Matter refers to substances that have mass and occupy space (or volume). The traditional definition of matter describes it as having three states: solid, liquid, and gas. These different states are caused by differences in the distances and angles between molecules or atoms, which result in differences in the energy that binds them. Solid structures are rigid or nearly rigid and have strong bonds. Molecules or atoms of liquids move around and have weak bonds, although they are not weak enough to readily break. Molecules or atoms of gases move almost independently of each other, are typically far apart, and do not form bonds. The current definition of matter describes it as having four states. The fourth is plasma, which is an ionized gas that has some electrons that are described as free because they are not bound to an atom or molecule.

All matter consists of atoms. Atoms consist of a nucleus and electrons. The nucleus consists of protons and neutrons. The properties of these are measurable; they have mass and an electrical charge. The nucleus is positively charged due to the presence of protons. Electrons are negatively charged and orbit the nucleus. The nucleus has considerably more mass than the surrounding electrons. Atoms can bond together to make molecules. Atoms that have an equal number of protons and electrons are electrically neutral. If the number of protons and electrons in an atom is not equal, the atom has a positive or negative charge and is an ion.

An element is matter with one particular type of atom. It can be identified by its atomic number, or the number of protons in its nucleus. There are approximately 117 elements currently known, 94 of which occur naturally on Earth. Elements from the periodic table include hydrogen, carbon, iron, helium, mercury, and oxygen. Atoms combine to form molecules. For example, two atoms of hydrogen (H) and one atom of oxygen (O) combine to form water (H_2O).

Compounds are substances containing two or more elements. Compounds are formed by chemical reactions and frequently have different properties than the original elements. Compounds are decomposed by a chemical reaction rather than separated by a physical one. Solutions are homogeneous mixtures composed of two or more substances that have become one. Mixtures contain two or more substances that are combined but have not reacted chemically with each other. Mixtures can be separated using physical methods, while compounds cannot.

A solution is a homogeneous mixture. A mixture is two or more different substances that are mixed together, but not combined chemically. Homogeneous mixtures are those that are uniform in their composition. Solutions consist of a solute (the substance that is dissolved) and a solvent (the substance that does the dissolving). An example is sugar water. The solvent is the water and the solute is the sugar. The intermolecular attraction between the solvent and the solute is called solvation. Hydration refers to solutions in which water is the solvent. Solutions are formed when the forces of the molecules of the solute and the solvent are as strong as the individual molecular forces of the solute and the solvent. An example is that salt (NaCl) dissolves in water to create a solution. The Na^+ and the Cl^- ions in salt interact with the molecules of water and vice versa to overcome the individual molecular forces of the solute and the solvent.

Elements are represented in upper case letters. If there is no subscript, it indicates there is only one atom of the element. Otherwise, the subscript indicates the number of atoms. In molecular formulas, elements are organized according to the Hill system. Carbon is first, hydrogen comes next, and the remaining elements are listed in alphabetical order. If there is no carbon, all elements are listed alphabetically. There are a couple of exceptions to these rules. First, oxygen is usually listed last in oxides. Second, in ionic compounds the positive ion is listed first, followed by the negative ion. In CO_2, for example, C indicates 1 atom of carbon and O_2 indicates 2 atoms of oxygen. The compound is carbon dioxide. The formula for ammonia (an ionic compound) is NH_3, which is one atom of nitrogen and three of hydrogen. H_2O is two atoms of hydrogen and one of oxygen. Sugar is $C_6H_{12}O_6$, which is 6 atoms of carbon, 12 of hydrogen, and 6 of oxygen.

An **atom** is one of the most basic units of matter. An atom consists of a central nucleus surrounded by electrons. The **nucleus** of an atom consists of protons and neutrons. It is positively charged, dense, and heavier than the surrounding electrons. The plural form of nucleus is nuclei. **Neutrons** are the uncharged atomic particles contained within the nucleus. The number of neutrons in a nucleus can be represented as "N." Along with neutrons, **protons** make up the nucleus of an atom. The number of protons in the nucleus determines the atomic number of an element. Carbon atoms, for example, have six protons. The atomic number of carbon is 6. **Nucleon** refers collectively to neutrons and protons. **Electrons** are atomic particles that are negatively charged and orbit the nucleus of an atom. The number of protons minus the number of electrons indicates the charge of an atom.

The **atomic number** of an element refers to the number of protons in the nucleus of an atom. It is a unique identifier. It can be represented as Z. Atoms with a neutral charge have an atomic number that is equal to the number of electrons. **Atomic mass** is also known as the mass number. The atomic mass is the total number of protons and neutrons in the nucleus of an atom. It is referred to as "A." The atomic mass (A) is equal to the number of protons (Z) plus the number of neutrons (N). This can be represented by the equation $A = Z + N$. The mass of electrons in an atom is basically insignificant because it is so small. **Atomic weight** may sometimes be referred to as "relative atomic mass," but should not be confused with atomic mass. Atomic weight is the ratio of the average mass per atom of a sample (which can include various isotopes of an element) to 1/12 of the mass of an atom of carbon-12.

Chemical properties are qualities of a substance which can't be determined by simply looking at the substance and must be determined through chemical reactions. Some chemical properties of elements include: atomic number, electron configuration, electrons per shell, electronegativity, atomic radius, and isotopes.

In contrast to chemical properties, **physical properties** can be observed or measured without chemical reactions. These include properties such as color, elasticity, mass, volume, and temperature. **Mass** is a measure of the amount of substance in an object. **Weight** is a measure of the gravitational pull of Earth on an object. **Volume** is a measure of the amount of space occupied. There are many formulas to determine volume. For example, the volume of a cube is the length of one side cubed (a^3) and the volume of a rectangular prism is length times width times height ($l \cdot w \cdot h$). The volume of an irregular shape can be determined by how much water it displaces. **Density** is a measure of the amount of mass per unit volume. The formula to find density is mass divided by volume ($D=m/V$). It is expressed in terms of mass per cubic unit, such as grams per cubic centimeter (g/cm^3). **Specific gravity** is a measure of the ratio of a substance's density compared to the density of water.

Both physical changes and chemical reactions are everyday occurrences. Physical changes do not result in different substances. For example, when water becomes ice it has undergone a physical change, but not a chemical change. It has changed its form, but not its composition. It is still H_2O.

Chemical properties are concerned with the constituent particles that make up the physicality of a substance. Chemical properties are apparent when chemical changes occur. The chemical properties of a substance are influenced by its electron configuration, which is determined in part by the number of protons in the nucleus (the atomic number). Carbon, for example, has 6 protons and 6 electrons. It is an element's outermost valence electrons that mainly determine its chemical properties. Chemical reactions may release or consume energy.

Periodic Table

The periodic table groups elements with similar chemical properties together. The grouping of elements is based on atomic structure. It shows periodic trends of physical and chemical properties and identifies families of elements with similar properties. It is a common model for organizing and understanding elements. In the periodic table, each element has its own cell that includes varying amounts of information presented in symbol form about the properties of the element. Cells in the table are arranged in rows (periods) and columns (groups or families). At minimum, a cell includes the symbol for the element and its atomic number. The cell for hydrogen, for example, which appears first in the upper left corner, includes an "H" and a "1" above the letter. Elements are ordered by atomic number, left to right, top to bottom.

In the periodic table, the groups are the columns numbered 1 through 18 that group elements with similar outer electron shell configurations. Since the configuration of the outer electron shell is one of the primary factors affecting an element's chemical properties, elements within the same group have similar chemical properties. Previous naming conventions for groups have included the use of Roman numerals and upper-case letters. Currently, the periodic table groups are: Group 1, alkali metals; Group 2, alkaline earth metals; Groups 3-12, transition metals; Group 13, boron family; Group 14; carbon family; Group 15, pnictogens; Group 16, chalcogens; Group 17, halogens; Group 18, noble gases.

In the periodic table, there are seven periods (rows), and within each period there are blocks that group elements with the same outer electron subshell (more on this in the next section). The number of electrons in that outer shell determines which group an element belongs to within a given block. Each row's number (1, 2, 3, etc.) corresponds to the highest number electron shell that is in use. For example, row 2 uses only electron shells 1 and 2, while row 7 uses all shells from 1-7.

Atomic radii will decrease from left to right across a period (row) on the periodic table. In a group (column), there is an increase in the atomic radii of elements from top to bottom. Ionic radii will be smaller than the atomic radii for metals, but the opposite is true for non-metals. From left to right, electronegativity, or an atom's likeliness of taking another atom's electrons, increases. In a group, electronegativity decreases from top to bottom. Ionization energy or the amount of energy needed to get rid of an atom's outermost electron, increases across a period and decreases down a group. Electron affinity will become more negative across a period but will not change much within a group. The melting point decreases from top to bottom in the metal groups and increases from top to bottom in the non-metal groups.

Electrons

Electrons are subatomic particles that orbit the nucleus at various levels commonly referred to as layers, shells, or clouds. The orbiting electron or electrons account for only a fraction of the atom's mass. They are much smaller than the nucleus, are negatively charged, and exhibit wave-like characteristics. Electrons are part of the lepton family of elementary particles. Electrons can occupy orbits that are varying distances away from the nucleus, and tend to occupy the lowest energy level they can. If an atom has all its electrons in the lowest available positions, it has a stable electron arrangement. The outermost electron shell of an atom in its uncombined state is known as the valence shell. The electrons there are called valence electrons, and it is their number that determines bonding behavior. Atoms tend to react in a manner that will allow them to fill or empty their valence shells.

There are seven electron shells. One is closest to the nucleus and seven is the farthest away. Electron shells can also be identified with the letters K, L, M, N, O, P, and Q. Traditionally, there were four subshells identified by the first letter of their descriptive name: s (sharp), p (principal), d (diffuse), and f (fundamental). The maximum number of electrons for each subshell is as follows: s is 2, p is 6, d is 10, and f is 14. Every shell has an s subshell, the second shell and those above also have a p subshell, the third shell and those above also have a d subshell, and so on. Each subshell contains atomic orbitals, which describes the wave-like characteristics of an electron or a pair of electrons expressed as two angles and the distance from the nucleus. Atomic orbital is a concept used to express the likelihood of an electron's position in accordance with the idea of wave-particle duality.

Electron configuration: This is a trend whereby electrons fill shells and subshells in an element in a particular order and with a particular number of electrons. The chemical properties of the elements reflect their electron configurations. Energy levels (shells) do not have to be completely filled before the next one begins to be filled. An example of electron configuration notation is $1s^22s^22p^5$, where the first number is the row (period), or shell. The letter refers to the subshell of the shell, and the number in superscript is the number of electrons in the subshell. A common shorthand method for electron configuration notation is to use a noble gas (in a bracket) to abbreviate the shells that elements have in common. For example, the electron configuration for neon is $1s^22s^22p^6$. The configuration for phosphorus is $1s^22s^22p^63s^23p^3$, which can be written as $[Ne]3s^23p^3$. Subshells are filled in the following manner: 1s, 2s, 2p, 3s, 3p, 4s, 3d, 4p, 5s, 4d, 5p, 6s, 4f, 5d, 6p, 7s, 5f, 6d, and 7p.

Most atoms are neutral since the positive charge of the protons in the nucleus is balanced by the negative charge of the surrounding electrons. Electrons are transferred between atoms when they come into contact with each other. This creates a molecule or atom in which the number of electrons does not equal the number of protons, which gives it a positive or negative charge. A negative ion is created when an atom gains electrons, while a positive ion is created when an atom loses electrons. An ionic bond is formed between ions with opposite charges. The resulting compound is neutral. Ionization refers to the process by which neutral particles are ionized into charged particles. Gases and plasmas can be partially or fully ionized through ionization.

Atoms interact by transferring or sharing the electrons furthest from the nucleus. Known as the outer or valence electrons, they are responsible for the chemical properties of an element. Bonds between atoms are created when electrons are paired up by being transferred or shared. If electrons are transferred from one atom to another, the bond is ionic. If electrons are shared, the bond is covalent. Atoms of the same element may bond together to form molecules or crystalline solids. When two or more different types of atoms bind together chemically, a compound is made. The physical properties of compounds reflect the nature of the interactions among their molecules. These interactions are determined by the structure of the molecule, including the atoms they consist of and the distances and angles between them.

Isotopes and Molecules

The number of protons in an atom determines the element of that atom. For instance, all helium atoms have exactly two protons, and all oxygen atoms have exactly eight protons. If two atoms have the same number of protons, then they are the same element. However, the number of neutrons in two atoms can be different without the atoms being different elements. Isotope is the term used to distinguish between atoms that have the same number of protons but a different number of neutrons. The names of isotopes have the element name with the mass number. Recall that the mass number is the number of protons plus the number of neutrons. For example, carbon-12 refers to an atom that has 6 protons, which makes it carbon, and 6 neutrons. In other words, 6 protons + 6 neutrons = 12. Carbon-13 has six protons and seven neutrons, and carbon-14 has six protons and eight neutrons. Isotopes can also be written with the mass number in superscript before the element symbol. For example, carbon-12 can be written as ^{12}C.

The important properties of water (H_2O) are high polarity, hydrogen bonding, cohesiveness, adhesiveness, high specific heat, high latent heat, and high heat of vaporization. It is essential to life as we know it, as water is one of the main if not the main constituent of many living things. Water is a liquid at room temperature. The high specific heat of water means it resists the breaking of its hydrogen bonds and resists heat and motion, which is why it has a relatively high boiling point and high vaporization point. It also resists temperature change. Water is peculiar in that its solid state floats in its liquid state. Most substances are denser in their solid forms. Water is cohesive, which means it is attracted to itself. It is also adhesive, which means it readily attracts other molecules. If water tends to adhere to another substance, the substance is said to be hydrophilic. Water makes a good solvent. Substances, particularly those with polar ions and molecules, readily dissolve in water.

Electrons in an atom can orbit different levels around the nucleus. They can absorb or release energy, which can change the location of their orbit or even allow them to break free from the atom. The outermost layer is the valence layer, which contains the valence electrons. The valence layer tends to have or share eight electrons. Molecules are formed by a chemical bond between atoms, a bond which occurs at the valence level. Two basic types of bonds are covalent and ionic. A covalent bond is formed when atoms share electrons. An ionic bond is formed when an atom transfers an electron to another atom. A hydrogen bond is a weak bond between a hydrogen atom of one molecule and an electronegative atom (such as nitrogen, oxygen, or fluorine) of another molecule. The Van der Waals force is a weak force between molecules. This type of force is much weaker than actual chemical bonds between atoms.

Reactions

Chemical reactions measured in human time can take place quickly or slowly. They can take fractions of a second or billions of years. The rates of chemical reactions are determined by how frequently reacting atoms and molecules interact. Rates are also influenced by the temperature and various properties (such as shape) of the reacting materials. Catalysts accelerate chemical reactions, while inhibitors decrease reaction rates. Some types of reactions release energy in the form of heat and light. Some types of reactions involve the transfer of either electrons or hydrogen ions between reacting ions, molecules, or atoms. In other reactions, chemical bonds are broken down by heat or light to form reactive radicals with electrons that will readily form new bonds. Processes such as the formation of ozone and greenhouse gases in the atmosphere and the burning and processing of fossil fuels are controlled by radical reactions.

Chemical equations describe chemical reactions. The reactants are on the left side before the arrow and the products are on the right side after the arrow. The arrow indicates the reaction or change. The coefficient, or stoichiometric coefficient, is the number before the element, and indicates the ratio of reactants to products in terms of moles. The equation for the formation of water from hydrogen and oxygen, for example, is $2H_2(g) + O_2(g) \rightarrow 2H_2O(l)$. The 2 preceding hydrogen and water is the coefficient, which means there are 2 moles of hydrogen and 2 of water. There is 1 mole of oxygen, which does not have to be indicated with the number 1. In parentheses, g stands for gas, l stands for liquid, s stands for solid, and aq stands for aqueous solution (a substance dissolved in water). Charges are shown in superscript for individual ions, but not for ionic compounds. Polyatomic ions are separated by parentheses so the ion will not be confused with the number of ions.

An unbalanced equation is one that does not follow the law of conservation of mass, which states that matter can only be changed, not created. If an equation is unbalanced, the numbers of atoms indicated by the stoichiometric coefficients on each side of the arrow will not be equal. Start by writing the formulas for each species in the reaction. Count the atoms on each side and determine if the number is equal. Coefficients must be whole numbers. Fractional amounts, such as half a molecule, are not possible. Equations can be balanced by multiplying the coefficients by a constant that will produce the smallest possible whole number coefficient. $H_2 + O_2 \rightarrow H_2O$ is an example of an unbalanced equation. The balanced equation is $2H_2 + O_2 \rightarrow 2H_2O$, which indicates that it takes two moles of hydrogen and one of oxygen to produce two moles of water.

One way to organize chemical reactions is to sort them into two categories: oxidation/reduction reactions (also called redox reactions) and metathesis reactions (which include acid/base reactions). Oxidation/reduction reactions can involve the transfer of one or more electrons, or they can occur as a result of the transfer of oxygen, hydrogen, or halogen atoms. The species that loses electrons is oxidized and is referred to as the reducing agent. The species that gains electrons is reduced and is referred to as the oxidizing agent. The element undergoing oxidation experiences an increase in its oxidation number, while the element undergoing reduction experiences a decrease in its oxidation number. Single replacement reactions are types of oxidation/reduction reactions. In a single replacement reaction, electrons are transferred from one chemical species to another. The transfer of electrons results in changes in the nature and charge of the species.

Single substitution, displacement, or replacement reactions are when one reactant is displaced by another to form the final product (A + BC → AB + C). Single substitution reactions can be cationic or anionic. When a piece of copper (Cu) is placed into a solution of silver nitrate ($AgNO_3$), the solution turns blue. The copper appears to be replaced with a silvery-white material. The equation is $2AgNO_3 + Cu \rightarrow Cu(NO_3)_2 + 2Ag$. When this reaction takes place, the copper dissolves and the silver in the silver nitrate solution precipitates (becomes a solid), thus resulting in copper nitrate and silver. Copper and silver have switched places in the nitrate.

Combination, or synthesis, reactions: In a combination reaction, two or more reactants combine to form a single product (A + B → C). These reactions are also called synthesis or addition reactions. An example is burning hydrogen in air to produce water. The equation is $2H_2(g) + O_2(g) \rightarrow 2H_2O(l)$. Another example is when water and sulfur trioxide react to form sulfuric acid. The equation is $H_2O + SO_3 \rightarrow H_2SO_4$.

Double displacement, double replacement, substitution, metathesis, or ion exchange reactions are when ions or bonds are exchanged by two compounds to form different compounds (AC + BD → AD + BC). An example of this is that silver nitrate and sodium chloride form two different products (silver chloride and sodium nitrate) when they react. The formula for this reaction is AgNO$_3$ + NaCl → AgCl + NaNO$_3$.

Double replacement reactions are metathesis reactions. In a double replacement reaction, the chemical reactants exchange ions but the oxidation state stays the same. One of the indicators of this is the formation of a solid precipitate. In acid/base reactions, an acid is a compound that can donate a proton, while a base is a compound that can accept a proton. In these types of reactions, the acid and base react to form a salt and water. When the proton is donated, the base becomes water and the remaining ions form a salt. One method of determining whether a reaction is an oxidation/reduction or a metathesis reaction is that the oxidation number of atoms does not change during a metathesis reaction.

A neutralization, acid-base, or proton transfer reaction is when one compound acquires H$^+$ from another. These types of reactions are also usually double displacement reactions. The acid has an H$^+$ that is transferred to the base and neutralized to form a salt.

Decomposition (or desynthesis, decombination, or deconstruction) reactions; in a decomposition reaction, a reactant is broken down into two or more products (A → B + C). These reactions are also called analysis reactions. Thermal decomposition is caused by heat. Electrolytic decomposition is due to electricity. An example of this type of reaction is the decomposition of water into hydrogen and oxygen gas. The equation is 2H$_2$O → 2H$_2$ + O$_2$. Decomposition is considered a chemical reaction whereby a single compound breaks down into component parts or simpler compounds. When a compound or substance separates into these simpler substances, the byproducts are often substances that are different from the original. Decomposition can be viewed as the opposite of combination reactions. Most decomposition reactions are endothermic. Heat needs to be added for the chemical reaction to occur. Separation processes can be mechanical or chemical, and usually involve re-organizing a mixture of substances without changing their chemical nature. The separated products may differ from the original mixture in terms of chemical or physical properties. Types of separation processes include filtration, crystallization, distillation, and chromatography. Basically, decomposition breaks down one compound into two or more compounds or substances that are different from the original; separation sorts the substances from the original mixture into like substances.

Endothermic reactions are chemical reactions that absorb heat and exothermic reactions are chemical reactions that release heat. Reactants are the substances that are consumed during a reaction, while products are the substances that are produced or formed. A balanced equation is one that uses reactants, products, and coefficients in such a way that the number of each type of atom (law of conservation of mass) and the total charge remains the same. The reactants are on the left side of the arrow and the products are on the right. The heat difference between endothermic and exothermic reactions is caused by bonds forming and breaking. If more energy is needed to break the reactant bonds than is released when they form, the reaction is endothermic. Heat is absorbed and the environmental temperature decreases. If more energy is released when product bonds form than is needed to break the reactant bonds, the reaction is exothermic. Heat is released and the environmental temperature increases.

The collision theory states that for a chemical reaction to occur, atoms or molecules have to collide with each other with a certain amount of energy. A certain amount of energy is required to breach the activation barrier. Heating a mixture will raise the energy levels of the molecules and the rate of reaction (the time it takes for a reaction to complete). Generally, the rate of reaction is doubled for every 10 degrees Celsius temperature increase. However, the increase needed to double a reaction rate increases as the temperature climbs. This is due to the increase in collision frequency that occurs as the temperature increases. Other factors that can affect the rate of reaction are surface area, concentration, pressure, and the presence of a catalyst.

The particles of an atom's nucleus (the protons and neutrons) are bound together by nuclear force, also known as residual strong force. Unlike chemical reactions, which involve electrons, nuclear reactions occur when two nuclei or nuclear particles collide. This results in the release or absorption of energy and products that are different from the initial particles. The energy released in a nuclear reaction can take various forms, including the release of kinetic energy of the product particles and the emission of very high energy photons known as gamma rays. Some energy may also remain in the nucleus. Radioactivity refers to the particles emitted from nuclei as a result of nuclear instability. There are many nuclear isotopes that are unstable and can spontaneously emit some kind of radiation. The most common types of radiation are alpha, beta, and gamma radiation, but there are several other varieties of radioactive decay.

Inorganic and Organic

The terms inorganic and organic have become less useful over time as their definitions have changed. Historically, inorganic molecules were defined as those of a mineral nature that were not created by biological processes. Organic molecules were defined as those that were produced biologically by a "life process" or "vital force." It was then discovered that organic compounds could be synthesized without a life process. Currently, molecules containing carbon are considered organic. Carbon is largely responsible for creating biological diversity, and is more capable than all other elements of forming large, complex, and diverse molecules of an organic nature. Carbon often completes its valence shell by sharing electrons with other atoms in four covalent bonds, which is also known as tetravalence.

The main trait of inorganic compounds is that they lack carbon. Inorganic compounds include mineral salts, metals and alloys, non-metallic compounds such as phosphorus, and metal complexes. A metal complex has a central atom (or ion) bonded to surrounding ligands (molecules or anions). The ligands sacrifice the donor atoms (in the form of at least one pair of electrons) to the central atom. Many inorganic compounds are ionic, meaning they form ionic bonds rather than share electrons. They may have high melting points because of this. They may also be colorful, but this is not an absolute identifier of an inorganic compound. Salts, which are inorganic compounds, are an example of inorganic bonding of cations and anions. Some examples of salts are magnesium chloride ($MgCl_2$) and sodium oxide (Na_2O). Oxides, carbonates, sulfates, and halides are classes of inorganic compounds. They are typically poor conductors, are very water soluble, and crystallize easily. Minerals and silicates are also inorganic compounds.

Two of the main characteristics of organic compounds are that they include carbon and are formed by covalent bonds. Carbon can form long chains, double and triple bonds, and rings. While inorganic compounds tend to have high melting points, organic compounds tend to melt at temperatures below 300° C. They also tend to boil, sublimate, and decompose below this temperature. Unlike inorganic compounds, they are not very water soluble. Organic molecules are organized into functional groups based on their specific atoms, which helps determine how they will react chemically. A few groups are alkanes, nitro, alkenes, sulfides, amines, and carbolic acids. The hydroxyl group (-OH) consists of alcohols. These molecules are polar, which increases their solubility. By some estimates, there are more than 16 million organic compounds.

Nomenclature refers to the manner in which a compound is named. First, it must be determined whether the compound is ionic (formed through electron transfer between cations and anions) or molecular (formed through electron sharing between molecules). When dealing with an ionic compound, the name is determined using the standard naming conventions for ionic compounds. This involves indicating the positive element first (the charge must be defined when there is more than one option for the valency) followed by the negative element plus the appropriate suffix. The rules for naming a molecular compound are as follows: write elements in order of increasing group number and determine the prefix by determining the number of atoms. Exclude mono for the first atom. The name for CO_2, for example, is carbon dioxide. The end of oxygen is dropped and "ide" is added to make oxide, and the prefix "di" is used to indicate there are two atoms of oxygen.

Acids and Bases

The potential of hydrogen (pH) is a measurement of the concentration of hydrogen ions in a substance in terms of the number of moles of H^+ per liter of solution. All substances fall between 0 and 14 on the pH scale. A lower pH indicates a higher H^+ concentration, while a higher pH indicates a lower H^+ concentration. Pure water has a neutral pH, which is 7. Anything with a pH lower than water (0-7) is considered acidic. Anything with a pH higher than water (7-14) is a base. Drain cleaner, soap, baking soda, ammonia, egg whites, and sea water are common bases. Urine, stomach acid, citric acid, vinegar, hydrochloric acid, and battery acid are acids. A pH indicator is a substance that acts as a detector of hydrogen or hydronium ions. It is halochromic, meaning it changes color to indicate that hydrogen or hydronium ions have been detected.

When they are dissolved in aqueous solutions, some properties of acids are that they conduct electricity, change blue litmus paper to red, have a sour taste, react with bases to neutralize them, and react with active metals to free hydrogen. A weak acid is one that does not donate all of its protons or disassociate completely. Strong acids include hydrochloric, hydriodic, hydrobromic, perchloric, nitric, and sulfuric. They ionize completely. Superacids are those that are stronger than 100 percent sulfuric acid. They include fluoroantimonic, magic, and perchloric acids. Acids can be used in pickling, a process used to remove rust and corrosion from metals. They are also used as catalysts in the processing of minerals and the production of salts and fertilizers. Phosphoric acid (H_3PO_4) is added to sodas and other acids are added to foods as preservatives or to add taste.

When they are dissolved in aqueous solutions, some properties of bases are that they conduct electricity, change red litmus paper to blue, feel slippery, and react with acids to neutralize their properties. A weak base is one that does not completely ionize in an aqueous solution, and usually has a low pH. Strong bases can free protons in very weak acids. Examples of strong bases are hydroxide compounds such as potassium, barium, and lithium hydroxides. Most are in the first and second groups of the periodic table. A superbase is extremely strong compared to sodium hydroxide and cannot be kept in an aqueous solution. Superbases are organized into organic, organometallic, and inorganic classes. Bases are used as insoluble catalysts in heterogeneous reactions and as catalysts in hydrogenation.

Some properties of salts are that they are formed from acid base reactions, are ionic compounds consisting of metallic and nonmetallic ions, dissociate in water, and are comprised of tightly bonded ions. Some common salts are sodium chloride (NaCl), sodium bisulfate, potassium dichromate ($K_2Cr_2O_7$), and calcium chloride ($CaCl_2$). Calcium chloride is used as a drying agent, and may be used to absorb moisture when freezing mixtures. Potassium nitrate (KNO_3) is used to make fertilizer and in the manufacture of explosives. Sodium nitrate ($NaNO_3$) is also used in the making of fertilizer. Baking soda (sodium bicarbonate) is a salt, as are Epsom salts [magnesium sulfate ($MgSO_4$)]. Salt and water can react to form a base and an acid. This is called a hydrolysis reaction.

A buffer is a solution whose pH remains relatively constant when a small amount of an acid or a base is added. It is usually made of a weak acid and its conjugate base (proton receiver) or one of its soluble salts. It can also be made of a weak base and its conjugate acid (proton donator) or one of its salts. A constant pH is necessary in living cells because some living things can only live within a certain pH range. If that pH changes, the cells could die. Blood is an example of a buffer. A pKa is a measure of acid dissociation or the acid dissociation constant. Buffer solutions can help keep enzymes at the correct pH. They are also used in the fermentation process, in dyeing fabrics, and in the calibration of pH meters. An example of a buffer is HC_2H_3O (a weak acid) and $NaC_2H_3O_2$ (a salt containing the $C_2H_3O_2^-$ ion).

General Concepts

Lewis formulas: These show the bonding or nonbonding tendency of specific pairs of valence electrons. Lewis dot diagrams use dots to represent valence electrons. Dots are paired around an atom. When an atom forms a covalent bond with another atom, the elements share the dots as they would electrons. Double and triple bonds are indicated with additional adjacent dots. Methane (CH_4), for instance, would be shown as a C with 2 dots above, below, and to the right and left and an H next to each set of dots. In structural formulas, the dots are single lines.

Kekulé diagrams: Like Lewis dot diagrams, these are two-dimensional representations of chemical compounds. Covalent bonds are shown as lines between elements. Double and triple bonds are shown as two or three lines and unbonded valence electrons are shown as dots.

Molar mass: This refers to the mass of one mole of a substance (element or compound), usually measured in grams per mole (g/mol). This differs from molecular mass in that molecular mass is the mass of one molecule of a substance relative to the atomic mass unit (amu).

Atomic mass unit (amu) is the smallest unit of mass, and is equal to 1/12 of the mass of the carbon isotope carbon-12. A mole (mol) is a measurement of molecular weight that is equal to the molecule's amu in grams. For example, carbon has an amu of 12, so a mole of carbon weighs 12 grams. One mole is equal to about 6.0221415×10^{23} elementary entities, which are usually atoms or molecules. This amount is also known as the Avogadro constant or Avogadro's number (N_A). Another way to say this is that one mole of a substance is the same as one Avogadro's number of that substance. One mole of chlorine, for example, is 6.0221415×10^{23} chlorine atoms. The charge on one mole of electrons is referred to as a Faraday.

The kinetic theory of gases assumes that gas molecules are small compared to the distances between them and that they are in constant random motion. The attractive and repulsive forces between gas molecules are negligible. Their kinetic energy does not change with time as long as the temperature remains the same. The higher the temperature is, the greater the motion will be. As the temperature of a gas increases, so does the kinetic energy of the molecules. In other words, gas will occupy a greater volume as the temperature is increased and a lesser volume as the temperature is decreased. In addition, the same amount of gas will occupy a greater volume as the temperature increases, but pressure remains constant. At any given temperature, gas molecules have the same average kinetic energy. The ideal gas law is derived from the kinetic theory of gases.

Charles's law: This states that gases expand when they are heated. It is also known as the law of volumes.

Boyle's law: This states that gases contract when pressure is applied to them. It also states that if temperature remains constant, the relationship between absolute pressure and volume is inversely proportional. When one increases, the other decreases. Considered a specialized case of the ideal gas law, Boyle's law is sometimes known as the Boyle-Mariotte law.

The ideal gas law is used to explain the properties of a gas under ideal pressure, volume, and temperature conditions. It is best suited for describing monatomic gases (gases in which atoms are not bound together) and gases at high temperatures and low pressures. It is not well-suited for instances in which a gas or its components are close to their condensation point. All collisions are perfectly elastic and there are no intermolecular attractive forces at work. The ideal gas law is a way to explain and measure the macroscopic properties of matter. It can be derived from the kinetic theory of gases, which deals with the microscopic properties of matter. The equation for the ideal gas law is $PV = nRT$, where "P" is absolute pressure, "V" is absolute volume, and "T" is absolute temperature. "R" refers to the universal gas constant, which is 8.3145 J/mol Kelvin, and "n" is the number of moles.

Physics

Thermodynamics

Thermodynamics is a branch of physics that studies the conversion of energy into work and heat. It is especially concerned with variables such as temperature, volume, and pressure. Thermodynamic equilibrium refers to objects that have the same temperature because heat is transferred between them to reach equilibrium. Thermodynamics takes places within three different types of systems; open, isolated, and closed systems. Open systems are capable of interacting with a surrounding environment and can exchange heat, work (energy), and matter outside their system boundaries. A closed system can exchange heat and work, but not matter. An isolated system cannot exchange heat, work, or matter with its surroundings. Its total energy and mass stay the same. In physics, surrounding environment refers to everything outside a thermodynamic system (system). The terms "surroundings" and "environment" are also used. The term "boundary" refers to the division between the system and its surroundings.

The laws of thermodynamics are generalized principles dealing with energy and heat.
- The zeroth law of thermodynamics states that two objects in thermodynamic equilibrium with a third object are also in equilibrium with each other. Being in thermodynamic equilibrium basically means that different objects are at the same temperature.
- The first law deals with conservation of energy. It states that neither mass nor energy can be destroyed; only converted from one form to another.
- The second law states that the entropy (the amount of energy in a system that is no longer available for work or the amount of disorder in a system) of an isolated system can only increase. The second law also states that heat is not transferred from a lower-temperature system to a higher-temperature one unless additional work is done.
- The third law of thermodynamics states that as temperature approaches absolute zero, entropy approaches a constant minimum. It also states that a system cannot be cooled to absolute zero.

Thermal contact refers to energy transferred to a body by a means other than work. A system in thermal contact with another can exchange energy with it through the process of heat transfer. Thermal contact does not necessarily involve direct physical contact. Heat is energy that can be transferred from one body or system to another without work being done. Everything tends to become less organized and less useful over time (entropy). In all energy transfers, therefore, the overall result is that the heat is spread out so that objects are in thermodynamic equilibrium and the heat can no longer be transferred without additional work.

The laws of thermodynamics state that energy can be exchanged between physical systems as heat or work, and that systems are affected by their surroundings. It can be said that the total amount of energy in the universe is constant. The first law is mainly concerned with the conservation of energy and related concepts, which include the statement that energy can only be transferred or converted, not created or destroyed. The formula used to represent the first law is $\Delta U = Q - W$, where ΔU is the change in total internal energy of a system, Q is the heat added to the system, and W is the work done by the system. Energy can be transferred by conduction, convection, radiation, mass transfer, and other processes such as collisions in chemical and nuclear reactions. As transfers occur, the matter involved becomes less ordered and less useful. This tendency towards disorder is also referred to as entropy.

The second law of thermodynamics explains how energy can be used. In particular, it states that heat will not transfer spontaneously from a cold object to a hot object. Another way to say this is that heat transfers occur from higher temperatures to lower temperatures. Also covered under this law is the concept that systems not under the influence of external forces tend to become more disordered over time. This type of disorder can be expressed in terms of entropy. Another principle covered under this law is that it is impossible to make a heat engine that can extract heat and convert it all to useful work. A thermal bottleneck occurs in machines that convert energy to heat and then use it to do work. These types of machines are less efficient than ones that are solely mechanical.

Conduction is a form of heat transfer that occurs at the molecular level. It is the result of molecular agitation that occurs within an object, body, or material while the material stays motionless. An example of this is when a frying pan is placed on a hot burner. At first, the handle is not hot. As the pan becomes hotter due to conduction, the handle eventually gets hot too. In this example, energy is being transferred down the handle toward the colder end because the higher speed particles collide with and transfer energy to the slower ones. When this happens, the original material becomes cooler and the second material becomes hotter until equilibrium is reached. Thermal conduction can also occur between two substances such as a cup of hot coffee and the colder surface it is placed on. Heat is transferred, but matter is not.

Convection refers to heat transfer that occurs through the movement or circulation of fluids (liquids or gases). Some of the fluid becomes or is hotter than the surrounding fluid, and is less dense. Heat is transferred away from the source of the heat to a cooler, denser area. Examples of convection are boiling water and the movement of warm and cold air currents in the atmosphere and the ocean. Forced convection occurs in convection ovens, where a fan helps circulate hot air.

Radiation is heat transfer that occurs through the emission of electromagnetic waves, which carry energy away from the emitting object. All objects with temperatures above absolute zero radiate heat.

Temperature is a measurement of an object's stored heat energy. More specifically, temperature is the average kinetic energy of an object's particles. When the temperature of an object increases and its atoms move faster, kinetic energy also increases. Temperature is not energy since it changes and is not conserved. Thermometers are used to measure temperature.

There are three main scales for measuring temperature. Celsius uses the base reference points of water freezing at 0 degrees and boiling at 100 degrees. Fahrenheit uses the base reference points of water freezing at 32 degrees and boiling at 212 degrees. Celsius and Fahrenheit are both relative temperature scales since they use water as their reference point. The Kelvin temperature scale is an absolute temperature scale. Its zero mark corresponds to absolute zero. Water's freezing and boiling points are 273.15 Kelvin and 373.15 Kelvin, respectively. Where Celsius and Fahrenheit are measured is degrees, Kelvin does not use degree terminology.

- Converting Celsius to Fahrenheit: $°F = \frac{9}{5}°C + 32$
- Converting Fahrenheit to Celsius: $°C = \frac{5}{9}(°F - 32)$
- Converting Celsius to Kelvin: $K = °C + 273.15$
- Converting Kelvin to Celsius: $°C = K - 273.15$

Heat capacity, also known as thermal mass, refers to the amount of heat energy required to raise the temperature of an object, and is measured in Joules per Kelvin or Joules per degree Celsius. The equation for relating heat energy to heat capacity is $Q = C\Delta T$, where Q is the heat energy transferred, C is the heat capacity of the body, and ΔT is the change in the object's temperature. Specific heat capacity, also known as specific heat, is the heat capacity per unit mass. Every element and compound has its own specific heat. For example, it takes different amounts of heat energy to raise the temperature of the same amounts of magnesium and lead by one degree. The equation for relating heat energy to specific heat capacity is $Q = mc\Delta T$, where m represents the mass of the object, and c represents its specific heat capacity.

Some discussions of energy consider only two types of energy: kinetic energy (the energy of motion) and potential energy (which depends on relative position or orientation). There are, however, other types of energy. Electromagnetic waves, for example, are a type of energy contained by a field. Another type of potential energy is electrical energy, which is the energy it takes to pull apart positive and negative electrical charges. Chemical energy refers to the manner in which atoms form into molecules, and this energy can be released or absorbed when molecules regroup. Solar energy comes in the form of visible light and non-visible light, such as infrared and ultraviolet rays. Sound energy refers to the energy in sound waves.

Energy is constantly changing forms and being transferred back and forth. An example of a heat to mechanical energy transformation is a steam engine, such as the type used on a steam locomotive. A heat source such as coal is used to boil water. The steam produced turns a shaft, which eventually turns the wheels. A pendulum swinging is an example of both a kinetic to potential and a potential to kinetic energy transformation. When a pendulum is moved from its center point (the point at which it is closest to the ground) to the highest point before it returns, it is an example of a kinetic to potential transformation. When it swings from its highest point toward the center, it is considered a potential to kinetic transformation. The sum of the potential and kinetic energy is known as the total mechanical energy. Stretching a rubber band gives it potential energy. That potential energy becomes kinetic energy when the rubber band is released.

Motion and Force

Mechanics is the study of matter and motion, and the topics related to matter and motion, such as force, energy, and work. Discussions of mechanics will often include the concepts of vectors and scalars. Vectors are quantities with both magnitude and direction, while scalars have only magnitude. Scalar quantities include length, area, volume, mass, density, energy, work, and power. Vector quantities include displacement, direction, velocity, acceleration, momentum, and force.

Motion is a change in the location of an object, and is the result of an unbalanced net force acting on the object. Understanding motion requires the understanding of three basic quantities: displacement, velocity, and acceleration.

Displacement
When something moves from one place to another, it has undergone *displacement*. Displacement along a straight line is a very simple example of a vector quantity. If an object travels from position $x = -5$ cm to $x = 5$ cm, it has undergone a displacement of 10 cm. If it traverses the same path in the opposite direction, its displacement is -10 cm. A vector that spans the object's displacement in the direction of travel is known as a displacement vector.

Velocity
There are two types of velocity to consider: *average velocity* and *instantaneous velocity*. Unless an object has a constant velocity or we are explicitly given an equation for the velocity, finding the instantaneous velocity of an object requires the use of calculus. If we want to calculate the *average velocity* of an object, we need to know two things: the displacement, or the distance it has covered, and the time it took to cover this distance. The formula for average velocity is simply the distance traveled divided by the time required. In other words, the average velocity is equal to the change in position divided by the change in time. Average velocity is a vector and will always point in the same direction as the displacement vector (since time is a scalar and always positive).

Acceleration
Acceleration is the change in the velocity of an object. On most test questions, the acceleration will be a constant value. Like position and velocity, acceleration is a vector quantity and will therefore have both magnitude and direction.

Most motion can be explained by Newton's three laws of motion:

Newton's first law
An object at rest or in motion will remain at rest or in motion unless acted upon by an external force. This phenomenon is commonly referred to as inertia, the tendency of a body to remain in its present state of motion. In order for the body's state of motion to change, it must be acted on by an unbalanced force.

Newton's second law
An object's acceleration is directly proportional to the net force acting on the object, and inversely proportional to the object's mass. It is generally written in equation form $F = ma$, where F is the net force acting on a body, m is the mass of the body, and a is its acceleration. Note that since the mass is always a positive quantity, the acceleration is always in the same direction as the force.

Newton's third law

For every force, there is an equal and opposite force. When a hammer strikes a nail, the nail hits the hammer just as hard. If we consider two objects, A and B, then we may express any contact between these two bodies with the equation $F_{AB} = -F_{BA}$, where the order of the subscripts denotes which body is exerting the force. At first glance, this law might seem to forbid any movement at all since every force is being countered with an equal opposite force, but these equal opposite forces are acting on different bodies with different masses, so they will not cancel each other out.

Energy

The two types of energy most important in mechanics are potential and kinetic energy. Potential energy is the amount of energy an object has stored within itself because of its position or orientation. There are many types of potential energy, but the most common is gravitational potential energy. It is the energy that an object has because of its height (h) above the ground. It can be calculated as $PE = mgh$, where m is the object's mass and g is the acceleration of gravity. Kinetic energy is the energy of an object in motion, and is calculated as $KE = mv^2/2$, where v is the magnitude of its velocity. When an object is dropped, its potential energy is converted into kinetic energy as it falls. These two equations can be used to calculate the velocity of an object at any point in its fall.

Work

Work can be thought of as the amount of energy expended in accomplishing some goal. The simplest equation for mechanical work (W) is $W = Fd$, where F is the force exerted and d is the displacement of the object on which the force is exerted. This equation requires that the force be applied in the same direction as the displacement. If this is not the case, then the work may be calculated as $W = Fd \cos(\theta)$, where θ is the angle between the force and displacement vectors. If force and displacement have the same direction, then work is positive; if they are in opposite directions, then work is negative; and if they are perpendicular, the work done by the force is zero.

As an example, if a man pushes a block horizontally across a surface with a constant force of 10 N for a distance of 20 m, the work done by the man is 200 N-m or 200 J. If instead the block is sliding and the man tries to slow its progress by pushing against it, his work done is -200 J, since he is pushing in the direction opposite the motion. If the man pushes vertically downward on the block while it slides, his work done is zero, since his force vector is perpendicular to the displacement vector of the block.

Friction

Friction is a force that arises as a resistance to motion where two surfaces are in contact. The maximum magnitude of the frictional force (f) can be calculated as $f = F_c \mu$, where F_c is the contact force between the two objects and μ is a coefficient of friction based on the surfaces' material composition. Two types of friction are static and kinetic. To illustrate these concepts, imagine a book resting on a table. The force of its weight (W) is equal and opposite to the force of the table on the book, or the normal force (N). If we exert a small force (F) on the book, attempting to push it to one side, a frictional force (f) would arise, equal and opposite to our force. At this point, it is a *static frictional force* because the book is not moving. If we increase our force on the book, we will eventually cause it to move. At this point, the frictional force opposing us will be a *kinetic frictional force*. Generally, the kinetic frictional force is lower than static frictional force (because the frictional coefficient for static friction is larger), which means that the amount of force needed to maintain the movement of the book will be less than what was needed to start it moving.

Gravitational force

Gravitational force is a universal force that causes every object to exert a force on every other object. The gravitational force between two objects can be described by the formula, $F = Gm_1m_2/r^2$, where m_1 and m_2 are the masses of two objects, r is the distance between them, and G is the gravitational constant, $G = 6.672 \times 10^{-11}$ N-m²/kg². In order for this force to have a noticeable effect, one or both of the objects must be extremely large, so the equation is generally only used in problems involving planetary bodies. For problems involving objects on the earth being affected by earth's gravitational pull, the force of gravity is simply calculated as $F = mg$, where g is 9.81 m/s² toward the ground.

Electrical force

Electrical force is a universal force that exists between any two electrically charged objects. Opposite charges attract one another and like charges repel one another. The magnitude of the force is directly proportional to the magnitude of the charges (q) and inversely proportional to the square of the distance (r) between the two objects: $F = kq_1q_2/r^2$, where $k = 9 \times 10^9$ N-m²/C². Magnetic forces operate on a similar principle.

Buoyancy

Archimedes's principle states that a buoyant (upward) force on a submerged object is equal to the weight of the liquid displaced by the object. Water has a density of one gram per cubic centimeter. Anything that floats in water has a lower density, and anything that sinks has a higher density. This principle of buoyancy can also be used to calculate the volume of an irregularly shaped object. The mass of the object (m) minus its apparent mass in the water (m_a) divided by the density of water (ρ_w), gives the object's volume: $V = (m-m_a)/\rho_w$.

Machines

Simple machines include the inclined plane, lever, wheel and axle, and pulley. These simple machines have no internal source of energy. More complex or compound machines can be formed from them. Simple machines provide a force known as a mechanical advantage and make it easier to accomplish a task. The inclined plane enables a force less than the object's weight to be used to push an object to a greater height. A lever enables a multiplication of force. The wheel and axle allows for movement with less resistance. Single or double pulleys allows for easier direction of force. The wedge and screw are forms of the inclined plane. A wedge turns a smaller force working over a greater distance into a larger force. The screw is similar to an incline that is wrapped around a shaft.

A certain amount of work is required to move an object. The amount cannot be reduced, but by changing the way the work is performed a mechanical advantage can be gained. A certain amount of work is required to raise an object to a given vertical height. By getting to a given height at an angle, the effort required is reduced, but the distance that must be traveled to reach a given height is increased. An example of this is walking up a hill. One may take a direct, shorter, but steeper route, or one may take a more meandering, longer route that requires less effort. Examples of wedges include doorstops, axes, plows, zippers, and can openers.

A lever consists of a bar or plank and a pivot point or fulcrum. Work is performed by the bar, which swings at the pivot point to redirect the force. There are three types of levers: first, second, and third class. Examples of a first-class lever include balances, see-saws, nail extractors, and scissors (which also use wedges). In a second-class lever the fulcrum is placed at one end of the bar and the work is performed at the other end. The weight or load to be moved is in between. The closer to the fulcrum the weight is, the easier it is to move. Force is increased, but the distance it is moved is decreased. Examples include pry bars, bottle openers, nutcrackers, and wheelbarrows. In a third-class lever the fulcrum is at one end and the positions of the weight and the location where the work is performed are reversed. Examples include fishing rods, hammers, and tweezers.

The center of a wheel and axle can be likened to a fulcrum on a rotating lever. As it turns, the wheel moves a greater distance than the axle, but with less force. Obvious examples of the wheel and axle are the wheels of a car, but this type of simple machine can also be used to exert a greater force. For instance, a person can turn the handles of a winch to exert a greater force at the turning axle to move an object. Other examples include steering wheels, wrenches, faucets, waterwheels, windmills, gears, and belts. Gears work together to change a force. The four basic types of gears are spur, rack and pinion, bevel, and worm gears. The larger gear turns slower than the smaller, but exerts a greater force. Gears at angles can be used to change the direction of forces.

A single pulley consists of a rope or line that is run around a wheel. This allows force to be directed in a downward motion to lift an object. This does not decrease the force required, just changes its direction. The load is moved the same distance as the rope pulling it. When a combination pulley is used, such as a double pulley, the weight is moved half the distance of the rope pulling it. In this way, the work effort is doubled. Pulleys are never 100% efficient because of friction. Examples of pulleys include cranes, chain hoists, block and tackles, and elevators.

Electrical Charges

A glass rod and a plastic rod can illustrate the concept of static electricity due to friction. Both start with no charge. A glass rod rubbed with silk produces a positive charge, while a plastic rod rubbed with fur produces a negative charge. The electron affinity of a material is a property that helps determine how easily it can be charged by friction. Materials can be sorted by their affinity for electrons into a triboelectric series. Materials with greater affinities include celluloid, sulfur, and rubber. Materials with lower affinities include glass, rabbit fur, and asbestos. In the example of a glass rod and a plastic one, the glass rod rubbed with silk acquires a positive charge because glass has a lower affinity for electrons than silk. The electrons flow to the silk, leaving the rod with fewer electrons and a positive charge. When a plastic rod is rubbed with fur, electrons flow to the rod and result in a negative charge.

The attractive force between the electrons and the nucleus is called the electric force. A positive (+) charge or a negative (-) charge creates a field of sorts in the empty space around it, which is known as an electric field. The direction of a positive charge is away from it and the direction of a negative charge is towards it. An electron within the force of the field is pulled towards a positive charge because an electron has a negative charge. A particle with a positive charge is pushed away, or repelled, by another positive charge. Like charges repel each other and opposite charges attract. Lines of force show the paths of charges. Electric force between two objects is directly proportional to the product of the charge magnitudes and inversely proportional to the square of the distance between the two objects. Electric charge is measured with the unit Coulomb (C). It is the amount of charge moved in one second by a steady current of one ampere (1C = 1A × 1s).

Insulators are materials that prevent the movement of electrical charges, while conductors are materials that allow the movement of electrical charges. This is because conductive materials have free electrons that can move through the entire volume of the conductor. This allows an external charge to change the charge distribution in the material. In induction, a neutral conductive material, such as a sphere, can become charged by a positively or negatively charged object, such as a rod.

The charged object is placed close to the material without touching it. This produces a force on the free electrons, which will either be attracted to or repelled by the rod, polarizing (or separating) the charge. The sphere's electrons will flow into or out of it when touched by a ground. The sphere is now charged. The charge will be opposite that of the charging rod.

Charging by conduction is similar to charging by induction, except that the material transferring the charge actually touches the material receiving the charge. A negatively or positively charged object is touched to an object with a neutral charge. Electrons will either flow into or out of the neutral object and it will become charged. Insulators cannot be used to conduct charges. Charging by conduction can also be called charging by contact. The law of conservation of charge states that the total number of units before and after a charging process remains the same. No electrons have been created. They have just been moved around. The removal of a charge on an object by conduction is called grounding.

Circuits

Electric potential, or electrostatic potential or voltage, is an expression of potential energy per unit of charge. It is measured in volts (V) as a scalar quantity. The formula used is $V = E/Q$, where V is voltage, E is electrical potential energy, and Q is the charge. Voltage is typically discussed in the context of electric potential difference between two points in a circuit. Voltage can also be thought of as a measure of the rate at which energy is drawn from a source in order to produce a flow of electric charge.

Electric current is the sustained flow of electrons that are part of an electric charge moving along a path in a circuit. This differs from a static electric charge, which is a constant non-moving charge rather than a continuous flow. The rate of flow of electric charge is expressed using the ampere (amp or A) and can be measured using an ammeter. A current of 1 ampere means that 1 coulomb of charge passes through a given area every second. Electric charges typically only move from areas of high electric potential to areas of low electric potential. To get charges to flow into a high potential area, you must to connect it to an area of higher potential, by introducing a battery or other voltage source.

Electric currents experience resistance as they travel through a circuit. Different objects have different levels of resistance. The ohm (Ω) is the measurement unit of electric resistance. The symbol is the Greek letter omega. Ohm's Law, which is expressed as $I = V/R$, states that current flow (I, measured in amps) through an object is equal to the potential difference from one side to the other (V, measured in volts) divided by resistance (R, measured in ohms). An object with a higher resistance will have a lower current flow through it given the same potential difference.

Movement of electric charge along a path between areas of high electric potential and low electric potential, with a resistor or load device between them, is the definition of a simple circuit. It is a closed conducting path between the high and low potential points, such as the positive and negative terminals on a battery. One example of a circuit is the flow from one terminal of a car battery to the other. The electrolyte solution of water and sulfuric acid provides work in chemical form to start the flow. A frequently used classroom example of circuits involves using a D cell (1.5 V) battery, a small light bulb, and a piece of copper wire to create a circuit to light the bulb.

Magnets

A magnet is a piece of metal, such as iron, steel, or magnetite (lodestone) that can affect another substance within its field of force that has like characteristics. Magnets can either attract or repel other substances. Magnets have two poles: north and south. Like poles repel and opposite poles (pairs of north and south) attract. The magnetic field is a set of invisible lines representing the paths of attraction and repulsion. Magnetism can occur naturally, or ferromagnetic materials can be magnetized. Certain matter that is magnetized can retain its magnetic properties indefinitely and become a permanent magnet. Other matter can lose its magnetic properties. For example, an iron nail can be temporarily magnetized by stroking it repeatedly in the same direction using one pole of another magnet. Once magnetized, it can attract or repel other magnetically inclined materials, such as paper clips. Dropping the nail repeatedly will cause it to lose its charge.

The motions of subatomic structures (nuclei and electrons) produce a magnetic field. It is the direction of the spin and orbit that indicate the direction of the field. The strength of a magnetic field is known as the magnetic moment. As electrons spin and orbit a nucleus, they produce a magnetic field. Pairs of electrons that spin and orbit in opposite directions cancel each other out, creating a net magnetic field of zero. Materials that have an unpaired electron are magnetic. Those with a weak attractive force are referred to as paramagnetic materials, while ferromagnetic materials have a strong attractive force. A diamagnetic material has electrons that are paired, and therefore does not typically have a magnetic moment. There are, however, some diamagnetic materials that have a weak magnetic field.

A magnetic field can be formed not only by a magnetic material, but also by electric current flowing through a wire. When a coiled wire is attached to the two ends of a battery, for example, an electromagnet can be formed by inserting a ferromagnetic material such as an iron bar within the coil. When electric current flows through the wire, the bar becomes a magnet. If there is no current, the magnetism is lost. A magnetic domain occurs when the magnetic fields of atoms are grouped and aligned. These groups form what can be thought of as miniature magnets within a material. This is what happens when an object like an iron nail is temporarily magnetized. Prior to magnetization, the organization of atoms and their various polarities are somewhat random with respect to where the north and south poles are pointing. After magnetization, a significant percentage of the poles are lined up in one direction, which is what causes the magnetic force exerted by the material.

Waves

Waves have energy and can transfer energy when they interact with matter. Although waves transfer energy, they do not transport matter. They are a disturbance of matter that transfers energy from one particle to an adjacent particle. There are many types of waves, including sound, seismic, water, light, micro, and radio waves.

The two basic categories of waves are mechanical and electromagnetic. Mechanical waves are those that transmit energy through matter. Electromagnetic waves can transmit energy through a vacuum. A transverse wave provides a good illustration of the features of a wave, which include crests, troughs, amplitude, and wavelength.

There are a number of important attributes of waves. Frequency is a measure of how often particles in a medium vibrate when a wave passes through the medium with respect to a certain point or node. Usually measured in Hertz (Hz), frequency might refer to cycles per second, vibrations per second, or waves per second. One Hz is equal to one cycle per second..

Period is a measure of how long it takes to complete a cycle. It is the inverse of frequency; where frequency is measure in cycles per second, period can be thought of as seconds per cycle, though it is measured in units of time only.

Speed refers to how fast or slow a wave travels. It is measured in terms of distance divided by time. While frequency is measured in terms of cycles per second, speed might be measured in terms of meters per second.

Amplitude is the maximum amount of displacement of a particle in a medium from its rest position, and corresponds to the amount of energy carried by the wave. High energy waves have greater amplitudes; low energy waves have lesser amplitudes. Amplitude is a measure of a wave's strength.

Rest position, also called equilibrium, is the point at which there is neither positive nor negative displacement.

Crest, also called the peak, is the point at which a wave's positive or upward displacement from the rest position is at its maximum. Trough, also called a valley, is the point at which a wave's negative or downward displacement from the rest position is at its maximum. A wavelength is one complete wave cycle. It could be measured from crest to crest, trough to trough, rest position to rest position, or any point of a wave to the corresponding point on the next wave.

Sound is a pressure disturbance that moves through a medium in the form of mechanical waves, which transfer energy from one particle to the next. Sound requires a medium to travel through, such as air, water, or other matter since it is the vibrations that transfer energy to adjacent particles, not the actual movement of particles over a great distance. Sound is transferred through the movement of atomic particles, which can be atoms or molecules. Waves of sound energy move outward in all directions from the source. Sound waves consist of compressions (particles are forced together) and rarefactions (particles move farther apart and their density decreases). A wavelength consists of one compression and one rarefaction. Different sounds have different wavelengths. Sound is a form of kinetic energy.

The electromagnetic spectrum is defined by frequency (f) and wavelength (λ). Frequency is typically measured in hertz and wavelength is usually measured in meters. Because light travels at a fairly constant speed, frequency is inversely proportional to wavelength, a relationship expressed by the formula $f = c/\lambda$, where c is the speed of light (about 300 million meters per second). Frequency multiplied by wavelength equals the speed of the wave; for electromagnetic waves, this is the speed of light, with some variance for the medium in which it is traveling. Electromagnetic waves include (from largest to smallest wavelength) radio waves, microwaves, infrared radiation (radiant heat), visible light, ultraviolet radiation, x-rays, and gamma rays. The energy of electromagnetic waves is carried in packets that have a magnitude inversely proportional to the wavelength. Radio waves have a range of wavelengths, from about 10^{-3} to 10^5 meters, while their frequencies range from 10^3 to about 10^{11} Hz.

Atoms and molecules can gain or lose energy only in particular, discrete amounts. Therefore, they can absorb and emit light only at wavelengths that correspond to these amounts. Using a process known as spectroscopy, these characteristic wavelengths can be used to identify substances.

Light is the portion of the electromagnetic spectrum that is visible because of its ability to stimulate the retina. It is absorbed and emitted by electrons, atoms, and molecules that move from one energy level to another. Visible light interacts with matter through molecular electron excitation (which occurs in the human retina) and through plasma oscillations (which occur in metals). Visible light is between ultraviolet and infrared light on the spectrum. The wavelengths of visible light cover a range from 380 nm (violet) to 760 nm (red). Different wavelengths correspond to different colors. The human brain interprets or perceives visible light, which is emitted from the sun and other stars, as color. For example, when the entire wavelength reaches the retina, the brain perceives the color white. When no part of the wavelength reaches the retina, the brain perceives the color black.

When light waves encounter an object, they are either reflected, transmitted, or absorbed. If the light is reflected from the surface of the object, the angle at which it contacts the surface will be the same as the angle at which it leaves, on the other side of the perpendicular. If the ray of light is perpendicular to the surface, it will be reflected back in the direction from which it came. When light is transmitted through the object, its direction may be altered upon entering the object. This is known as refraction. The degree to which the light is refracted depends on the speed at which light travels in the object. Light that is neither reflected nor transmitted will be absorbed by the surface and stored as heat energy. Nearly all instances of light hitting an object will involve a combination of two or even all three of these.

When light waves are refracted, or bent, an image can appear distorted. Sound waves and water waves can also be refracted. Diffraction refers to the bending of waves around small objects and the spreading out of waves past small openings. The narrower the opening, the greater the level of diffraction will be. Larger wavelengths also increase diffraction. A diffraction grating can be created by placing a number of slits close together, and is used more frequently than a prism to separate light. Different wavelengths are diffracted at different angles. The particular color of an object depends upon what is absorbed and what is transmitted or reflected. For example, a leaf consists of chlorophyll molecules, the atoms of which absorb all wavelengths of the visible light spectrum except for green, which is why a leaf appears green. Certain wavelengths of visible light can be absorbed when they interact with matter. Wavelengths that are not absorbed can be transmitted by transparent materials or reflected by opaque materials.

The various properties of light have numerous real life applications. For example, polarized sunglasses have lenses that help reduce glare, while non-polarized sunglasses reduce the total amount of light that reaches the eyes. Polarized lenses consist of a chemical film of molecules aligned in parallel. This allows the lenses to block wavelengths of light that are intense, horizontal, and reflected from smooth, flat surfaces. The "fiber" in fiber optics refers to a tube or pipe that channels light. Because of the composition of the fiber, light can be transmitted greater distances before losing the signal. The fiber consists of a core, cladding, and a coating. Fibers are bundled, allowing for the transmission of large amounts of data.

Common Subject: Mechanical Comprehension

In order to succeed on the mechanical concepts section of the exam, you will need to be familiar with basic concepts in physics and mechanics. Don't worry: the exam does not dwell on obscure theories or require you to make complicated calculations. The equations that are included in this section of the book are meant to illustrate the relationships of physics, not to show you how to solve numerical problems. You do, however, need to understand the essential properties of physics, and how they apply to real-life situations. In order to help you along, we have included a full primer on all of the concepts that may come up on this section of the exam. Important terms and concepts are placed in bold. Finally, although we have tried to make this section of the guidebook as easy to read as possible, we still recommend that you take your time and avoid reading in a hurry. You may need to read some of this information a few times before fully absorbing it. Whenever possible, try to imagine some everyday examples for the concepts we discuss; after all, applying the theories of physics to the materials of everyday life is one way to define mechanics.

Kinematics

To begin, we will look at the basics of physics. At its heart, physics is just a set of explanations for the ways in which matter and energy behave. There are three key concepts used to describe how matter moves:
1. Displacement
2. Velocity
3. Acceleration

Displacement

Concept: where and how far an object has gone
Calculation: final position – initial position

When something changes its location from one place to another, it is said to have undergone displacement. If a golf ball is hit across a sloped green into the hole, the displacement only takes into account the final and initial locations, not the path of the ball.

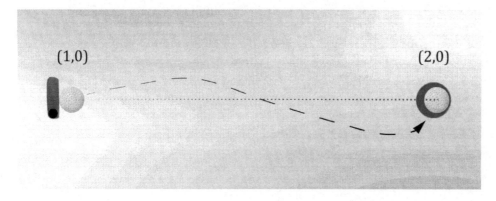

Displacement along a straight line is a very simple example of a vector quantity: that is, it has both a magnitude and a direction. Direction is as important as magnitude in many measurements. If we

can determine the original and final position of the object, then we can determine the total displacement with this simple equation:
$$\text{Displacement} = \text{final position} - \text{original position}$$

The hole (final position) is at the Cartesian coordinate location (2, 0) and the ball is hit from the location (1, 0). The displacement is:
$$\text{Displacement} = (2,0) - (1,0)$$
$$\text{Displacement} = (1,0)$$

The displacement has a magnitude of 1 and a direction of the positive x direction.

Velocity

Concept: the rate of moving from one position to another
Calculation: change in position / change in time

Velocity answers the question, "How quickly is an object moving?" For example, if a car and a plane travel between two cities which are a hundred miles apart, but the car takes two hours and the plane takes one hour, the car has the same displacement as the plane, but a smaller velocity.

In order to solve some of the problems on the exam, you may need to assess the velocity of an object. If we want to calculate the average velocity of an object, we must know two things. First, we must know its displacement. Second, we must know the time it took to cover this distance. The formula for average velocity is quite simple:
$$\text{average velocity} = \frac{\text{displacement}}{\text{change in time}}$$
Or
$$\text{average velocity} = \frac{\text{final position} - \text{original position}}{\text{final time} - \text{original time}}$$

To complete the example, the velocity of the plane is calculated to be:
$$\text{plane average velocity} = \frac{100 \text{ miles}}{1 \text{ hour}} = 100 \text{ miles per hour}$$

The velocity of the car is less:
$$\text{car average velocity} = \frac{100 \text{ miles}}{2 \text{ hours}} = 50 \text{ miles per hour}$$

Often, people confuse the words *speed* and *velocity*. There is a significant difference. The average velocity is based on the amount of displacement, a vector. Alternately, the average speed is based on the distance covered or the path length. The equation for speed is:

$$\text{average speed} = \frac{\text{total distance traveled}}{\text{change in time}}$$

Notice that we used total distance and *not* change in position, because speed is path-dependent.

If the plane traveling between cities had needed to fly around a storm on its way, making the distance traveled 50 miles greater than the distance the car traveled, the plane would still have the same total displacement as the car.

The calculation for the speed: For this reason, average speed can be calculated:

$$\text{plane average speed} = \frac{150 \text{ miles}}{1 \text{ hour}} = 150 \text{ miles per hour}$$

$$\text{car average speed} = \frac{100 \text{ miles}}{2 \text{ hours}} = 50 \text{ miles per hour}$$

Acceleration

Concept: how quickly something changes from one velocity to another
Calculation: change in velocity / change in time

Acceleration is the rate of change of the velocity of an object. If a car accelerates from zero velocity to 60 miles per hour (88 feet per second) in two seconds, the car has an impressive acceleration. But if a car performs the same change in velocity in eight seconds, the acceleration is much lower and not as impressive.

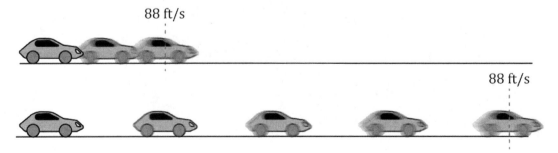

To calculate average acceleration, we may use the equation:
$$\text{average acceleration} = \frac{\text{change in velocity}}{\text{change in time}}$$

The acceleration of the cars is found to be:
$$\text{Car \#1 average acceleration} = \frac{88 \text{ feet per second}}{2 \text{ seconds}} = 44 \frac{\text{feet}}{\text{second}^2}$$
$$\text{Car \#2 average acceleration} = \frac{88 \text{ feet per second}}{8 \text{ seconds}} = 11 \frac{\text{feet}}{\text{second}^2}$$

Acceleration will be expressed in units of distance divided by time squared; for instance, meters per second squared or feet per second squared.

Projectile Motion

A specific application of the study of motion is projectile motion. Simple projectile motion occurs when an object is in the air and experiencing only the force of gravity. We will disregard drag for this topic. Some common examples of projectile motion are thrown balls, flying bullets, and falling rocks. The characteristics of projectile motion are:
1. The horizontal component of velocity doesn't change
2. The vertical acceleration due to gravity affects the vertical component of velocity

Because gravity only acts downwards, objects in projectile motion only experience acceleration in the y direction (vertical). The horizontal component of the object's velocity does not change in flight. This means that if a rock is thrown out off a cliff, the horizontal velocity, (think the shadow if the sun is directly overhead) will not change until the ball hits the ground.

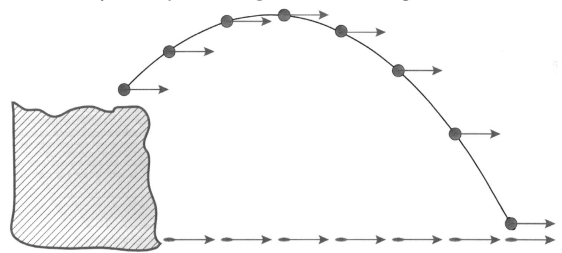

The velocity in the vertical direction is affected by gravity. Gravity imposes an acceleration of $g = 9.8 \frac{m}{s^2}$ or $32 \frac{ft}{s^2}$ downward on projectiles. The vertical component of velocity at any point is equal to:

vertical velocity = original vertical velocity − g × time

When these characteristics are combined, there are three points of particular interest in a projectile's flight. At the beginning of a flight, the object has a horizontal component and a vertical component giving it a large speed. At the top of a projectile's flight, the vertical velocity equals zero, making the top the slowest part of travel. When the object passes the same height as the launch, the vertical velocity is opposite of the initial vertical velocity making the speed equal to the initial speed.

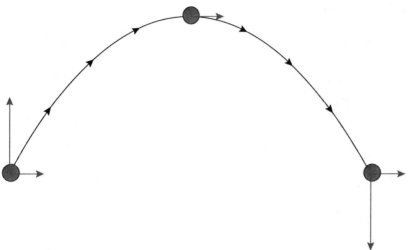

If the object continues falling below the initial height from which it was launched (e.g., it was launched from the edge of a cliff), it will have an even greater velocity than it did initially from that point until it hits the ground.

Rotational Kinematics

Concept: increasing the radius increases the linear speed
Calculation: linear speed = radius × rotational speed

Another interesting application of the study of motion is rotation. In practice, simple rotation is when an object rotates around a point at a constant speed. Most questions covering rotational kinematics will provide the distance from a rotating object to the center of rotation (radius) and ask about the linear speed of the object. A point will have a greater linear speed when it is farther from the center of rotation.

If a potter is spinning his wheel at a constant speed of one revolution per second, the clay six inches away from the center will be going faster than the clay three inches from the center. The clay directly in the center of the wheel will not have any linear velocity.

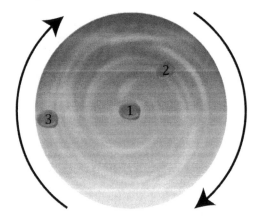

To find the linear speed of rotating objects using radians, we use the equation:

$$linear\ speed = (rotational\ speed\ [in\ radians]) \times (radius)$$

Using degrees, the equation is:

$$linear\ speed = (rotational\ speed\ [in\ degrees]) \times \frac{\pi\ radians}{180\ degrees} \times (radius)$$

To find the speed of the pieces of clay we use the known values (rotational speed of 1 revolution per second, radii of 0 inches, 3 inches, and 6 inches) and the knowledge that one revolution = 2 pi.

$$clay\ \#1\ speed = \left(2\pi \frac{rad}{s}\right) \times (0\ inches) = 0 \frac{inches}{second}$$

$$clay\ \#2\ speed = \left(2\pi \frac{rad}{s}\right) \times (3\ inches) = 18.8 \frac{inches}{second}$$

$$clay\ \#3\ speed = \left(2\pi \frac{rad}{s}\right) \times (6\ inches) = 37.7 \frac{inches}{second}$$

Cams

In the study of motion, a final application often tested is the cam. A cam and follower system allows mechanical systems to have timed, specified, and repeating motion. Although cams come in varied forms, tests focus on rotary cams. In engines, a cam shaft coordinates the valves for intake and exhaust. Cams are often used to convert rotational motion into repeating linear motion.

Cams rotate around one point. The follower sits on the edge of the cam and moves along with the edge. To understand simple cams, count the number of bumps on the cam. Each bump will cause the follower to move outwards.

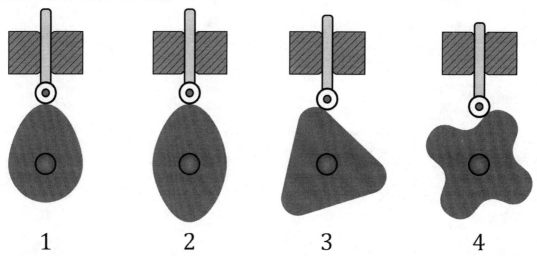

Another way to consider cams is to unravel the cam profile into a straight object. The follower will then follow the top of the profile.

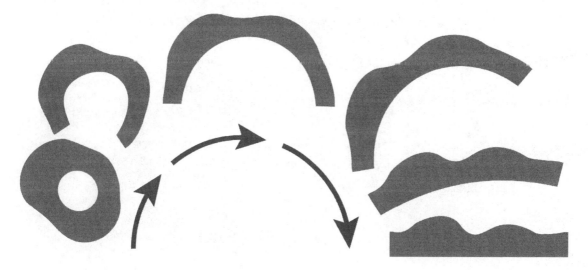

Kinetics

Newton's three laws of mechanics

The questions on the exam may require you to demonstrate familiarity with the concepts expressed in Newton's three laws of motion which relate to the concept of force.

Newton's first law – A body at rest will tend to remain at rest, while a body in motion will tend to remain in motion, unless acted upon by an external force.

Newton's second law – The acceleration of an object is directly proportional to the force being exerted on it and inversely proportional to its mass.

Newton's third law – For every force, there is an equal and opposite force.

First Law

Concept: Unless something interferes, an object won't start or stop moving

Although intuition supports the idea that objects do not start moving until a force acts on them, the idea of an object continuing forever without any forces can seem odd. Before Newton formulated his laws of mechanics, general thought held that some force had to act on an object continuously in order for it to move at a constant velocity. This seems to make sense: when an object is briefly pushed, it will eventually come to a stop. Newton, however, determined that unless some other force acted on the object (most notably friction or air resistance), it would continue in the direction it was pushed at the same velocity forever.

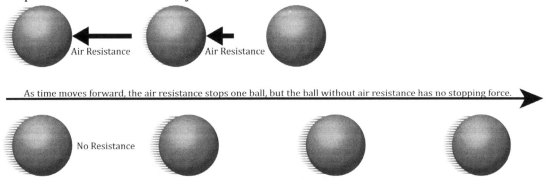

Second Law

Concept: Acceleration increases linearly with force.

Although Newton's second law can be conceptually understood as a series of relationships describing how an increase in one factor will decrease another factor, the law can be understood best in equation format:

$$Force = mass \times acceleration$$

Or

$$Acceleration = \frac{force}{mass}$$

Or

$$Mass = \frac{force}{acceleration}$$

Each of the forms of the equation allows for a different look at the same relationships. To examine the relationships, change one factor and observe the result. If a steel ball, with a diameter of 6.3 cm, has a mass of 1 kg and an acceleration of 1 m/s², then the net force on the ball will be 1 Newton.

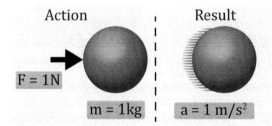

Third Law

Concept: Nothing can push or pull without being pushed or pulled in return.

When any object exerts a force on another object, the other object exerts the opposite force back on the original object. To observe this, consider two spring-based fruit scales, both tipped on their sides as shown with the weighing surfaces facing each other. If fruit scale #1 is pressing fruit scale #2 into the wall, it exerts a force on fruit scale #2, measurable by the reading on scale #2. However, because fruit scale #1 is exerting a force on scale #2, scale #2 is exerting a force on scale #1 with an opposite direction, but the same magnitude.

Force

Concept: a push or pull on an object
Calculation: $Force = mass \times acceleration$

A force is a vector which causes acceleration of a body. Force has both magnitude and direction. Furthermore, multiple forces acting on one object combine in vector addition. This can be demonstrated by considering an object placed at the origin of the coordinate plane. If it is pushed along the positive direction of the *x*-axis, it will move in this direction; if the force acting on it is in the positive direction of the *y*-axis, it will move in that direction. However, if both forces are applied at the same time, then the object will move at an angle to both the *x* and *y* axes, an angle determined by the relative amount of force exerted in each direction. In this way, we may see that the resulting force is a vector sum; that is, a net force that has both magnitude and direction.

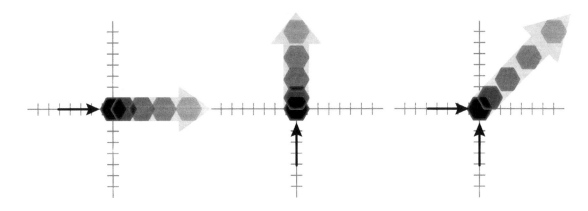

Mass

Concept: the amount of matter

Mass can be defined as the quantity of matter in an object. If we apply the same force to two objects of different mass, we will find that the resulting acceleration is different. Newton's Second Law of Motion describes the relationship between mass, force, and acceleration in the equation: ***Force = mass × acceleration***. In other words, the acceleration of an object is directly proportional to the force being exerted on it and inversely proportional to its mass.

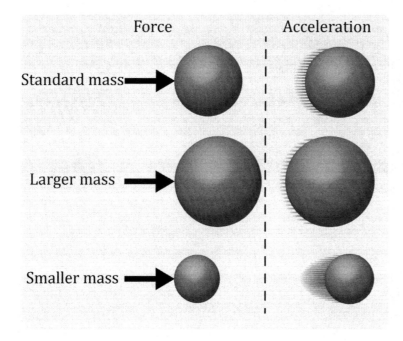

Gravity

Gravity is a force which exists between all objects with matter. Gravity is a pulling force between objects meaning that the forces on the objects point toward the opposite object. When Newton's third law is applied to gravity, the force pairs from gravity are shown to be equal in magnitude and opposite in direction.

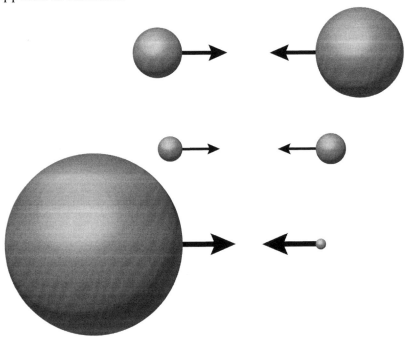

Weight

Weight is sometimes confused with mass. While mass is the amount of matter, weight is the force exerted by the earth on an object with matter by gravity. The earth pulls every object of mass toward its center while every object of mass pulls the earth toward its center. The object's pull on the earth is equal in magnitude to the pull which the earth exerts, but, because the mass of the earth is very large in comparison (5.97×10^{24} kg), only the object appears to be affected by the force.

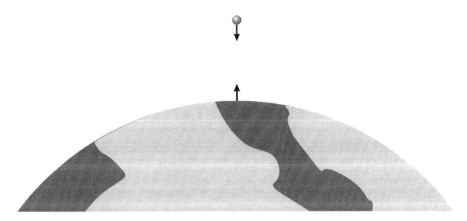

The gravity of earth causes a constant acceleration due to gravity (g) at a specific altitude. For most earthbound applications the acceleration due to gravity is 32.2 ft/s² or 9.8 m/s² in a downward direction. The equation for the force of gravity (weight) on an object is the equation from Newton's Second Law with the constant acceleration due to gravity (g).

$$Force = mass \times acceleration$$
$$Weight = mass \times acceleration\ due\ to\ gravity$$
$$W = m \times g$$

The SI (International Standard of Units) unit for weight is the Newton $\left(\frac{kg \times m}{s^2}\right)$. The English Engineering unit system uses the pound, or lb, as the unit for weight and force $\left(\frac{slug \times ft}{s^2}\right)$. Thus, a 2 kg object under the influence of gravity would have a weight of:

$$W = m \times g$$
$$W = 2\ kg \times 9.8\ \frac{m}{s^2} = 18.6\ N\ downwards$$

Normal force

Concept: the force perpendicular to a contact surface

The word "normal" is used in mathematics to mean perpendicular, and so the force known as normal force should be remembered as the perpendicular force exerted on an object that is resting on some other surface. For instance, if a box is resting on a horizontal surface, we may say that the normal force is directed upwards through the box (the opposite, downward force is the weight of the box). If the box is resting on a wedge, the normal force from the wedge is not vertical but is perpendicular to the wedge edge.

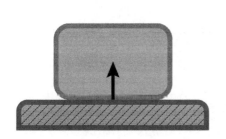
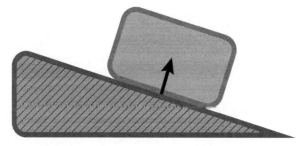

Tension

Concept: the pulling force from a cord

Another force that may come into play on the exam is called tension. Anytime a cord is attached to a body and pulled so that it is taut, we may say that the cord is under tension. The cord in tension applies a pulling tension force on the connected objects. This force is pointed away from the body and along the cord at the point of attachment. In simple considerations of tension, the cord is assumed to be both without mass and incapable of stretching. In other words, its only role is as the connector between two bodies. The cord is also assumed to pull on both ends with the same magnitude of tension force.

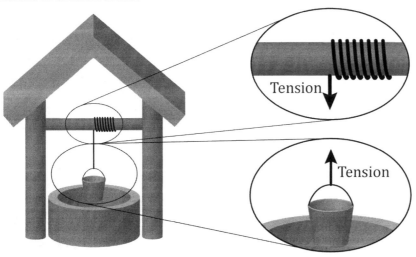

Friction

Concept: Friction is a resistance to motion between contacting surfaces

In order to illustrate the concept of friction, let us imagine a book resting on a table. As it sits, the force of its weight is equal to and opposite of the normal force. If, however, we were to exert a force on the book, attempting to push it to one side, a frictional force would arise, equal and opposite to our force. This kind of frictional force is known as static frictional force.

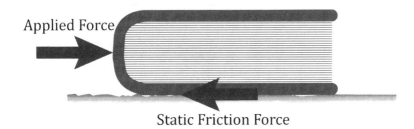

As we increase our force on the book, however, we will eventually cause it to accelerate in the direction of our force. At this point, the frictional force opposing us will be known as kinetic friction. For many combinations of surfaces, the magnitude of the kinetic frictional force is lower than that of the static frictional force, and consequently, the amount of force needed to maintain the movement of the book will be less than that needed to initiate the movement.

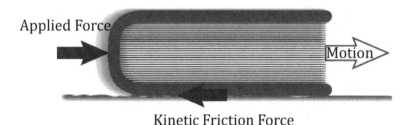

Rolling Friction

Occasionally, a question will ask you to consider the amount of friction generated by an object that is rolling. If a wheel is rolling at a constant speed, then the point at which it touches the ground will not slide, and there will be no friction between the ground and the wheel inhibiting movement. In fact, the friction at the point of contact between the wheel and the ground is static friction necessary to propulsion with wheels. When a vehicle accelerates, the static friction between the wheels and ground allows the vehicle to achieve acceleration. Without this friction, the vehicle would spin its wheels and go nowhere.

Although the static friction does not impede movement for the wheels, a combination of frictional forces can resist rolling motion. One such frictional force is bearing friction. Bearing friction is the kinetic friction between the wheel and an object it rotates around, such as a stationary axle.

Most questions will consider bearing friction the only force stopping a rotating wheel. There are many other factors that affect the efficiency of a rolling wheel such as deformation of the wheel, deformation of the surface, and force imbalances, but the net resulting friction can be modeled as a simple kinetic rolling friction. Rolling friction or rolling resistance is the catch-all friction for the combination of all the losses which impede wheels in real life.

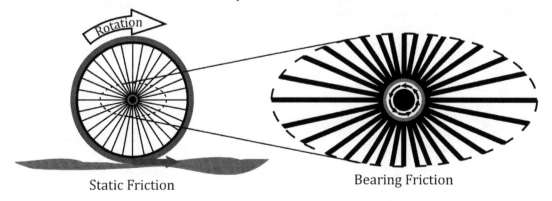

Drag force

Friction can also be generated when an object is moving through air or liquid. A drag force occurs when a body moves through some fluid (either liquid or gas) and experiences a force that opposes the motion of the body. The drag force is greater if the air or fluid is thicker or is moving in the direction opposite to the object. Obviously, the higher the drag force, the greater amount of positive force required to keep the object moving forward.

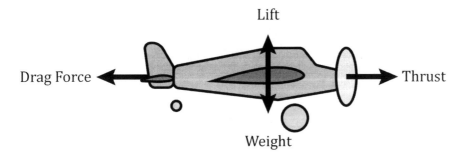

Balanced Forces

An object is in equilibrium when the sum of all forces acting on the object is zero. When the forces on an object sum to zero, the object does not accelerate. Equilibrium can be obtained when forces in the y-direction sum to zero, forces in the x-direction sum to zero, or forces in both directions sum to zero.

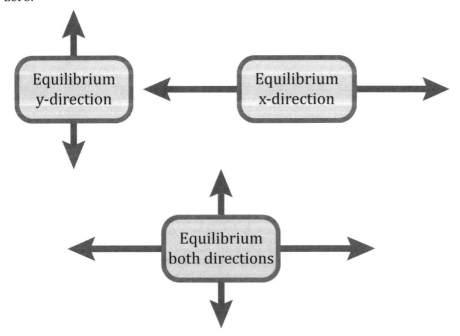

In most cases, a problem will provide one or more forces acting on object and ask for a force to balance the system. The force will be the opposite of the current force or sum of current forces.

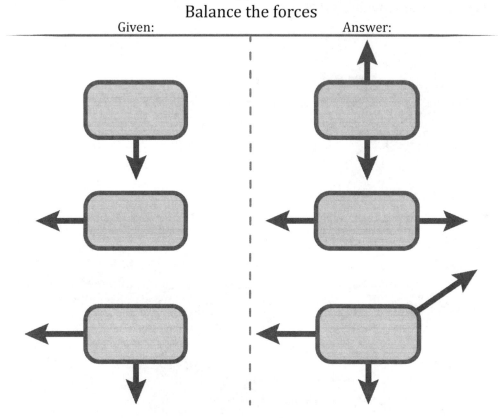

Rotational Kinetics

Many equations and concepts in linear kinematics and kinetics transfer to rotation. For example, angular position is an angle. Angular velocity, like linear velocity, is the change in the position (angle) divided by the time. Angular acceleration is the change in angular velocity divided by time. Although most tests will not require you to perform angular calculations, they will expect you to understand the angular version of force: torque.

Concept: Torque is a twisting force on an object
Calculation: $Torque = radius \times force$

Torque, like force, is a vector and has magnitude and direction. As with force, the sum of torques on an object will affect the angular acceleration of that object. The key to solving problems with torque is understanding the lever arm. A better description of the torque equation is:

Torque = force × the distance perpedicular to the force to the center of rotation

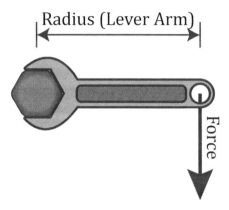

Because torque is directly proportional to the radius, or lever arm, a greater lever arm will result in a greater torque with the same amount of force. The wrench on the right has twice the radius and, as a result, twice the torque.

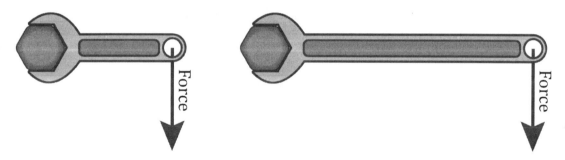

Alternatively, a greater force also increases torque. The wrench on the right has twice the force and twice the torque.

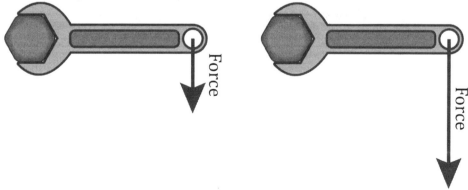

Work/Energy

Work

Concept: Work is the transfer of energy from one object to another
Calculation: Work = force × displacement

The equation for work in one dimension is fairly simple:
$$Work = Force \times displacement$$
$$W = F \times d$$

In the equation, the force and the displacement are the magnitude of the force exerted and the total change in position of the object on which the force is exerted, respectively. If force and displacement have the same direction, then the work is positive. If they are in opposite directions, however, the work is negative.

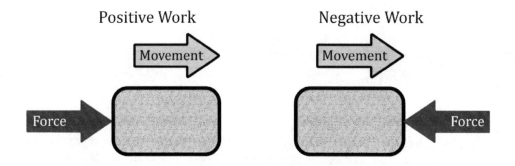

For two-dimensional work, the equation is a bit more complex:
$$Work = Force \times displacement \times cos(\theta \text{ between displacement and force})$$
$$W = F \times d \times cos(\theta)$$

The angle in the equation is the angle between the direction of the force and the direction of the displacement. Thus, the work done when a box is pulled at a 20 degree angle with a force of 100 lb for 20 ft will be less than the work done when a differently weighted box is pulled horizontally with a force of 100 lb for 20 ft.

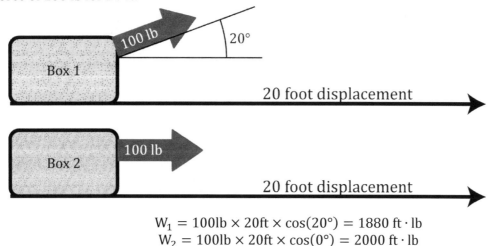

$$W_1 = 100\text{lb} \times 20\text{ft} \times \cos(20°) = 1880 \text{ ft} \cdot \text{lb}$$
$$W_2 = 100\text{lb} \times 20\text{ft} \times \cos(0°) = 2000 \text{ ft} \cdot \text{lb}$$

The unit ft · lb is the unit for both work and energy.

Energy

Concept: the ability of a body to do work on another object

Energy is a word that has found a million different uses in the English language, but in physics it refers to the measure of a body's ability to do work. In physics, energy may not have a million meanings, but it does have many forms. Each of these forms, such as chemical, electric, and nuclear, is the capability of an object to perform work. However, for the purpose of most tests, mechanical energy and mechanical work are the only forms of energy worth understanding in depth. Mechanical energy is the sum of an object's kinetic and potential energies. Although they will be introduced in greater detail, these are the forms of mechanical energy:

Kinetic Energy – energy an object has by virtue of its motion
Gravitational Potential Energy – energy by virtue of an object's height
Elastic Potential Energy – energy stored in compression or tension

Neglecting frictional forces, mechanical energy is conserved.

As an example, imagine a ball moving perpendicular to the surface of the earth, with its weight the only force acting on it. As the ball rises, the weight will be doing work on the ball, decreasing its speed and its kinetic energy, and slowing it down until it momentarily stops. During this ascent, the potential energy of the ball will be rising. Once the ball begins to fall back down, it will lose potential energy as it gains kinetic energy. Mechanical energy is conserved throughout; the potential energy of the ball at its highest point is equal to the kinetic energy of the ball at its lowest point prior to impact.

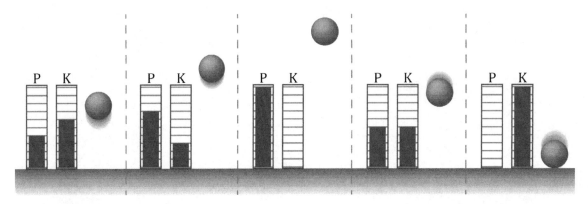

In systems where friction and air resistance are not negligible, we observe a different sort of result. For example, imagine a block sliding across the floor until it comes to a stop due to friction. Unlike a compressed spring or a ball flung into the air, there is no way for this block to regain its energy with a return trip. Therefore, we cannot say that the lost kinetic energy is being stored as potential energy. Instead, it has been dissipated and cannot be recovered. The total mechanical energy of the block-floor system has been not conserved in this case but rather reduced. The total energy of the system has not decreased, since the kinetic energy has been converted into thermal energy, but that energy is no longer useful for work.

Energy, though it may change form, will be neither created nor destroyed during physical processes. However, if we construct a system and some external force performs work on it, the result may be slightly different. If the work is positive, then the overall store of energy is increased; if it is negative, however, we can say that the overall energy of the system has decreased.

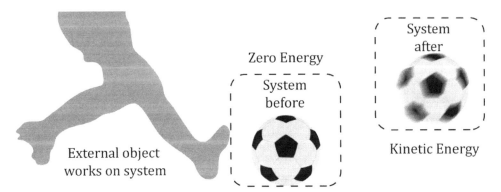

Kinetic energy

The kinetic energy of an object is the amount of energy it possesses by reason of being in motion. Kinetic energy cannot be negative. Changes in kinetic energy will occur when a force does work on an object, such that the motion of the object is altered. This change in kinetic energy is equal to the amount of work that is done. This relationship is commonly referred to as the work-energy theorem.

One interesting application of the work-energy theorem is that of objects in a free fall. To begin with, let us assert that the force acting on such an object is its weight, equal to its mass times g (the force of gravity). The work done by this force will be positive, as the force is exerted in the direction in which the object is traveling. Kinetic energy will, therefore, increase, according to the work-kinetic energy theorem.

If the object is dropped from a great enough height, it eventually reaches its terminal velocity, where the drag force is equal to the weight, so the object is no longer accelerating and its kinetic energy remains constant.

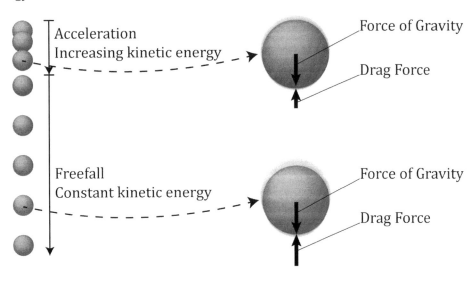

Gravitational Potential Energy

Gravitational potential energy is simply the potential for a certain amount of work to be done by one object on another using gravity. For objects on earth, the gravitational potential energy is equal to the amount of work which the earth can act on the object. The work which gravity performs on objects moving entirely or partially in the vertical direction is equal to the force exerted by the earth (weight) times the distance traveled in the direction of the force (height above the ground or reference point):

Work from gravity = weight × height above the ground

Thus, the gravitational potential energy is the same as the potential work.

Gravitational Potential Energy = weight × height

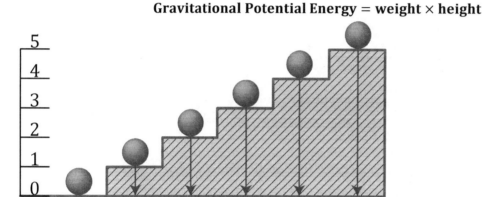

Elastic Potential Energy

Elastic potential energy is the potential for a certain amount of work to be done by one object on another using elastic compression or tension. The most common example is the spring. A spring will resist any compression or tension away from its equilibrium position (natural position). A small buggy is pressed into a large spring. The spring contains a large amount of elastic potential energy. If the buggy and spring are released, the spring will push exert a force on the buggy for a distance. This work will put kinetic energy into the buggy. The energy can be imagined as a liquid poured from one container into another. The spring pours its elastic energy into the buggy, which receives the energy as kinetic energy.

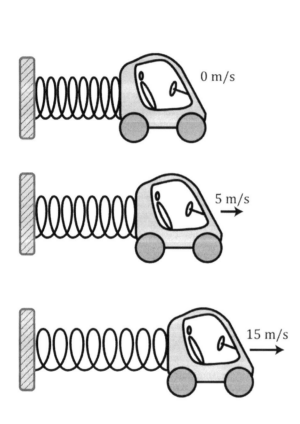
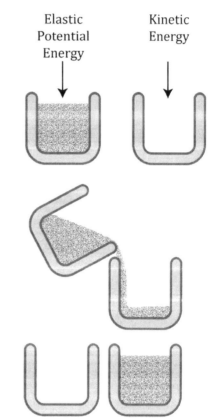

Power

Concept: the rate of work
Calculation: work/time

On occasion, you may need to demonstrate an understanding of power, as it is defined in applied physics. Power is the rate at which work is done. Power, like work and energy, is a scalar quantity. Power can be calculated by dividing the amount of work performed by the amount of time in which the work was performed.

$$\text{Power} = \frac{\text{work}}{\text{time}}$$

If more work is performed in a shorter amount of time, more power has been exerted. Power can be expressed in a variety of units. The preferred metric expression is one of watts or joules per seconds. For engine power, it is often expressed in horsepower.

Machines

Simple machines

Concept: Tools which transform forces to make tasks easier.

As their job is to transform forces, simple machines have an input force and an output force or forces. Simple machines transform forces in two ways: direction and magnitude. A machine can change the direction of a force, with respect to the input force, like a single stationary pulley which only changes the direction of the output force. A machine can also change the magnitude of the force like a lever.

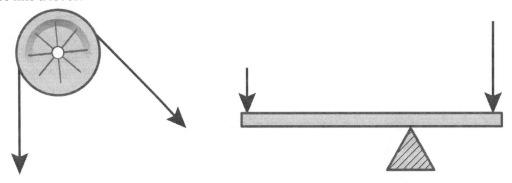

Simple machines include the inclined plane, the wedge, the screw, the pulley, the lever, and the wheel.

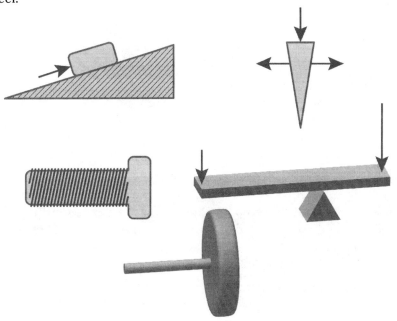

Mechanical Advantage

Concept: the amount of change a simple machine provides to the magnitude of a force
Calculation: output force/input force

Mechanical advantage is the measure of the output force divided by the input force. Thus, mechanical advantage measures the change performed by a machine. Machines cannot create energy, only transform it. Thus, in frictionless, ideal machines, the input work equals the output work.

$$Work_{input} = Work_{output}$$
$$force_{input} \times distance_{input} = force_{output} \times distance_{output}$$

This means that a simple machine can increase the force of the output by decreasing the distance which the output travels or it can increase the distance of the output by decreasing the force at the output.

By moving parts of the equation for work, we can arrive at the equation for mechanical advantage.

$$\textbf{Mechanical Advantage} = \frac{\textbf{force}_{\textbf{output}}}{\textbf{force}_{\textbf{input}}} = \frac{\textbf{distance}_{\textbf{input}}}{\textbf{distance}_{\textbf{output}}}$$

If the mechanical advantage is greater than one, the output force is greater than the input force and the input distance is greater than the output distance. Conversely, if the mechanical advantage is less than one, the input force is greater than the output force and the output distance is greater than the input distance. In equation form this is:

If Mechanical Advantage > 1:
$$force_{input} < force_{output} \text{ and } distance_{output} < distance_{input}$$

If Mechanical Advantage < 1:
$$force_{input} > force_{output} \text{ and } distance_{output} > distance_{input}$$

Inclined plane

The inclined plane is perhaps the most common of the simple machines. It is simply a flat surface that elevates as you move from one end to the other; a ramp is an easy example of an inclined plane. Consider how much easier it is for an elderly person to walk up a long ramp than to climb a shorter but steeper flight of stairs; this is because the force required is diminished as the distance increases. Indeed, the longer the ramp, the easier it is to ascend.

On the exam, this simple fact will most often be applied to moving heavy objects. For instance, if you have to move a heavy box onto the back of a truck, it is much easier to push it up a ramp than to lift it directly onto the truck bed. The longer the ramp, the greater the mechanical advantage, and the easier it will be to move the box. The mechanical advantage of an inclined plane is equal to the slant length divided by the rise of the plane.

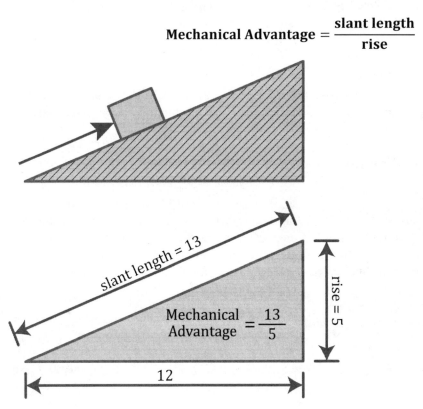

As you solve this kind of problem, however, remember that the same amount of work is being performed whether the box is lifted directly or pushed up a twenty-foot ramp; a simple machine only changes the force and the distance.

Wedge

A wedge is a variation on the inclined plane, in which the wedge moves between objects or parts and forces them apart. The unique characteristic of a wedge is that, unlike an inclined plane, it is designed to move. Perhaps the most familiar use of the wedge is in splitting wood. A wedge is driven into the wood by hitting the flat back end. The thin end of a wedge is easier to drive into the wood since it has less surface area and, therefore, transmits more force per area. As the wedge is driven in, the increased width helps to split the wood.

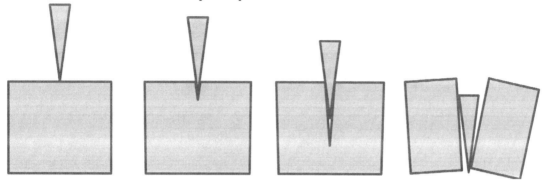

The exam may require you to select the wedge that has the highest mechanical advantage. This should be easy: the longer and thinner the wedge, the greater the mechanical advantage. The equation for mechanical advantage is:

$$\text{Mechanical Advantage} = \frac{\text{Length}}{\text{Width}}$$

Screw

A screw is simply an inclined plane that has been wound around a cylinder so that it forms a sort of spiral.

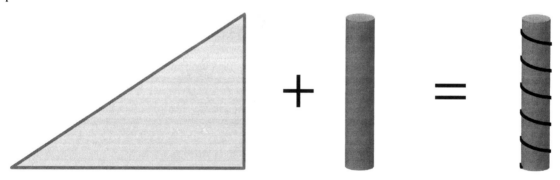

When it is placed into some medium, as for instance wood, the screw will move either forward or backward when it is rotated. The principle of the screw is used in a number of different objects, from jar lids to flashlights. On the exam, you are unlikely to see many questions regarding screws, though you may be presented with a given screw rotation and asked in which direction the screw will move. However, for consistency's sake, the equation for the mechanical advantage is a modification of the inclined plane's equation. Again, the formula for an inclined plane is:

$$Mechanical\ Advantage = \frac{slant\ length}{rise}$$

Because the rise of the inclined plane is the length along a screw, length between rotations = rise. Also, the slant length will equal the circumference of one rotation = 2πr.

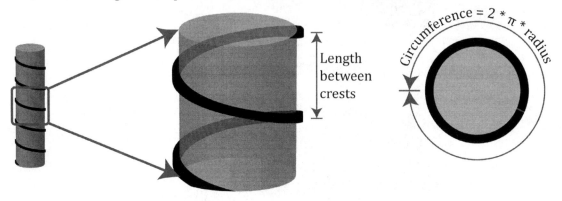

$$\text{Mechanical Advantage} = \frac{2 \times \pi \times \text{radius}}{\text{length between crests}}$$

Lever

The lever is the most common kind of simple machine. See-saws, shovels, and baseball bats are all examples of levers. There are three classes of levers which are differentiated by the relative orientations of the fulcrum, resistance, and effort. The fulcrum is the point at which the lever rotates; the effort is the point on the lever where force is applied; the resistance is the part of the lever that acts in response to the effort.

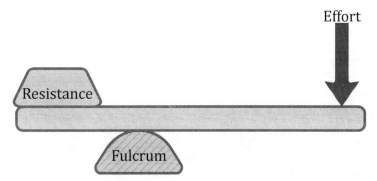

The mechanical advantage of a lever depends on the distances of the effort and resistance from the fulcrum. Mechanical advantage equals:

$$\text{Mechanical Advantage} = \frac{\text{effort distance}}{\text{resistance distance}}$$

For each class of lever, the location of the important distances changes:

First Class Lever

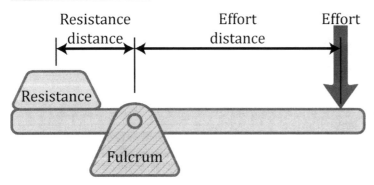

Second Class Lever

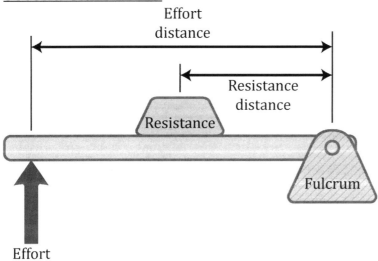

Third Class Lever

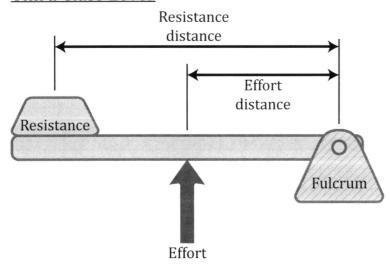

In a first class lever, the fulcrum is between the effort and the resistance. A seesaw is a good example of a first class lever: when effort is applied to force one end up, the other end goes down, and vice versa. The shorter the distance between the fulcrum and the resistance, the easier it will be to move the resistance. As an example, consider whether it is easier to lift another person on a see-saw when they are sitting close to the middle or all the way at the end. A little practice will show you that it is much more difficult to lift a person the farther away he or she is on the see-saw.

In a second class lever, the resistance is in-between the fulcrum and the effort. While a first class lever is able to increase force and distance through mechanical advantage, a second class lever is only able to increase force. A common example of a second class lever is the wheelbarrow: the force exerted by your hand at one end of the wheelbarrow is magnified at the load. Basically, with a second class lever you are trading distance for force; by moving your end of the wheelbarrow a bit farther, you produce greater force at the load.

Third class levers are used to produce greater distance. In a third class lever, the force is applied in between the fulcrum and the resistance. A baseball bat is a classic example of a third class lever; the bottom of the bat, below where you grip it, is considered the fulcrum. The end of the bat, where the ball is struck, is the resistance. By exerting effort at the base of the bat, close to the fulcrum, you are able to make the end of the bat fly quickly through the air. The closer your hands are to the base of the bat, the faster you will be able to make the other end of the bat travel.

Pulley

The pulley is a simple machine in which a rope is carried by the rotation of a wheel. Another name for a pulley is a block. Pulleys are typically used to allow the force to be directed from a convenient location. For instance, imagine you are given the task of lifting a heavy and tall bookcase. Rather than tying a rope to the bookcase and trying to lift it up, it would make sense to tie a pulley system to a rafter above the bookcase and run the rope through it, so that you could pull down on the rope and lift the bookcase. Pulling down allows you to incorporate your weight (normal force) into the act of lifting, thereby making it easier.

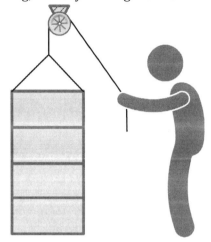

If there is just one pulley above the bookcase, you have created a first-class lever which will not diminish the amount of force that needs to be applied to lift the bookcase. There is another way to use a pulley, however, that can make the job of lifting a heavy object considerably easier. First, tie the rope directly to the rafter. Then, attach a pulley to the top of the bookcase and run the rope through it. If you can then stand so that you are above the bookcase, you will have a much easier time lifting this heavy object. Why? Because the weight of the bookcase is now being distributed: half of it is acting on the rafter, and half of it is acting on you. In other words, this arrangement allows you to lift an object with half the force. This simple pulley system, therefore, has a mechanical advantage of 2. Note that in this arrangement, the unfixed pulley is acting like a second-class lever. The price you pay for your mechanical advantage is that whatever distance you raise your end of the rope, the bookcase will only be lifted half as much.

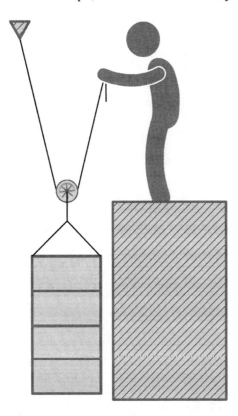

Of course, it might be difficult for you to find a place high enough to enact this system. If this is the case, you can always tie another pulley to the rafter and run the rope through it and back down to the floor. Since this second pulley is fixed, the mechanical advantage will remain the same.

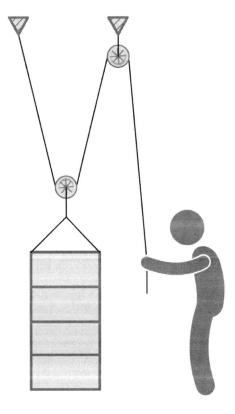

There are other, slightly more complex ways to obtain an even greater mechanical advantage with a system of pulleys. On the exam, you may be required to determine the pulley and tackle (rope) arrangement that creates the greatest mechanical advantage. The easiest way to determine the answer is to count the number of ropes that are going to and from the unfixed pulley; the more ropes coming and going, the greater the mechanical advantage.

Wheel and axle

Another basic arrangement that makes use of simple machines is called the wheel and axle. When most people think of a wheel and axle, they immediately envision an automobile tire. The steering wheel of the car, however, operates on the same mechanical principle, namely that the force required to move the center of a circle is much greater than the force require to move the outer rim of a circle. When you turn the steering wheel, you are essentially using a second-class lever by increasing the output force by increasing the input distance. The force required to turn the wheel from the outer rim is much less than would be required to turn the wheel from its center. Just imagine how difficult it would be to drive a car if the steering wheel was the size of a saucer!

Conceptually, the mechanical advantage of a wheel is easy to understand. For instance, all other things being equal, the mechanical advantage created by a system will increase along with the radius. In other words, a steering wheel with a radius of 12 inches has a greater mechanical advantage than a steering wheel with a radius of ten inches; the same amount of force exerted on the rim of each wheel will produce greater force at the axis of the larger wheel.

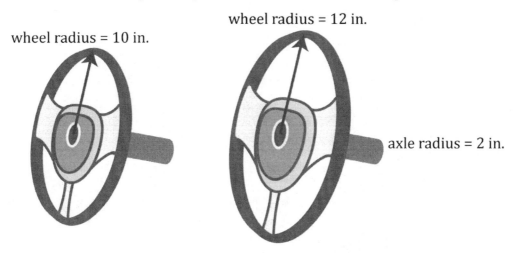

The equation for the mechanical advantage of a wheel and axle is:
$$\text{Mechanical Advantage} = \frac{\text{radius}_{\text{wheel}}}{\text{radius}_{\text{axle}}}$$

Thus, the mechanical advantage of the steering wheel with a larger radius will be:
$$Mechanical\ Advantage = \frac{12\ inches}{2\ inches} = 6$$

Gears

The exam may ask you questions involving some slightly more complex mechanisms. It is very common, for instance, for there to be a couple of questions concerning gears. Gears are a system of interlocking wheels that can create immense mechanical advantages. The amount of mechanical advantage, however, will depend on the gear ratio; that is, on the relation in size between the gears.

When a small gear is driving a big gear, the speed of the big gear is relatively slow; when a big gear is driving a small gear, the speed of the small gear is relatively fast.

The equation for the mechanical advantage is:

$$\text{Mechanical Advantage} = \frac{\text{Torque}_{output}}{\text{Torque}_{input}} = \frac{r_{output}}{r_{input}} = \frac{\#\text{ of teeth}_{output}}{\#\text{ of teeth}_{input}}$$

Note that mechanical advantage is greater than 1 when the output gear is larger. In these cases, the output velocity (ω) will be lower. The equation for the relative speed of a gear system is:

$$\frac{\omega_{input}}{\omega_{output}} = \frac{r_{output}}{r_{input}}$$

$$\text{Mechanical Advantage} = \frac{teeth_{output}}{teeth_{input}} = \frac{20}{10} = 2$$

$$\text{Mechanical Advantage} = \frac{teeth_{output}}{teeth_{input}} = \frac{16}{8} = 2$$

Uses of Gears

Gears are used to change direction of output torque, change location of output torque, change amount of output torque, and change angular velocity of output.

Change output direction

Change torque location

Change torque amount

Change output velocity

Gear Ratios

A gear ratio is a measure of how much the speed and torque are changing in a gear system. It is the ratio of output speed to input speed. Because the number of teeth is directly proportional to the speed in meshing gears, a gear ratio can also be calculated using the number of teeth on the gears. When the driving gear has 30 teeth and the driven gear has 10 teeth, the gear ratio is 3:1.

$$Gear\ Ratio = \frac{\#\ of\ teeth_{driving}}{\#\ of\ teeth_{driven}} = \frac{30}{10} = \frac{3}{1} = 3:1$$

This means that the smaller, driven gear rotates 3 times for every 1 rotation of the driving gear.

The Hydraulic Jack

The hydraulic jack is a simple machine using two tanks and two pistons to change the amount of an output force.

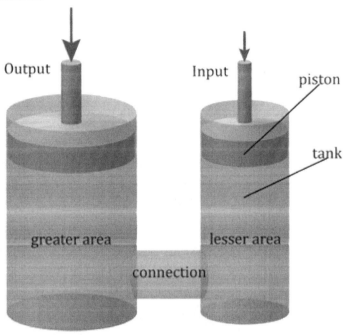

Since fluids are effectively incompressible, when you apply pressure to one part of a contained fluid, that pressure will have to be relieved in equal measure elsewhere in the container. Supposed the input piston has half the surface area of the output piston (10 in² compared to 20 in²), and it is being pushed downward with 50 pounds of force. The pressure being applied to the fluid is $50\ lb \div 10\ in^2 = 5\frac{lb}{in^2}$ or 5 psi. When that 5 psi of pressure is applied to the output piston, it pushes that piston upward with a force of $5\frac{lb}{in^2} \times 20\ in^2 = 100\ lb$.

The hydraulic jack functions similarly to a first class lever, but with the important factor being the area of the pistons rather than the length of the lever arms. Note that the mechanical advantage is based on the relative areas, not the relative radii, of the pistons. The radii must be squared to compute the relative areas.

$$\text{Mechanical Advantage} = \frac{\text{Force}_{output}}{\text{Force}_{input}} = \frac{\text{area}_{output}}{\text{area}_{input}} = \frac{\text{radius}_{output}^2}{\text{radius}_{input}^2}$$

Pulleys and Belts

Another system involves two pulleys connected by a drive belt (a looped band that goes around both pulleys). The operation of this system is similar to that of gears, with the exception that the pulleys will rotate in the same direction, while interlocking gears will rotate in opposite directions. A smaller pulley will always spin faster than a larger pulley, though the larger pulley will generate more torque.

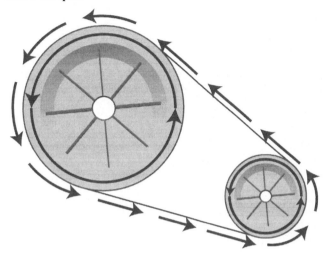

The speed ratio between the pulleys can be determined by comparing their radii; for instance, a 4-inch pulley and a 12-inch pulley will have a speed ratio of 3:1.

Momentum/Impulse

Linear momentum

Concept: how much a body will resist stopping
Calculation: $momentum = mass \times velocity$

In physics, linear momentum can be found by multiplying the mass and velocity of an object:
$$\text{Momentum} = \text{mass} \times \text{velocity}$$

Momentum and velocity will always be in the same direction. Newton's second law describes momentum, stating that the rate of change of momentum is proportional to the force exerted and is in the direction of the force. If we assume a closed and isolated system (one in which no objects leave or enter, and upon which the sum of external forces is zero), then we can assume that the momentum of the system will neither increase nor decrease. That is, we will find that the momentum is a constant. The law of conservation of linear momentum applies universally in physics, even in situations of extremely high velocity or with subatomic particles.

Collisions

This concept of momentum takes on new importance when we consider collisions. A collision is an isolated event in which a strong force acts between each of two or more colliding bodies for a brief period of time. However, a collision is more intuitively defined as one or more objects hitting each other.

When two bodies collide, each object exerts a force on the opposite member. These equal and opposite forces change the linear momentum of the objects. However, when both bodies are considered, the net momentum in collisions is conserved.

There are two types of collisions: elastic and inelastic. The difference between the two lies in whether kinetic energy is conserved. If the total kinetic energy of the system is conserved, the collision is elastic. Visually, elastic collisions are collisions in which objects bounce perfectly. If some of the kinetic energy is transformed into heat or another form of energy, the collision is inelastic. Visually, inelastic collisions are collisions in which the objects do not bounce perfectly or even stick to each other.

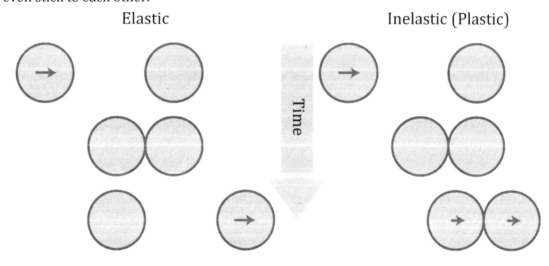

If the two bodies involved in an elastic collision have the same mass, then the body that was moving will stop completely, and the body that was at rest will begin moving at the same velocity as the projectile was moving before the collision.

Fluids

Fluids

Concept: liquids and gasses

A few of the questions on the exam will probably require you to consider the behavior of fluids. It sounds obvious, perhaps, but fluids can best be defined as substances that flow. A fluid will conform, slowly or quickly, to any container in which it is placed. Both liquids and gasses are considered to be fluids. Fluids are essentially those substances in which the atoms are not arranged in any permanent, rigid way. In ice, for instance, atoms are all lined up in what is known as a crystalline lattice, while in water and steam the only intermolecular arrangements are haphazard connections between neighboring molecules.

Flow Rates

When liquids flow in and out of containers with certain rates, the change in volume is the volumetric flow in minus the volumetric flow out. Volumetric flow is essentially the amount of volume moved past some point divided by the time it took for the volume to pass.

$$\text{Volumetric flow rate} = \frac{\text{volume moved}}{\text{time for the movement}}$$

If the flow into a container is greater than the flow out, the container will fill with the fluid. However, if the flow out of a container is greater than the flow into a container, the container will drain of the fluid.

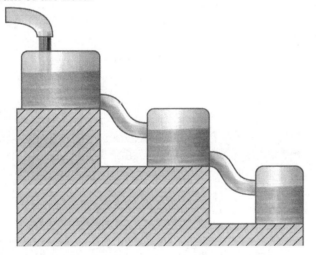

Density

Concept: how much mass is in a specific volume of a substance
Calculation: $density = \rho = \frac{mass}{volume}$

Density is essentially how much stuff there is in a volume or space. The density of a fluid is generally expressed with the symbol ρ (the Greek letter rho.) The density may be found with the simple equation:

$$\text{density} = \rho = \frac{\text{mass}}{\text{volume}}$$

Density is a scalar property, meaning that it has no direction component.

Pressure

Concept: The amount of force applied per area
Calculation: $Pressure = \frac{force}{area}$

Pressure, like fluid density, is a scalar and does not have a direction. The equation for pressure is concerned only with the magnitude of that force, not with the direction in which it is pointing. The SI unit of pressure is the Newton per square meter, or Pascal.

$$\text{Pressure} = \frac{\text{force}}{\text{area}}$$

As every deep-sea diver knows, the pressure of water becomes greater the deeper you go below the surface; conversely, experienced mountain climbers know that air pressure decreases as they gain a higher altitude. These pressures are typically referred to as hydrostatic pressures because they involve fluids at rest.

Pascal's principle

The exam may also require you to demonstrate some knowledge of how fluids move. Anytime you squeeze a tube of toothpaste, you are demonstrating the idea known as Pascal's principle. This principle states that a change in the pressure applied to an enclosed fluid is transmitted undiminished to every portion of the fluid as well as to the walls of the containing vessel.

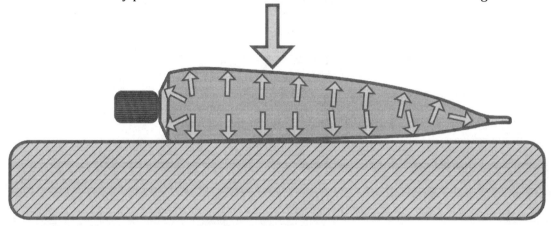

Buoyant force

If an object is submerged in water, it will have a buoyant force exerted on it in the upward direction. Often, of course, this buoyant force is much too small to keep an object from sinking to the bottom. Buoyancy is summarized in Archimedes' principle; a body wholly or partially submerged in a fluid will be buoyed up by a force equal to the weight of the fluid that the body displaces.

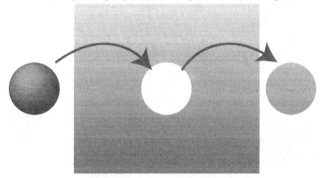 

If the buoyant force is greater than the weight of an object, the object will go upward. If the weight of the object is greater than the buoyant force, the object will sink. When an object is floating on the surface, the buoyant force has the same magnitude as the weight.

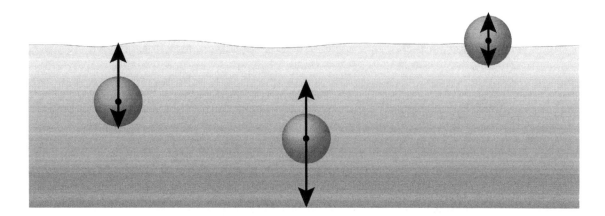

Even though the weight of a floating object is precisely balanced by a buoyant force, these forces will not necessarily act at the same point. The weight will act from the center of mass of the object, while the buoyancy will act from the center of mass of the hole in the water made by the object (known as the center of buoyancy). If the floating object is tilted, then the center of buoyancy will shift and the object may be unstable. In order to remain in equilibrium, the center of buoyancy must always shift in such a way that the buoyant force and weight provide a restoring torque, one that will restore the body to its upright position. This concept is, of course, crucial to the construction of boats which must always be made to encourage restoring torque.

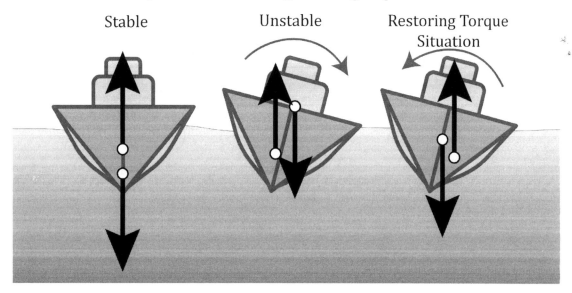

Ideal fluids

Because the motion of actual fluids is extremely complex, the exam usually assumes ideal fluids when they set up their problems. Using ideal fluids in fluid dynamics problems is like discounting friction in other problems. Therefore, when we deal with ideal fluids, we are making four assumptions. It is important to keep these in mind when considering the behavior of fluids on the exam. First, we are assuming that the flow is steady; in other words, the velocity of every part of the fluid is the same. Second, we assume that fluids are incompressible, and, therefore, have a consistent density. Third, we assume that fluids are nonviscous, meaning that they flow easily and without resistance. Fourth, we assume that the flow of ideal fluids is irrotational: that is, particles in the fluid will not rotate around a center of mass.

Bernoulli's principle

When fluids move, they do not create or destroy energy; this modification of Newton's second law for fluid behavior is called Bernoulli's principle. It is essentially just a reformulation of the law of conservation of mechanical energy for fluid mechanics.

The most common application of Bernoulli's principle is that pressure and speed are inversely related, assuming constant altitude. Thus if the elevation of the fluid remains constant and the speed of a fluid particle increases as it travels along a streamline, the pressure will decrease. If the fluid slows down, the pressure will increase.

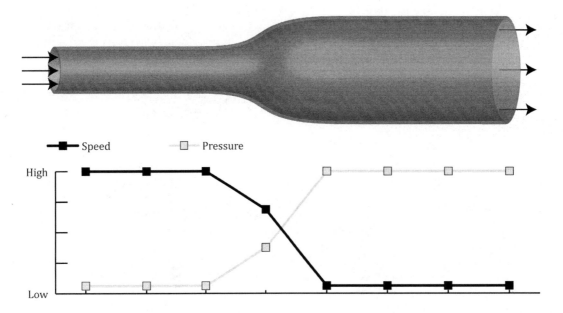

Heat Transfer

Heat Transfer

Heat is a type of energy. Heat transfers from the hot object to the cold object through the three forms of heat transfer: conduction, convection, and radiation.

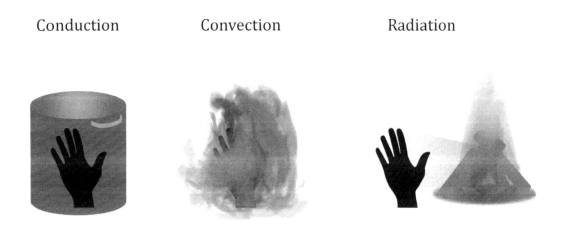

Conduction is the transfer of heat by physical contact. When you touch a hot pot, the pot transfers heat to your hand by conduction.

Convection is the transfer of heat by the movement of fluids. When you put your hand in steam, the steam transfers heat to your hand by convection.

Radiation is the transfer of heat by electromagnetic waves. When you put your hand near a campfire, the fire heats your hand by radiation.

Phase Changes

Materials exist in four phases or states: solid, liquid, gas, and plasma. However, as most tests will not cover plasma, we will focus on solids, liquids, and gases. The solid state is the densest in almost all cases (water is the most notable exception), followed by liquid, and then gas.

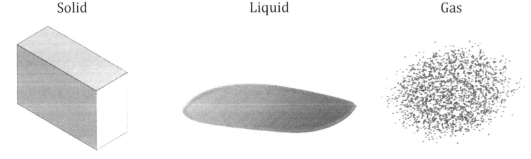

The impetus for phase change (changing from one phase to another) is heat. When a solid is heated, it will change into a liquid. The same process of heating will change a liquid into a gas.

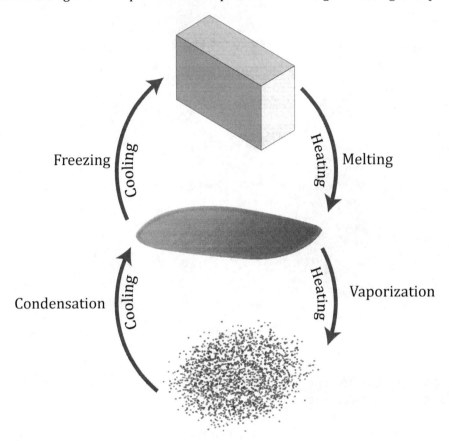

Optics

Optics

Lenses change the way light travels. Lenses are able to achieve this by the way in which light travels at different speeds in different mediums. The essentials to optics with lenses deal with concave and convex lenses. Concave lenses make objects appear smaller, while convex lenses make objects appear larger.

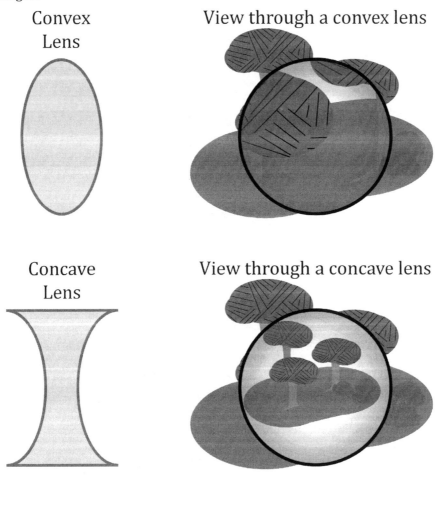

Electricity

Electric Charge

Much like gravity, electricity is an everyday observable phenomenon which is very complex, but may be understood as a set of behaviors. As the gravitational force exists between objects with mass, the electric force exists between objects with electrical charge. In all atoms, the protons have a positive charge, while the electrons have a negative charge. An imbalance of electrons and protons in an object results in a net charge. Unlike gravity, which only pulls, electrical forces can push objects apart as well as pulling them together.

Similar electric charges repel each other. Opposite charges attract each other.

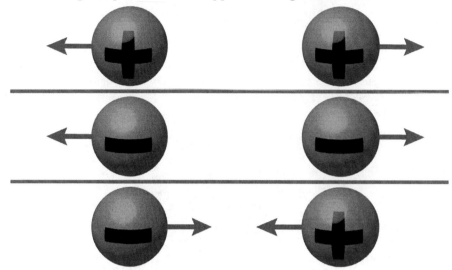

Current

Electrons (and electrical charge with it) move through conductive materials by switching quickly from one atom to another. This electrical flow can manipulate energy like mechanical systems.

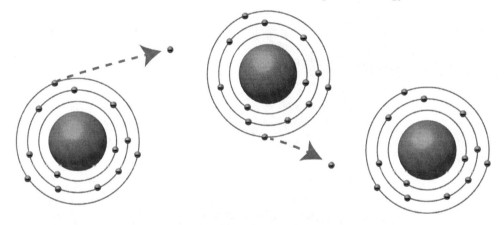

The term for the rate at which the charge flows through a conductive material is current. Because each electron carries a specific charge, current can be thought of as the number of electrons passing a point in a length of time. Current is measured in Amperes (A), each unit of which is approximately 6.24×10^{18} electrons per second.

Electric current carries energy much like moving balls carry energy.

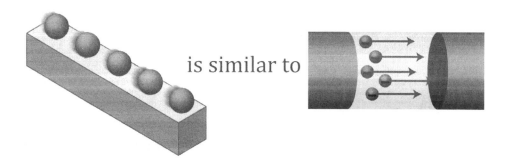

Voltage

Voltage is the potential for electric work. Voltage is the push behind electrical work. Voltage is similar to gravitational potential energy.

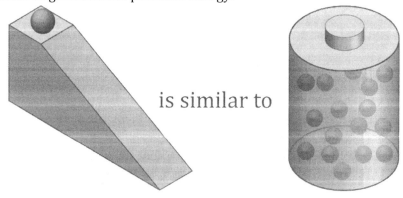

Anything used to generate a voltage, such as a battery or a generator, is called a voltage source. Voltage is conveniently measured in Volts (V).

Resistance

Resistance is the amount of pressure to slow electrical current. Electrical resistance is much like friction, resisting flow and dissipating energy.

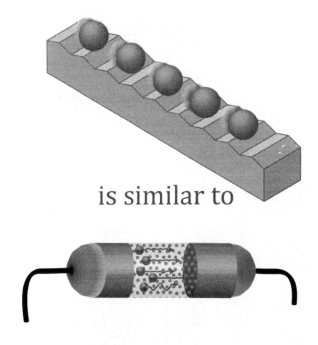

Different objects have different resistances. A resistor is an electrical component designed to have a specific resistance, measured in Ohms (Ω).

Basic Circuits

A circuit is a closed loop through which current can flow. A simple circuit contains a voltage source and a resistor. The current flows from the positive side of the voltage source through the resistor to the negative side of the voltage source.

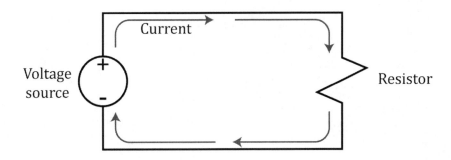

If we plot the voltage of a simple circuit, the similarities to gravitational potential energy appear.

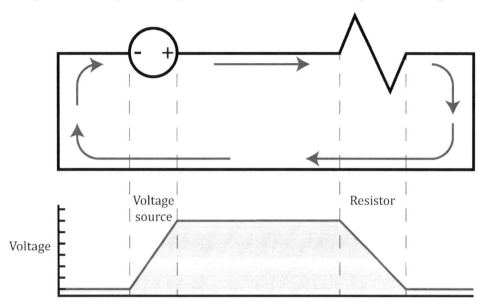

If we consider the circuit to be a track, the electrons would be balls, the voltage source would be a powered lift, and the resistor would be a sticky section of the track. The lift raises the balls, increasing their potential energy. This potential energy is expended as the balls roll down the sticky section of the track.

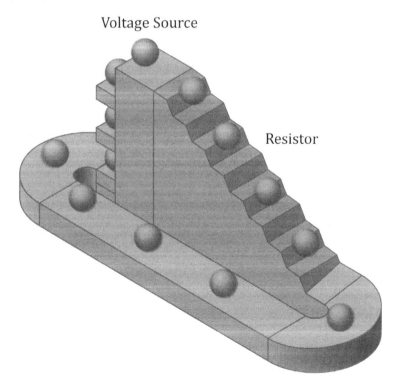

Ohm's Law

A principle called Ohm's Law explains the relationship between the voltage, current, and resistance. The voltage drop over a resistance is equal to the amount of current times the resistance:

Voltage (V) = current (I) × resistance (R)

We can gain a better understanding of this equation by looking at a reference simple circuit and then changing one variable at a time to examine the results.

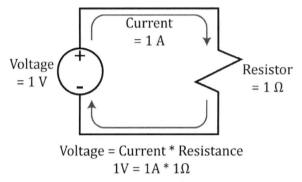

Voltage = Current * Resistance
1V = 1A * 1Ω

Increased Resistance	Increased Current	Increased Voltage

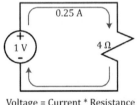 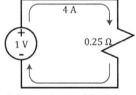 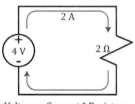

Voltage = Current * Resistance   Voltage = Current * Resistance   Voltage = Current * Resistance
1V = 0.25A * 4Ω   1V = 4A * 0.25Ω   4V = 2A * 2Ω

Series Circuits

A series circuit is a circuit with two or more resistors on the same path. The same current runs through both resistors. However, the total voltage drop splits between the resistors. The resistors in series can be added together to make an equivalent basic circuit.

$$R_{equiv} = R_1 + R_2$$

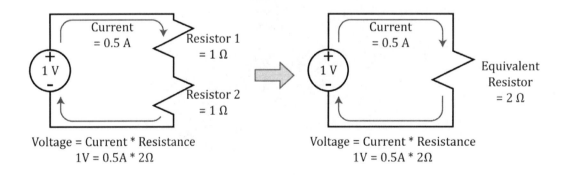

Voltage = Current * Resistance   Voltage = Current * Resistance
1V = 0.5A * 2Ω   1V = 0.5A * 2Ω

Parallel Circuits

A parallel circuit is a circuit with two or more resistors on different, parallel paths. Unlike the series circuit, the current splits between the different paths in a parallel circuit. Resistors in parallel can be reduced to an equivalent circuit, but not by simply adding the resistances. The inverse of the equivalent resistance of parallel resistors is equal to the sum of the inverses of the resistance of each leg of the parallel circuit. In equation form that means:

$$\frac{1}{R_{equiv}} = \frac{1}{R_1} + \frac{1}{R_2}$$

Or when solved for equivalent resistance:

$$R_{equiv} = \frac{1}{\frac{1}{R_1} + \frac{1}{R_2}}$$

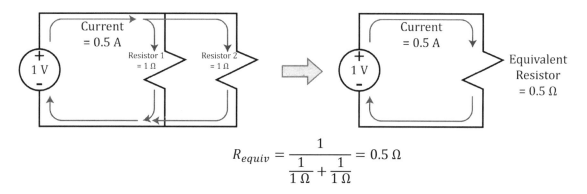

$$R_{equiv} = \frac{1}{\frac{1}{1\,\Omega} + \frac{1}{1\,\Omega}} = 0.5\,\Omega$$

Electrical Power

Electrical power, or the energy output over time, is equal to the current resulting from a voltage source times the voltage of that source:

Power (P) = current (I) × voltage (V)

Thanks to Ohm's Law, we can write this relation in two other ways:

$$P = I^2 R$$

$$P = \frac{V^2}{R}$$

For instance, if a circuit is composed of a 9 Volt battery and a 3 Ohm resistor, the power output of the battery will be:

$$Power = \frac{V^2}{R} = \frac{9^2}{3} = 27\,Watts$$

AC vs. DC

Up until this point, current has been assumed to flow in one direction. One directional flow is called Direct Current (DC). However, there is another type of electric current: Alternating Current (AC). Many circuits use AC power sources, in which the current flips back and forth rapidly between directions.

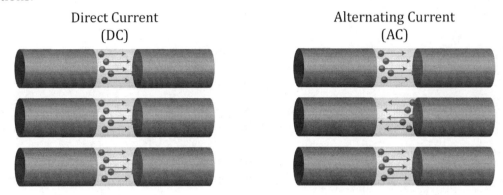

Capacitors

Capacitors are electrical components which store voltage. Capacitors are made from two conductive surfaces separated from each other by a space and/or insulation. Capacitors resist changes to voltage. Capacitors don't stop AC circuits (although they do affect the current flow), but they do stop DC circuits, acting as open circuits.

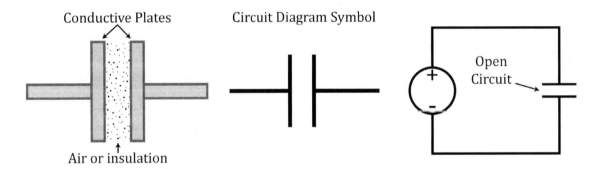

Inductors

Inductors are electrical components which effectively store current. Inductors use the relationship between electricity and magnetism to resist changes in current by running the current through coils of wire. Inductors don't stop DC circuits, but they do resist AC circuits as AC circuits utilize changing currents.

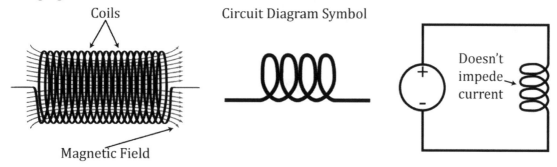

Diodes

Diodes are electrical components which limit the flow of electricity to one direction. If current flows through a diode in the intended direction, the diode will allow the flow. However, a diode will stop current if it runs the wrong way.

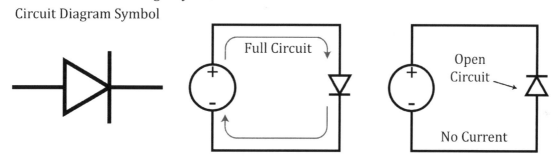

Magnetism

Magnetism

Magnetism is an attraction between opposite poles of magnetic materials and a repulsion between similar poles of magnetic materials. Magnetism can be natural or induced with the use of electric currents. Magnets almost always exist with two polar sides: north and south. A magnetic force exists between two poles on objects. Different poles attract each other. Like poles repel each other.

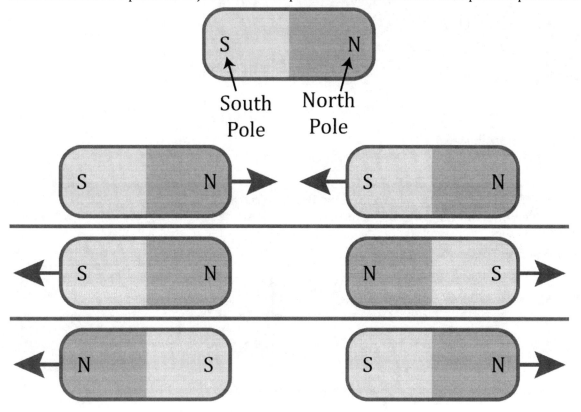

ASVAB: Assembling Objects

This section of the test measures your ability to imagine the way an object will look when its pieces are joined together properly. The questions that make up this section will begin with a picture of five separate parts. Each of these parts will be marked with one or more letters. These letters will be assigned to places on the part. Sometimes the letter will appear directly over part of the object, and sometimes a line will be drawn from the letter to the appropriate spot on the object. If there is a dotted line drawn from the letter to the object, this means that the letter corresponds to a place on the side of the object that cannot be seen. Your task will be to imagine how the object will look when all of the parts are connected such that the letters touch one another. In other words, all of the parts marked with an *A* will need to touch; all of the parts marked with a *B* will need to touch, and so on. You will be given five assembled objects from which to choose your answer. This format will be the same for all twenty questions in the assembly section.

The format of the assembly questions, then, is fairly straightforward. However, you will need to be able to work efficiently. This means approaching the assembly questions with a strategy in mind. There are a couple of different ways to proceed. Perhaps the best way to approach assembly questions is to proceed in order and work methodically. In other words, begin by mentally connecting the *A*s; once you have this assembly in mind, move on to the *B*s, and so forth. One mistake that many test-takers make is to attempt the entire assembly all at once. Many of these problems will involve four or five parts, and it will be too easy to get confused if you do not work in order. Indeed, the test administrator will be sure to include a few close-but-incorrect answers designed to trip up people who work too fast. The best way to proceed through an assembly question is to firmly decide on the arrangement of the object one piece at a time.

In tandem with this strategy, you can use the process of elimination to quickly reduce the number of possible answers. After you have mentally connected the *A*s, for instance, you can go through and rule out all of those answers that do not have the *A*s connected properly. You can then proceed through the Bs, ruling out a few more answers, until you are left with a single possible right answer, which you can then take a moment to confirm.

There are a few things to keep in mind when considering possible solutions to an assembly problem. To begin with, remember that the pieces can be turned in any way. They may be rotated, spun, and flipped. They may not, however, be folded, bent, or twisted. Furthermore, they may not change in size; each piece should be the same size in the answer as it is in the original picture. Occasionally, the makers of the exam will place the pieces in the correct configuration but will drastically alter the sizes of one or more pieces. This automatically invalidates the answer. It is best to imagine the pieces as solid three-dimensional objects, which can be manipulated in all directions but cannot have their fundamental size and shape altered.

Similarly, make sure that the answer you select has the appropriate number of pieces. If there are five different components in the original drawing, there must be five connected parts in the answer. It is common for the exam to leave out a piece or include an extra piece in one or more of the possible answers; even if these answers are right in all other respects, they cannot be correct if they include more or fewer pieces than the original drawing.

Once you have worked through the problem in this systematic manner and have selected your answer, go back and double-check your work. To begin with, make sure that all of the letters are joined properly in the answer you have chosen. Then, make sure that your answer has the right number of pieces. Finally, make sure that the pieces in the assembled object are similar in size to the pieces in the original drawing. If all of these factors check out, you can be comfortable that you have selected the right answer.

This methodical way of working through assembly problems may seem too time-consuming given the limits of the exam, but with a little bit of practice you can speed through the process within the time given. In fact, by adhering to this organized way of solving assembly problems, you will actually save time, since you will never get confused or lost in a problem and have to go back to the beginning. All it takes to master these problems is a little strategic knowledge and a little preparation. Having covered the strategy, let's now take a look at a few ways to prepare for this section of the exam.

There are a few common activities you can use to prepare for the assembly section of the exam. For instance, even though jigsaw puzzles do not offer an exact replication of the items you will encounter on the exam, they still exercise your spatial reasoning skills. Envisioning which pieces of the puzzle will fit together, and how the resulting arrangement will look, is a great way to hone your assembly skills. For an even greater challenge, try turning all the pieces of the puzzle face down, so that you cannot use the picture to guide your work: this will force you to rely more on your sense of orientation and arrangement.

Another way to prepare for this section of the exam is to take apart a small piece of machinery and study its configuration. Yard sales and junk bins are great places to find old appliances and electronics equipment. With the help of a screwdriver, wrench, and pair of pliers, you should be able to take apart most any appliance with ease. Once you have broken the item down into a pile of parts, see if you can put those parts back together into a functional whole. Even better, have someone else take apart the item and then see if you can put it back together. Of course, it is not recommended that you try this with valuable or expensive pieces of hardware. Also, some electronic appliances contain small batteries which should not be handled by non-professionals; always obey any warnings listed on the equipment. Nevertheless, taking apart and putting together small machines can be a fantastic way to improve your assembly skills.

Finally, there is a wealth of spatial intelligence exercises to be found on the Internet. Just by entering "spatial intelligence" into a search engine, you should receive listings for dozens of simple, free puzzles and games that strengthen your ability to visualize and perform basic assemblies. Some of these programs are so sophisticated that they allow you to manipulate three-dimensional objects on your computer screen! While for many people working on a computer is no substitute for direct contact with an object, these on-line exercises are a clean, fast way to stretch your mental muscles.

Whatever method you choose, be sure to prepare for at least a few hours before sitting for the examination. For most students, the most difficult thing about the assembly questions is getting comfortable with the format and learning how to approach the problem. By remembering the strategies discussed above, and utilizing some of the suggested practice exercises, you can make sure that you will be ready to attack assembly problems immediately.

AFOQT: Verbal Analogies

What are analogies questions?

Analogies are pairs of terms that have a common relationship. Analogies questions are presented in the format, "A is to B as C is to D," meaning that terms A and B are related to one another in the same or similar way that terms C and D are related to each other. Terms A and B do not have to be related to terms C and D at all, though they usually will be.

Usually in the question, you will be given terms A, B, and C, and will just have to supply term D from the choices given. Occasionally, you may be given only terms A and B, and you will have to select a pair of terms for C and D.

What sort of relationships will there be?

Below are some examples of the types of analogies that may appear on the exam. Most of the questions you encounter will be relatively simple relationships, but here is an extensive list of the types of analogies that might show up.

Characteristic

Some characteristic analogies will focus on a characteristic of something else.
- Dog : Paw – The foot of a dog is its paw.
- Lady : Lovely – A lady has a lovely personality.
- Outrageous : Lies – Lies can be described as being outrageous.

Some characteristic analogies will focus on something that is NOT a characteristic of something else.
- Desert : Humidity – A desert does not have humidity.
- Job : Unemployed – A person without a job is unemployed.
- Quick : Considered – A quick decision is often not very considered.

Source
- Casting : Metal – A casting is made from metal.
- Forest : Trees – A forest is composed of trees.
- Slogans : Banners – A slogan is printed on banners.

Location
- Eiffel Tower : Paris – The Eiffel Tower is a structure in Paris.
- Welsh : Wales – The Welsh are the inhabitants of Wales.
- Pound : England – The pound is the monetary unit of England.

Sequential
- One : Two – These are consecutive numbers.
- Birth : Death – These are the first and last events of a life or project.
- Spring : Summer – The season of spring immediately precedes summer.

Cause/Effect
- Storm : Hail – Hail can be caused by a storm.
- Heat : Fire – Heat results from a fire.
- Monotony : Boredom – Boredom is a consequence of monotony.

Creator/Creation
- Carpenter : House – A carpenter builds a house.
- Painter : Portrait – A painter makes a portrait.
- Burroughs : Tarzan – Edgar Rice Burroughs wrote the novel Tarzan.

Provider/Provision
- Job : Salary – A job provides a salary.
- Therapist : Treatment – A therapist treats patients.
- Army : Defense – An army enables national defense.

Object/Function
- Pencil : Write – A pencil is used to write.
- Pressure : Barometer – A barometer measures pressure.
- Frown : Unhappy – A frown shows unhappiness.

User/Tool
- Carpenter : Hammer – A carpenter uses a hammer.
- Teacher : Chalk – A teacher uses chalk.
- Farmer : Tractor – A farmer drives a tractor.

Whole/Part
- Door : House – A door is part of a house.
- State : Country – A country is made up of states.
- Day : Month – A month consists of many days.

Grammatical Transformation
- Ran : Run – These are different tenses of the same verb.
- *Die : Dice – These are singular and plural forms.*
- We : Our – These are pronouns related to groups.

Translation
- Satan : Lucifer – These are both names for the devil.
- Bon Voyage : Farewell – These are the French and English words for goodbye.
- Japan : Nippon – These are two names for the same country.

Category
- Door : Window – Both a door and a window are parts of a house.
- Thigh : Shin – Both a thigh and a shin are parts of a leg.
- Measles : Mumps – Both measles and mumps are types of diseases.

Synonym or Definition

These are analogies in which both terms have a similar meaning.
- Chase : Pursue – Both of these terms mean to "go after".
- Achieve : Accomplish – Both of these terms refer to the successful attainment of a goal.
- Satiate : Satisfy – Both of these terms mean to gratify a desire.

Antonym or Contrast

These are analogies in which both terms have an opposite meaning.
- Disguise : Reveal – To disguise something is not to reveal it, but to conceal it.
- Peace : War – Peace is a state in which there is no war.
- Forget : Remember – The word "remember" means not to forget something.

Intensity

These are analogies in which either one term expresses a higher degree of something than the other term.
- Exuberant : Happy – To be exuberant is to be extremely happy.
- Break : Shatter – To shatter is to strongly break.
- Deluge : Rain – A deluge is a heavy rain.

What strategies can I use?

A huge vocabulary is not necessary to succeed on analogies questions (though it certainly doesn't hurt). In most cases though, you can determine the answer even if you don't recognize all the words. The strategies listed here will help you develop the ability to recognize basic relationships and apply simple steps and methods to solving them.

Determine the Relationship

Don't focus on the meanings, but rather the relationship between the two words.
To understand the relationship, first create a sentence that links the two words and puts them into perspective. The sentence that you use to connect the words can be simple at first.
- Example:
  - Wood : Fire
  - *Wood* feeds a *fire*.

Then go through each answer choice and replace the words with the answer choices. If the question is easy, then that may be all that is necessary. If the question is hard, you might have to fine-tune your sentence.
- Example:
  - Fire : Wood :: Cow : (a. grass, *b.* farmer)

Using the initial sentence, you would state "Grass feeds a cow." This is correct, but then so is the next answer choice "Farmer feeds a cow." So which is right? Modify the sentence to be more specific.
- Example: "Wood feeds a fire and is consumed."

This modified sentence makes answer choice B incorrect and answer choice A clearly correct, because while "Grass feeds a cow and is consumed" is correct, "farmer feeds a cow and is consumed" is definitely wrong.

If your initial sentence seems correct with more than one answer choice, then keep modifying it until only one answer choice makes sense.

Similar Choices

If you don't know the word, don't worry. Look at the answer choices and just use them. Remember that three of the answer choices will always be wrong. If you can find a common relationship between any three answer choices, then you know they are all wrong. Find the answer choice that does not have a common relationship to the other answer choices and it will be the correct answer.
- Example:
  - Tough : Rugged :: Hard : (a. soft, b. easy, c. delicate, d. rigid)

In this example the first three choices are all opposites of the term "hard". Even if you don't know that rigid means the same as hard, you know it must be correct, because the other three all had the same relationship. They were all opposites, so they must all be wrong. The one that has a different relationship from the other three must be correct. So don't worry if you don't know a word. Focus on the answer choices that you do understand and see if you can identify common relationships. Even identifying two word pairs with the same relationship (for example, two word pairs that are both opposites) will allow you to eliminate those two answer choices, for they are both wrong.

A simple way to remember this is that if you have two or more answer choices that have the exact same relationship, then they are both or all wrong.
- Example: (a. neat, b. orderly)

Since the two answer choices above are synonyms and therefore have the same relationship with the matching term, then you know that they both must be wrong, because they both can't be correct, and for all intents and purposes they are the same word.

Be sure to read all of the choices. You may find an answer choice that seems right at first, but continue reading and you may find a better choice.

Difficult words are usually synonyms or antonyms (opposites). Whenever you have extremely difficult words that you don't understand, look at the answer choices. Try and identify whether two or more of the answer choices are either synonyms or antonyms. Remember that if you can find two word pairs that have the same relationship (for example, they are both synonyms) then you can eliminate them both.

Eliminate Answers

Eliminate choices as soon as you realize they are wrong, but be careful! Make sure you consider all of the possible answer choices. Don't worry if you are stuck between two that seem right. By eliminating the other two possible choices, your odds are now 50/50. Rather than wasting too much time, play the odds. You are guessing, but guessing wisely, because you've been able to knock out some of the answer choices that you know are wrong. If you are eliminating choices and realize that the answer choice you are left with is also obviously wrong, don't panic. Start over and consider each choice again. There may easily be something that you missed the first time and will realize on the second pass.

Word Types

The correct answer choice will contain words that are the same type of word as those in the word pair.
- Example:
  - Artist : Paintbrush

In this example, an artist is a person, while a paintbrush is an object. The correct answer will have one word that describes a person and another word that describes an object.
- Example:
  - Hedge : Gardener :: Rock : (a. wind, b. sculptor)

In this example, you could create the sentence, "Gardener cuts away at hedges." Both answer choices seem correct with this sentence, "Wind cuts away at rocks" through the process of erosion, and "sculptor cuts away at rocks" using a hammer and chisel. The difference is that a gardener is a person, as is a sculptor, while the wind is a thing, which makes answer choice B correct.

Nearly and Perfect Opposites

When you have determined which pair of terms you should work with, and know that the provided pair is an opposite, then you must find the opposite of the remaining unmatched term. Nearly opposite may often be more correct, because the goal is to test your understanding of the nuances, or little differences, between words. A perfect opposite may not exist, so don't be concerned if your answer choice is not a complete opposite. Focus upon edging closer to the word. Eliminate the words that you know aren't correct first. Then narrow your search. Cross out the words that are the most similar to the main word until you are left with the one that is the least similar.

Prefixes

Take advantage of every clue that the word might include. Prefixes and suffixes can be a huge help. Usually they allow you to determine a basic meaning. Pre- means before, post- means after, pro – is positive, de- is negative. From these prefixes and suffixes, you can get an idea of the general meaning of the word and look for its opposite. Beware though of any traps. Just because con is the opposite of pro, doesn't necessarily mean congress is the opposite of progress!

Positive vs. Negative

Many words can be easily determined to be a positive word or a negative word. Words such as despicable, and gruesome, bleak are all negative. Words such as ecstatic, praiseworthy, and magnificent are all positive. You will be surprised at how many words can be considered as either positive or negative. If you recognize a positive/negative relationship between the given pair of terms, then focus in on the answer choices that would duplicate that positive/negative relationship with the remaining term.

Word Strength

When analyzing a word, determine how strong it is. For example, stupendous and good are both positive words. However, stupendous is a much stronger positive adjective than good. Also, towering or gigantic are stronger words than tall or large. Search for an answer choice with either the same or opposite strength (depending on the relationship of the matched terms) to the remaining term.

Type and Topic

Another key is what type of word is the unmatched term. If the unmatched term is an adjective describing height, then look for the answer choice to be an adjective describing height as well. Match both the type and topic of the main word. The type refers the parts of speech, whether the word is an adjective, adverb, or verb. The topic refers to what the definition of the word includes, such as descriptive sizes (large, small, gigantic, etc).

Form a Sentence

Many words seem more natural in a sentence. *Specious* reasoning, *irresistible* force, and *uncanny* resemblance are just a few of the word combinations that usually go together. When faced with an uncommon word that you barely understand, try to put the word in a sentence that makes sense. It will help you to understand the word's meaning and make it easier to determine its relationship. Once you have a good descriptive sentence that utilizes a main term or answer choice properly, plug in the answer choice or main term and see if a solid relationship can be established.

Use Logic

Ask yourself questions about each answer choice to see if they are logical.
- Example:
  - Aromas : Smelt :: Poundings : (a. seen, *b.* heard)

Would poundings be "seen"? Or would pounding be "heard"? It can logically be deduced that poundings are heard.

The Trap of Familiarity

Don't just choose a word because you recognize it. On difficult questions, you may only recognize one or two words. There won't be any made up words on the test, so don't think that just because you only recognize one word means that word must be correct. If you don't recognize three words, then focus on the one that you do recognize. Is it correct? Try your best to determine if it fits the sentence you've created that shows the relationship between terms. If it does, that is great, but if it doesn't, eliminate it. Each word you eliminate increases your chances of getting the question correct.

Tough Questions

If you are stumped on a problem or it appears too hard or too difficult, don't waste time. Move on! Remember though, if you can quickly check for obviously incorrect answer choices, your chances of guessing correctly are greatly improved. Before you completely give up, at least try to knock out a couple of possible answers. Eliminate what you can and then guess at the remainder before moving on.

Read Carefully

Understand the analogy. Read the terms and answer choices carefully. Don't miss the question because you misread the terms. There are only a few words in each question, so you can spend time reading them carefully. Yet a happy medium must be attained, so don't waste too much time. You must read carefully, but efficiently.

Brainstorm

If you get stuck on a difficult analogy, spend a few seconds quickly brainstorming. Run through the complete list of possible relationships. Break down each answer choice into all of the potential combinations with the two possible analogous terms. Since there are four answer choices and each answer choice could form a pair with one of two terms, then there are only eight possible relationships to test. Look at each relationship and see if it would make sense. Test with sentences to determine if any relationship can be established. By systematically going through all possibilities, you may find something that you would otherwise overlook.

AFOQT: Situational Judgment

What do situational judgment questions look like?

All situational judgment questions will have a similar format. Test takers will be given a scenario which requires some action to be taken to solve a problem that has come up, usually involving interpersonal and/or official relationships between an officer and his subordinates and/or his superiors. The test taker will then be shown several possible actions that could be taken, and will be told to select both the **most effective** and the **least effective** of the actions listed.

What are situational judgment questions testing?

This is a newer section of the AFOQT, added to make the test a better predictor of success as an Air Force officer. These questions are primarily focused on testing a person in the areas of judgment and self-sufficient decision making abilities. In order to do well on this section, test takers will need to show that they can lead subordinates and solve problems independently by using their core competencies of resource management, communication, innovation, mentoring, leadership, professionalism, and integrity.

How can I prepare?

Situational judgment questions aren't the kind of questions that lend themselves easily to preparation, as there really isn't any material for a person to review and memorize. However, you should keep in mind the qualities listed above (resource management, communication, innovation, mentoring, leadership, professionalism, and integrity) when answering questions. Your answers should reflect these qualities as much as possible. Also, you should avoid choosing any answer which involves going to a superior for advice or help unless there are no other viable options, or discussing a person's shortcomings behind their back no matter their rank. Officers are expected to be resourceful men and women of character.

AFOQT: Physical Science

What do physical science questions look like?

Physical science questions primarily test your understanding of scientific terms. You won't be asked to perform complex physics calculations or balance a chemical reaction. The purpose of this section is to make sure you paid attention in high school science and retained some of the general concepts.

How can I prepare?

The best way to prepare for these questions is to brush up on your science terminology and concepts. We've included a glossary of terms here to give you head start, but to be more thorough, a good idea would be to find a high school physical science textbook and look through the full glossary in there.

A

Absolute zero: The lowest possible temperature (-273.15°C).
Atmospheric pressure: The pressure exerted by the gases in the air. Units of measurement are kilopascals (kPa), atmospheres (atm), millimeters of mercury (mm Hg) and Torr. Standard atmospheric pressure is 100 kPa, 1atm, 760 mm Hg or 760 Torr.
Atom: The smallest particle of an element; a nucleus and its surrounding electrons.
Atomic mass: The mass of an atom measured in atomic mass units (amu). An atomic mass unit is equal to one-twelfth of the atom of carbon-12. Atomic mass is now more generally used instead of atomic weight. Example: the atomic mass of chlorine is about 35 amu.
Atomic number: Also known as proton number, it is the number of electrons or the number of protons in an atom. Example: the atomic number of gold is 79.
Atomic weight: A common term used to mean the average molar mass of an element. This is the mass per mole of atoms. Example: the atomic weight of chlorine is about 35 g/mol.

B

Boiling point: The temperature at which a substance undergoes a phase change from a liquid to a gas.

C

Celsius scale (°C): A temperature scale on which the freezing point of water is at 0 degrees and the normal boiling point at standard atmospheric pressure is 100 degrees.
Change of state: A change between two of the three states of matter, solid, liquid and gas. Example: when water evaporates it changes from a liquid to a gaseous state.
Compound: A chemical consisting of two or more elements chemically bonded together. Example: Calcium can combine with carbon and oxygen to make calcium carbonate ($CaCO_3$), a compound of all three elements.
Condensation: The formation of a liquid from a gas. This is a change of state, also called a phase change.
Conduction: (i) the exchange of heat (heat conduction) by contact with another object, or (ii) allowing the flow of electrons (electrical conduction).
Convection: The exchange of heat energy with the surroundings produced by the flow of a fluid due to being heated or cooled.

D

Decay (radioactive decay): The way that a radioactive element changes into another element due to loss of mass through radiation. Example: uranium 238 decays with the loss of an alpha particle to form thorium 234.

Density: The mass per unit volume (e.g. g/cm^3).

Diffusion: The slow mixing of one substance with another until the two substances are evenly mixed. Mixing occurs because of differences in concentration within the mixture. Diffusion works rapidly with gases, very slowly with liquids.

Dissolve: To break down a substance in a solution without causing a reaction.

E

Electrical potential: The energy produced by an electrochemical cell and measured by the voltage or electromotive force (emf).

Electron: A tiny, negatively charged particle that is part of an atom. The flow of electrons through a solid material such as a wire produces an electric current.

Element: A substance that cannot be decomposed into simpler substance by chemical means. Examples: calcium, iron, gold.

Explosive: A substance which, when a shock is applied to it, decomposes very rapidly, releasing a very large amount of heat and creating a large volume of gases as a shock wave.

F

Fluid: Able to flow; either a liquid or a gas.

Freezing point: The temperature at which a substance undergoes a phase change from a liquid to a solid. It is the same temperature as the melting point.

G

Gamma rays: Waves of radiation produced as the nucleus of a radioactive element rearranges itself into a tighter cluster of protons and neutrons. Gamma rays carry enough energy to damage living cells.

Gas/gaseous phase: A form of matter in which the molecules form no definite shape and are free to move about to uniformly fill any vessel they are put in. A gas can easily be compressed into a much smaller volume.

Group: A vertical column in the Periodic Table. There are eight groups in the table. Their numbers correspond to the number of electrons in the outer shell of the atoms in the group. Example: Group 2 contains beryllium, magnesium, calcium, strontium, barium and radium.

H

Half-life: The time it takes for the radiation coming from a sample of a radioactive element to decrease by half.

Heat: The energy that is transferred when a substance is at a different temperature to that of its surroundings.

Heat capacity: The ratio of the heat supplied to a substance, compared with the rise in temperature that is produced.

Heat of combustion: The amount of heat given off by a mole of a substance during combustion. This heat is a property of the substance and is the same no matter what kind of combustion is involved. Example: heat of combustion of carbon is 94.05 kcal (x 4.18 kJ/kcal = 393.1 kJ).

I

Ion: An atom, or group of atoms, that has gained or lost one or more electrons and so developed an electrical charge. Ions behave differently from electrically neutral atoms and molecules. They can move in an electric field, and they can also bind strongly to solvent molecules such as water. Positively charged ions are called cations; negatively charged ions are called anions. Ions can carry an electrical current through solutions.

Isotope: One of two or more atoms of the same element that have the same number of protons in their nucleus (atomic number), but which have a different number of neutrons (atomic mass). Example: carbon-12 and carbon-14.

K

Kinetic energy: The energy an object has by virtue of its being in motion.

L

Latent heat: The amount of heat that is absorbed or released during the process of changing state between gas, liquid or solid. For example, heat is absorbed when a substance melts and it is released again when the substance solidifies.

Liquid/liquid phase: A form of matter that has a fixed volume but no fixed shape.

M

Mass: The amount of matter in an object. In everyday use the word weight is often used (somewhat incorrectly) to mean mass.

Matter: Anything that has mass and takes up space.

Melting point: The temperature at which a substance changes state from a solid phase to a liquid phase. It is the same as freezing point.

Metal: A class of elements that is a good conductor of electricity and heat, has a metallic luster, is malleable and ductile, forms cations and has oxides that are bases. Metals are formed as cations held together by a sea of electrons. A metal may also be an alloy of these elements. Example: sodium, calcium, gold.

Mixture: A material that can be separated into two or more substances using physical means. Example: a mixture of copper (II) sulfate and cadmium sulfide can be separated by filtration.

Mole: 1 mole is the amount of a substance which contains Avogadro's number (about 6×10^{23}) of particles. Example: 1 mole of carbon-12 weighs exactly 12 g.

Molecule: A group of two or more atoms held together by chemical bonds. Example: O_2.

N

Neutron: A particle inside the nucleus of an atom that is neutral and has no charge.

Newton (N): The unit of force required to give one kilogram an acceleration of one meter per second every second (1 m/s^2).

Noble gases: The members of Group 8 of the Periodic Table: helium, neon, argon, krypton, xenon, and radon. These gases are almost entirely unreactive.

Nucleus: The small, positively charged particle at the centre of an atom. The nucleus is responsible for most of the mass of an atom.

P

Period: A row in the Periodic Table.

Periodic Table: A chart organizing elements by atomic number and chemical properties into groups and periods.

Phase: A particular state of matter. A substance may exist as a solid, liquid or gas and may change between these phases with addition or removal of energy. Examples: ice, liquid and vapor are the three phases of water. Ice undergoes a phase change to water when heat energy is added.
Photon: A parcel of light energy.
Potential energy: The energy an object has by virtue of its position or orientation, most commonly its height above some reference point, or amount of compression as with a spring.
Pressure: The force per unit area measured in Pascals.
Proton: A positively charged particle in the nucleus of an atom that balances out the charge of the surrounding electrons.

R

Radiation: The exchange of energy with the surroundings through the transmission of waves or particles of energy. Radiation is a form of energy transfer that can happen through space; no intervening medium is required (as would be the case for conduction and convection).

S

Solid/solid phase: A rigid form of matter which maintains its shape, whatever its container.

AFOQT: Table Reading

What are these questions testing?

The table reading section tests your ability to quickly and accurately locate information stored in a table. The questions require you to find a particular number in a table given a set of coordinates.

What is the question format?

The questions will be given in groups of around 5 questions, where each group will refer to a table of numbers, with column and row headers. Each question will give an ordered pair of numbers that indicate the location in the table where the correct answer can be found. The ordered pair is given in the form (x, y), where x is the column number and y is the row number.

The questions will be presented like this:

1. (3, -2)

a. 56
b. 39
c. 64
d. 11
e. 92

	-3	-2	-1	0	1	2	3
3	41	39	84	77	35	42	37
2	75	57	95	16	93	16	15
1	34	54	50	89	26	19	94
0	66	89	65	23	13	42	20
-1	15	97	86	76	76	58	92
-2	80	92	78	52	90	11	56
-3	88	81	61	79	35	64	52

To answer the question, look at the ordered pair. It indicates that the number you are looking for is in the column labeled 3, and the row labeled -2. In this table, that number is 56, answer choice A.

What strategies can I use to answer the questions quickly and accurately?

The best way to approach these questions is methodically. Take the first number in the ordered pair and find it on the column headers. Keep your finger on that number while you take the second number of the ordered pair and locate it on the row headers. Put another finger on that number. Drag the first finger straight down the column until you get to the row that your other finger is on. The number at the intersection of the indicated column and row is your answer.

If you find yourself staring at the table for too long trying to be sure that you've selected the right number, you may find it helpful to draw lines on the graph, or take two pieces of scratch paper and line up their edges with the column and row numbers given in the question so that the number you are looking for appears at the corner where the two pieces of paper come together.

What are the common mistakes to avoid?

Since the process for answering these questions is very straightforward, most errors are the result of trying to go too quickly. The more you practice these sorts of questions, the faster you will be able to accurately answer them. Once you've practiced for a while, you'll be able to get a feel for how fast you can go while accurately answering all the questions. On test day, force yourself to go no faster than this.

One common mistake made on these questions is taking the ordered pair in the question to be ordinal coordinates rather than numbers referencing the column and row labels. For example, suppose a question asks for the ordered pair (1, 2). Under the pressure of the test, many people will instinctively go to the first column and second row, and take that number. Don't succumb to pressure on the test. Just follow the procedure and work through the questions at your ideal speed.

AFOQT: Instrument Comprehension

What are these questions testing?

These questions are designed to test your familiarity with and understanding of common instruments.

What is the question format?

The question format will vary depending on the types of instruments that the test covers. However, this section of the book will focus on the compass and the artificial horizon, two instruments commonly used in airplanes.

Test questions related to these instruments will illustrate a compass and an artificial horizon inside an airplane cockpit. Based on the readings of both of these instruments, you will have to determine the position and orientation of the airplane.

How do I read the compass and artificial horizon instruments?

The compass, a relatively intuitive instrument with which many people are familiar, shows which direction a person or vehicle is facing. When a person is facing north, for example, the needle on the compass points toward the "N." If the person is facing a direction between south and southeast, the needle will point between "S" and "SE."

The artificial horizon is an instrument that shows how the nose and wings of a plane are tilted. For most people, the artificial horizon is less intuitive and less familiar than the compass. However, if you imagine yourself actually flying in a plane, the artificial horizon becomes easier to read and understand.

The artificial horizon has two components that illustrate how the nose of an airplane is tilted with respect to the ground: the miniature wings and the horizon bar. The miniature wings represent the actual wings of the aircraft, and the horizon bar represents the horizon, the imaginary line that divides the ground and the sky from the pilot's point of view. When the miniature wings are level with the horizon bar, the plane is level. When the miniature wings are above the horizon bar, the plane is tilted upward, and when the miniature wings are below the horizon bar, the plane is tilted downward. These categories of nose tilt are shown in the drawing below.

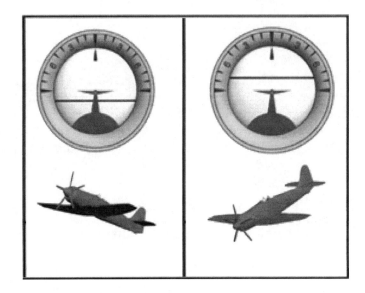

To illustrate how the wings are tilted from side to side, the artificial horizon instrument also has a dial with degree marks representing the bank angle. A needle on the dial indicates the exact bank angle, and the horizon bar is tilted accordingly, as shown in the picture below. If the left wing of the plane is tilted downward, the needle will be to the right of the center of the dial; if the right wing is tilted downward; the needle will be to the left of the center. Note that the tilted horizon bar reflects the pilot's point of view: if the left wing of the plane is tilted downward, the horizon will appear to be tilted in the opposite direction.

To answer the questions, you will have to use information from both the compass and artificial horizon to determine how the plane is oriented. If the plane is flying north, it will appear to fly into the page in the illustrations.

For example, based on the compass and artificial horizon shown below, which of the answer choices represents the orientation of the plane?

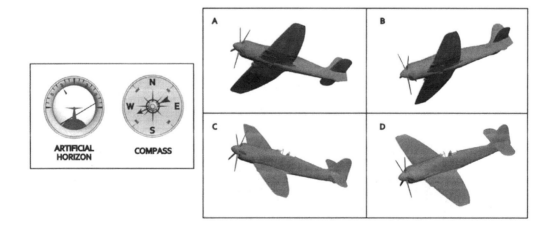

ANSWER: The answer is A. First, notice that the compass is pointing in a west-southwest direction. If a north-flying plane is facing into the page, then a westbound plane will be facing left. The compass indicates that the plane is flying somewhere between west and southwest, so the illustration will show a plane that appears to be facing left and just slightly out of the page.

Second, notice that the miniature wings in the artificial horizon are above the horizon line, and the needle on the dial is to the left of the center. From this information, you know that 1) the nose of the plane is tilted upward, and 2) the left wing of the plane is tilted upward, and the right wing is tilted downward. Because only the plane illustrated in choice A fits this description, it is the correct answer.

How can I improve my ability to read the compass and artificial horizon?

Most people don't encounter these instruments on an everyday basis, so the best way to improve your ability to read them is simply to do the practice question in this section. However, although the artificial horizon is not commonly found outside of aircraft, you might want to practice using a real compass if you are still having trouble with these questions. You can find a reasonably-priced compass at most outdoor or sporting goods stores.

AFOQT: Block Counting

What are these questions testing?

These questions are designed to test your spatial, geometric, and logical abilities.

What is the question format?

The test will show a drawing of a three-dimensional arrangement of blocks with the same size and shape, and ask you to identify how many other blocks a particular block is touching. Typically, the blocks are arranged in irregular shapes, and some of the blocks are hidden. You will have to use spatial intuition and reasoning to determine how many blocks are touching the block in question.

How do I know how many blocks are touching the particular block?

First, it is important to know which blocks qualify as "touching" the other block. If at least part of a face of one block touches at least part of a face of another block, those blocks are considered to touch each other. However, if a block shares only a corner or an edge with another block, the two blocks do not touch.

The example below illustrates the difference between touching and non-touching blocks. Block A is touching blocks 1, 2, and 3 because part of a face of block A is touching each of these blocks. However, blocks 4 and 5 do *not* touch block A because they only contact block A at an edge; that is, they share no area with any faces of block A. Therefore, block A is touching three blocks in this picture.

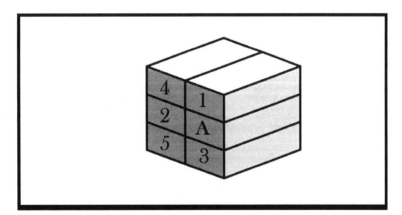

Sometimes, the blocks may be positioned so that certain blocks are hidden from view. In these cases, you will have to use basic spatial intuition and logic to determine the number of blocks touching a particular block.

In the example below, try to figure out how many blocks are touching block A.

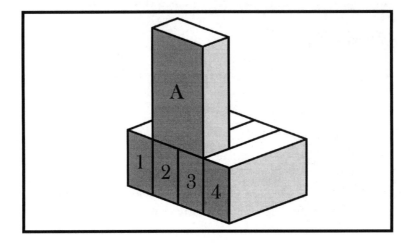

Answer: There are two blocks touching block A. Even though blocks 1 and 4 each share an edge with block A, they do not "touch" block A, as the word is used in the context of the test. Only blocks 2 and 3 actually touch a face of block A, so block A is only touching two blocks.

Are there any ways to make it easier to count the number of touching blocks?

Remember that for a block to count as "touching," it much contact a face of that block. Therefore, it might help to count the number of blocks that are touching each face of the block in question.

Try to apply this strategy to the example below, in order to count the number of blocks touching block A.

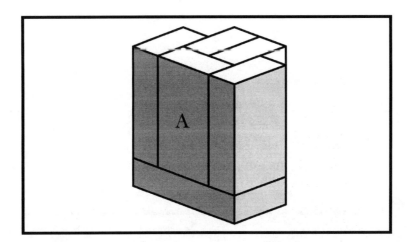

Answer: The back face of block A is touching two blocks. The left face of block A is touching one block, as is the right side. The bottom face is touching one block. Therefore, block A is touching a total of five blocks.

If you are still having trouble with the practice questions, it might help to try to re-create the example problems with a set of rectangular blocks. Practicing with physical blocks will make it easier to visualize hidden blocks when the blocks are drawn on paper.

AFOQT: Aviation Information

What are these questions testing?

The aviation information section tests your knowledge of basic aviation information. This includes a variety of things including aircraft terminology, the basic physics involved in flight, and common airport information.

How can I improve my ability to answer these questions?

Since these are all knowledge based questions, you can improve your success rate here by reading up on aircraft operation and airport information. The section below is an overview of the basics in these areas.

Fixed-wing aircraft

There are six basic components of a fixed-wing aircraft: wings; fuselage; tail assembly; landing gear; powerplant; and flight controls and control surfaces.

<u>Wings</u>
The **wings** are the primary airfoils of the plane. An airfoil is anything designed to produce lift when it moves through the air. The leading edge of an airfoil is thicker and rounder than the trailing edge, and the top surface of the airfoil has a greater curve than the bottom. The result is that air flows more quickly over the top of the wing, and the greater air pressure beneath pushes the wing, and thus the plane, upwards.

The wings connect to either side of the fuselage. Planes are designated as high-, mid-, and low-wing, depending on where the wings are attached. The wings themselves are described as either cantilever or semi-cantilever. A cantilever wing has sufficient internal support structures to keep it steady in its location. A semi-cantilever wing, on the other hand, requires additional external support structures. The trailing edge of a wing typically has two control surfaces attached by means of a hinge: flaps run from the fuselage to the middle of the wing, and ailerons run from the middle of the wing to the tip. By raising and lowering the flaps and ailerons, the pilot can roll the plane. The plane will roll when the ailerons and flaps are pointed in opposite directions. When the plane is cruising, however, these control surfaces are aligned with the rest of the wing. During takeoff and landing, both surfaces are extended, which increases the lift.

The distance from one wingtip to the other is the wingspan, and the distance from the leading edge to the trailing edge is called the chord. The chord line runs through the wing from leading edge to trailing edge: it divides the wing into upper and lower surfaces. The mean camber line runs along the inside of the wing, such that the parts of the wing above and below it are equal in thickness. The camber is the curvature of the airfoil: if an airfoil is heavily curved, it has a high camber. The thickness of a wing is measured at its greatest point. The shape of the wings when viewed from overhead is known as the planform.

When the wings are not attached parallel to the horizontal plane, the angle they make with the horizontal plane is called the dihedral angle. A positive dihedral wing angle (wings angling above the horizontal plane) keeps the plane stable when it rolls, as it will encourage the plane to return to its original position. This does diminish the maneuverability of the plane, which is why the wings of fighter jets are usually horizontal or even pointed slightly downwards (anhedral).

The shape of the wings has a major influence on the handling, maneuverability, and speed of the plane. Today's planes generally have a straight, sweep, or delta shape. Straight wings may be rectangular, elliptical (rounded), or tapered. They are commonly found on sailplanes, gliders, and other low-speed aircraft.

A swept wing provides better handling at high speeds, but makes the plane slightly less stable at low speeds. Since most modern aircraft are designed to operate at high speeds, this is the most commonly used wing style. Wings may be swept forward or back, though forward-swept wings are rarely seen. In general, a higher angle of sweep is used for planes that are meant to travel faster and be more maneuverable; however, more extreme sweeps require much greater speeds for takeoff and landing.

The delta wing shape is triangular, so that the leading edge of the wing has a high sweep angle while the trailing edge is mostly, if not completely, straight. A delta shape enables the plane to travel but also requires very high takeoff and landing speeds. Many of the earliest supersonic aircraft used the delta wing shape, as did the space shuttles.

Fuselage

The **fuselage** is the main body of the airplane. The basic features of the fuselage are the cockpit, cabin, cargo area, and attachment points for external components, like the wings and landing gear. Some planes designed for specific purposes may not have all of these components; for instance, a fighter jet will not have a cabin for passengers or a cargo area, since it needs to be light and maneuverable. The fuselage may be described as either truss or monocoque, depending on whether its strength is created by triangular arrangements of steel or aluminum tubing or by bulkheads, stringers and formers. A stringer is a support structure that runs the length of the fuselage, while a former runs perpendicular.

Tail assembly

The **tail assembly**, or empennage, includes the vertical and horizontal stabilizers, elevators, rudders, and trim tabs. The stabilizers are fixed (non-adjustable) surfaces that extend from the back end of the fuselage. The elevators are positioned along the trailing edges of the horizontal stabilizers; the pilot can move them to raise or lower the nose of the plane. The rudders are connected to the trailing edge of the vertical stabilizer, and are used to move the nose of the plane to the left or right, typically in combination with the ailerons. The trim tabs, finally, are movable surfaces that extend off the trailing edges of the rudder, elevators, and ailerons, and are used to make smaller adjustments.

Landing gear

The **landing gear** usually consists of three sets of wheels used for takeoffs and landings, though some planes have special non-wheel landing gear for landing on snow or water. Landing gear is commonly retractable, meaning that it is pulled up inside the plane during flight to reduce drag. In a typical arrangement, wheel sets are positioned either under each wing or on the sides of the fuselage, with the third wheel set being under the nose or the tail. Having the third wheel set under the nose, known as **tricycle** arrangement, is the most common arrangement on modern aircraft, but having the third wheel set under the tail is still known as **conventional** arrangement. Whether located under the tail or the nose, the third wheel will typically be able to rotate so that the plane can turn while traveling on the ground. The addition of extra wheels to each set allows the plane to handle a greater weight.

Powerplant
In aviation, the **powerplant** is the part of the plane that supplies the thrust. A jet engine operates by compressing the air that comes in the front, burning it along with fuel, and then blasting it out the back. There are different methods for compressing the air, but most jet engines do so by slowing it down with a set of small rotating blades. This greatly increases the air pressure at the front of the engine. The compressed air is then forced into a different section, where it is mixed with fuel and burned. As it then expands, it is pushed at great force through a series of turbines, the turning of which moves the compressor blades at the front of the engine, supplying both power and air. The exhausted air then passes out the back of the engine, which propels the plane forward. Some jet engines have afterburners, which feed extra fuel into the area between the turbines and the rear exhaust, increasing forward thrust.

In a propeller plane, on the other hand, the powerplant is the propellers and the engine. The propellers have tilted blades, which push air backwards and thereby push the plane forward. There are two types of propeller: fixed-pitch or variable-pitch. The blade angle of a fixed-pitch propeller cannot be adjusted by the pilot. Variable-pitch propellers allow the pilot, usually indirectly via the plane's control systems, to adjust the pitch of the propeller blades to alter the amount of thrust being generated. Some variable-pitch propellers are designed to operate only at a single rotational speed, allowing the engine to be much simpler and more efficient, so the amount of thrust is controlled entirely by the pitch of the blades. These are known as constant-speed propellers.

The engines of a propeller plane turn the crankshafts, which turn the propellers. The engines also are responsible for powering the plane's electrical system. The location of the engines on a propeller plane may vary. Single-engine planes typically have their engines in front of the fuselage, while multi-engine planes usually have their engines underneath the wings. Some multi-engine planes have engines in both locations.

Flight envelope

During flight, there are four forces a pilot must manage: lift, gravity, thrust, and drag. These forces act downward (gravity), upward (lift), forward (thrust), and backward (drag). The collective input of these forces is known as the flight envelope.

Gravity
The weight of a plane is the primary force that must be overcome for flight to take place. The force of **gravity** on a given object is the same, regardless of orientation, though it varies slightly with large changes in altitude.

Aviation experts distinguish between different types of weight. The basic weight includes the aircraft and any internal or external equipment that will remain a part of the plane during flight. The operating weight is the basic weight plus the crew and any other nonexpendable items not included in the basic weight. The gross weight is the total weight of the aircraft and all contents at any given time. The weight of the airplane when it has no usable fuel is called the zero fuel weight.

Lift
In order to overcome gravity, the plane must generate **lift**. Lift is the upward force of air pressure on the aircraft, primarily the wings, that allows it to achieve and maintain altitude. In order to generate lift, the plane typically must be traveling forward at considerable speed.

If the wing tilts too far back the airflow may stop over the wing's upper surface, which will result in a rapid loss of altitude and often control of the plane. This is known as a stall, and it may be avoided by decreasing the angle of attack, so that normal airflow over the top of the wing is not interrupted.

Thrust
The speed required for generating lift is provided by the aircraft's **thrust**. It ensures that the aircraft is able to continue moving forward at sufficient speed to generate lift. As was discussed in the previous section, thrust is generated by the powerplant of the aircraft, usually one or more jet engines or propellers.

Drag
An aircraft's thrust is countered by **drag**, the resistance to forward movement provided by the air that the aircraft is traveling through. At anything above normal walking or running speeds, air resistance is a noticeable hindrance to motion, and it only increases as airspeed goes up. The primary implication that this has on aircraft is that the faster the aircraft goes, the more thrust is required just to overcome the drag and maintain a constant speed.

There are two types of drag: profile drag and induced drag. Profile drag is the drag that exists when any object moves through the air. It is the result of the plane pushing air aside as it moves. Profile drag can be minimized by designing the aircraft to have a better wind profile. Induced drag, on the other hand, is drag that results from the wings generating lift. Part of the process of generating lift involves the wings redirecting the oncoming air downward (think Newton's third law), and this causes additional drag.

Atmospheric conditions
The flight envelope is significantly affected by the atmospheric conditions, primarily the density of the surrounding air and the speed and direction of any wind. The density is turn determined by the temperature, pressure, and humidity of the air. Lower temperatures, higher pressures, and lower humidity are all associated with higher density air. Denser air will produce greater lift, but will also produce more drag. Air pressure is most closely associated with altitude. In general, pressure decreases with altitude, so as you go higher up, the pressure of the surrounding air decreases.

With regard to wind, flying into the wind (headwind) has a similar type of impact to flying in denser air, though of much greater magnitude. In a headwind, the aircraft will have a higher speed relative to the surrounding air, which means it will experience greater drag and lift forces. Similarly, if the aircraft is flying the same direction as the wind (tailwind), it will have a lower speed relative to the surrounding air, and will experience reduced drag and lift.

Flight concepts and terminology

Flight attitude

The flight attitude is described in terms of three axes, all of which meet at the plane's center of mass.

The **longitudinal** axis is the axis that extends from the center forward toward the nose and rearward toward the tail. The **lateral** axis extends from the center out to the right and left, perpendicular to the longitudinal axis. Typically the lateral axis passes through (over/under) the wings. Both of these axes are in the horizontal plane when the aircraft is level. The **vertical** axis meanwhile extends straight upward and downward from the aircraft's center, perpendicular to the other two axes. The motion of the aircraft can be described in relation to these axes: Rotation about the longitudinal axis is called roll; rotation about the lateral axis is called pitch; rotation about the vertical axis is called yaw. In turn, these three types of motion are controlled by three sets of flight control surfaces. Roll is controlled by the ailerons, pitch by the elevators, and yaw by the rudder. This information is summarized in the table below, and is expanded upon in the following section.

Axis	Motion	Control surface
Longitudinal	Roll	Ailerons
Lateral	Pitch	Elevators
Vertical	Yaw	Rudder

Flight controls

Flight controls are divided into primary and secondary groups.

Primary

The primary **flight control surfaces** are the ailerons, rudder, and elevator.

The **ailerons** are responsible for the roll, or movement around the longitudinal axis. The ailerons extend from the trailing edges of the wings as shown in the figure below and can be manipulated by the pilot to cause the wing to either dip below or elevate above the horizontal plane.

The joystick (or control wheel) controls the roll of the aircraft. By pushing the stick (or turning the wheel) to the left, the pilot raises the left aileron and lowers the right aileron, causing the left wing to dip and the right wing to elevate.

The **elevators** control the plane's pitch, or movement around the lateral axis. They are attached to the trailing edges of the horizontal stabilizers at the rear of the aircraft. Depending on the design of the plane, there may be one elevator that extends across the length of the horizontal stabilizer, or there may be two elevators, divided by the vertical stabilizer, as shown in the figure below. When the elevators are undivided, they are sometimes referred to as a stabilator.

The joystick also controls the pitch of the aircraft. By pulling the stick back, the pilot raises the elevators, causing the tail of the plane to experience downward force, thus raising the nose of the plane. Pushing the stick forward will have the opposite effect on the elevators and will result in the nose of the plane dropping as the tail is pushed upward.

The **rudder** is a large flap attached by a hinge to the vertical stabilizer. It controls the motion of the plane around its vertical axis. The rudder can swing to the right or the left, causing the plane to turn (yaw) in either direction.

The rudder is controlled with two pedals: when the pilot pushes on the right pedal, the rudder swings out to the right, causing leftward pressure on the tail of the aircraft. This results in the nose of the plane turning to the right. Similarly, if the pilot pushes on the left pedal, the rudder will swing to the left, causing the tail to move right, and the nose to turn left.

The pilot also controls the amount of power or thrust being produced by the engines by manipulating the **throttle**. It is considered a primary flight control because the pilot must manage the thrust to ensure that the plane will be able to accomplish its intended maneuvers. With all three of the primary control surfaces, it is important to remember that the speed of the aircraft relative to the surrounding air determines the magnitude of the aircraft's response to the control. A plane that is traveling at 300 mph will roll much more quickly than one that is traveling at 200 mph in response to the same amount of aileron manipulation. The same is true of the other two types of motion.

Flight maneuvers usually involve the use of multiple controls. To make a proper turn, for instance, the pilot will need to employ the rudder, ailerons, and elevators. The bank is established by raising and lowering the ailerons, and the rudder pedals counteract any adverse yaw that occurs. Adverse yaw is the drifting of the nose caused by the extra drag on the downward-pointing aileron. Also, because extra lift is needed during a turn, the pilot must increase the angle of attack by applying downward elevator pressure. The amount of back elevator pressure required will be in proportion to the sharpness of the turn. This will be discussed in greater detail in the section on flight maneuvers.

Secondary
The secondary flight control surfaces include the flaps and leading edge devices, spoilers, and trim systems.

The **flaps** are connected to the trailing edges of the wings; they are raised or lowered to adjust the lift or drag. The retractable flaps on modern airplanes make it possible to cruise at a high speed and land at a low speed. On the opposite end of the wing, leading-edge devices accomplish much the same purpose. There are a number of different **leading-edge devices**: fixed slats, moveable slats, and leading edge flaps.

Spoilers are attached to the wings of some airplanes in order to diminish the lift and increase the drag. Spoilers can also be useful for roll control, in part because they reduce adverse yaw. This is accomplished by raising the spoiler on the side of the turn. This reduces the lift and creates more drag on that side, which causes that wing to drop and the plane to bank and yaw to that side. If both of the spoilers are raised at the same time, the plane can descend without increasing its speed. Raising the spoilers also improves the performance of the brakes, because they eliminate lift and push the plane down onto its wheels.

Trim systems exist mainly to ease the work of the pilot. They are attached to the trailing edges of one or more of the primary control surfaces. Small aircraft often have a single trim tab attached to the elevator. This tab is adjusted with a small wheel or crank, and its position is displayed in the cockpit. When the tab is deflected upwards, the trailing edge of the elevator is forced downward and the tail is pushed up, which lowers the nose of the plane.

Typically, a pilot first will achieve the desired pitch, power, attitude, and configuration, and then use the trim tabs to resolve the remaining control pressures. There are control pressures generated by any change in the flight condition, so trimming is necessary after any change. Trimming is complete when the pilot has eliminated any heaviness in the nose or tail of the plane.

Flight maneuvers

The four basic maneuvers in flight are straight-and-level flight, turning, climbing, and descending. As the name suggests, **straight-and-level flight** involves keeping the aircraft headed in a particular direction at a particular altitude. Maintaining straight-and-level flight requires frequent adjustment, much the same way as the driver of a car has to make frequent adjustments to maintain a straight path on a windy day or when driving on a rough uneven road.

Making a smooth **turn** requires the use of all four primary controls: the throttle is set to achieve a speed suitable to the desired type of turn, the ailerons bank the wings and the elevators raise the nose to establish the rate of turn, and the rudder is employed to counter any undesired yaw resulting from the effects of the other controls or to introduce desired yaw.

There are three classes of turn: shallow, medium, and deep. A shallow turn has a bank of less than 20 degrees. At angles this shallow, most planes will tend to try to stabilize themselves back to a level angle, so the pilot must maintain some pressure on the stick to ensure that the plane doesn't pull out of the bank prematurely. A medium turn has a bank of roughly 20 to 45 degrees. Most planes will tend to stay in a medium bank until the pilot makes an adjustment. Finally, a steep turn is one in which the bank is greater than 45 degrees. For angles this steep, most planes will tend to try to increase the banking angle even further unless the pilot counters that tendency by maintaining some pressure on the stick in a stabilizing direction.

While the pilot is getting the plane to the desired bank angle, he will also be pulling back on the stick to ensure that the nose of the plane does not dip during the bank. This also serves to increase the rate at which the plane turns its heading. In general, the steeper the bank, the more sharply the pilot must pull back on the stick to maintain altitude. Because the lowered aileron on the raised wing generally creates more drag than the raised aileron on the lowered wing, the airplane tends to yaw in the direction opposite to the turn. For this reason, the pilot must at the same time apply rudder pressure in the direction of the turn.

To initiate a **climb**, an aircraft's nose is angled upward so that it gains altitude. Several things that remain constant while the aircraft is flying level change when the nose of the plane is raised. The two most significant are the effective angle of gravity and the angle of attack of the wings.

When the aircraft is level, **gravity** acts entirely in direction of the vertical axis of the plane. When the plane angles upward, the force of gravity, which still acts straight down like before, now has a component in the longitudinal direction, since the rear of the plane is now pointed slightly toward the ground. Additionally, since raising the noise of the aircraft increases the angle of attack of the wings, the amount of **drag** the aircraft experiences goes up considerably during a climb. This means that, in order to maintain flight, the thrust must now overcome both an increased amount of drag and part of the force of gravity.

If the nose of the aircraft is raised to quickly or without a sufficient increase in thrust to account for the changing flight conditions, the aircraft may **stall**. The most common cause of a stall is the plane not generating enough thrust to maintain air speed, which means that the lift being generated by the wings is not sufficient to keep the plane in the air, so the plane ceases to fly in a practical sense and instead begins to fall. To correct a stall, the pilot must angle the nose of the aircraft steeply downward and increase the throttle to generate enough airspeed in the forward direction so that the control surfaces are effective in controlling the flight of the plane again, and pull out of the dive. As should be apparent from this description, recovering from a stall involves significant loss of altitude, which makes stalling at low altitudes extremely dangerous.

There are a few different styles of controlled **descent**, but they all involve manipulation of the same two factors: **pitch** and **thrust**. By angling the nose of the plane downward, the pilot reduces the angle of attack of the wings and consequently reduces the amount of lift generated by the wings. This causes the plane to lose altitude. Similarly, by pulling back on the throttle, the pilot reduces the amount of thrust being generated, which in turn reduces the plane's air speed and the amount of lift generated by the wings, also resulting in a loss of altitude.

A **glide** is a controlled descent in which little or no engine power is used, and the plane drifts downward at a regular pace. The pilot manages a glide by balancing the forces of lift and gravity as they act on the plane.

When a pilot is executing a **landing**, the nose of the plane will actually be angled upward, but the throttle will be pulled way back to ensure that the plane continues its descent all the way to the ground.

Helicopters

In many ways, the operation of a helicopter is based on the same fundamentals as airplane flight. A helicopter is subject to the same four fundamental forces of lift, weight, thrust, and drag. Unlike an airplane, however, a helicopter applies most of its thrust vertically. When a helicopter flies at a constant speed in a stable horizontal path, the lift is equal to the weight and the forward thrust is equal to the drag. The helicopter will increase its horizontal speed if the thrust is greater than the drag, and will increase its altitude if the lift is greater than the weight. If the helicopter is hovering (i.e., not moving at all), there is no drag or forward thrust; only gravity and vertical thrust or lift, which are balanced.

The manner in which a helicopter generates lift is considerably different from that of an airplane. Whereas a plane derives its lift from the natural flow of air over the wing, the helicopter spins its "wing" rapidly and at a variable angle, giving it a variety of options for angles of attack. Because the main rotor of the helicopter is being torqued with such great force, it exerts the same amount of torque back on the fuselage of the helicopter but in the opposite direction (Newton's third law again). This necessitates a tail rotor to provide the force required to the keep the fuselage from spinning around while in flight. This function of the tail rotor is called **torque control**. Manipulation of the tail rotor is also used to change the heading of the helicopter.

Helicopter controls
Piloting a helicopter requires the use of three controls: the cyclic (stick), the collective, and the directional control system. The cyclic controls the longitudinal and lateral movement of the helicopter by adjusting the tilt of the main rotor. Moving the stick forward tilts the rotor forward, which in turn pushes the helicopter forward.

The collective is a tube running up from the cockpit floor to the left of the pilot. It has a handle that may be raised or lowered to affect the pitch, as well as a throttle that wraps around the handle and can be used to alter the engine torque. The collective controls the angle of the main rotor blades. If the handle is pulled up, the leading edge of the rotor blade lifts relative to the trailing edge.

The directional control system is a pair of pedals the pilot uses to alter the pitch of the tail rotor blades. Pressing one or the other of the pedals will cause the tail rotor to exert more or less force on the fuselage, which will in turn affect the heading of the helicopter.

A helicopter pilot must use all three of the controls at the same time. The cyclic and collective adjust the action of the main rotor, which must be compensated for with adjustment to the tail rotor. For instance, if the speed of the main rotor increases during a climb, the pilot will need to increase the amount of force generated by the tail rotor to ensure the fuselage does not begin to rotate.

If the helicopter loses engine power for some reason, the pilot will need to rely on autorotation, or the spinning of the rotors that is generated by airflow rather than the engine. The amount of torque on the fuselage will be smaller during autorotation, but it will still be enough to require the use of the tail rotor.

<u>Unique forces</u>
A helicopter generates some other forces that distinguish it from an airplane. **Translational lift** is extra lift a helicopter experiences when traveling in a forward direction.

The **Coriolis force** is another physical phenomenon related to helicopters. The Coriolis force is the increase in rotational speed that occurs when the weight of a spinning object moves closer to the rotation center. In the case of a helicopter, having a greater portion of the weight closer to the base of the blade will cause the rotor to move faster, or to require less power to move at the same rotational speed.

If the main rotor increases the flow of air over the rear part of the main rotor disc, than the rear part will have a smaller angle of attack. The result of this will be less lift in the rear part of the rotor disc. This is called the **transverse flow effect**. However, when a force is applied to a spinning disc, the effects will occur ninety degrees later. This phenomenon is known as gyroscopic precession.

Airport information

At an **airport**, the areas controlled by the aircraft traffic controller are called the movement (or maneuvering) areas. These include the runways and taxiways. Runways may be composed of all different materials, ranging from grass and dirt to asphalt and concrete. At a general aviation airport, the runways may be as little as 800 feet long and 26 feet wide, while an international airport may have runways that are 18,000 feet long and 260 feet across. The markings on a runway are white, but are usually outlined in black so that they may be better seen. Taxiways and areas not meant to be traveled by aircraft are marked in yellow.

There are three basic types of runway: visual, nonprecision instrument, and precision instrument. **Visual runways** are typical of small airports: they have no markings, though the boundaries and center lines may be indicated in some way. They are called visual runways because the pilot must be able to see the ground in order to land. It is not possible to land a plane on a visual runway simply with the use of instruments.

With a **nonprecision instrument runway**, a pilot may be able to make his approach using instruments. Specifically, this sort of runway can provide feedback on the horizontal position of the plane as it nears. Nonprecision instrument runways are commonly found at small and medium airports. These runways may have threshold markings, centerlines, and designators. These runways may also have a special mark, called an aiming point, between 1000 and 1500 feet long along the centerline of the runway.

Medium and large airports will have **precision instrument runways**, which give the pilot feedback on both horizontal and vertical position when the plane is on instrument approach. A precision instrument runway includes thresholds, designators, centerlines, aiming points, blast pads, stopways, and touchdown zone marks every 500 feet from the 500 foot to the 3000 foot mark.

Runways are named according to their **direction** on the compass, ranging from 01 to 36. So, for instance, due south would be runway 18 ("one-eight"), and due west would be runway 27 ("two-seven"). In North America, the runways are named in accordance with geographic (grid) north, rather than magnetic north. Of course, a runway may have two names, one for each direction in which it is used. The same runway may be referred to as runway 05 ("zero-five") or runway 23 ("two-three") depending on the direction it is being used on a given day. In most cases, fixed-wing aircraft take off and land against the wind, because the extra amount of air over the wing will increase lift (and reduce the required ground speed).

In the event that **multiple runways** travel in the same direction, they will be distinguished from each other by their relative positions according to an observer on approach from the appropriate direction: left or right runway if there are only two; left, right, or center runway if there are three. Of course, a runway that is on the right when travelling in one direction will be on the left when it is being used in the opposite direction.

In most cases, **runway lights** are operated by the airport control tower. There are a number of different components to a runway lighting system. A Runway Centerline Lighting System is a line of white lights mounted every fifty feet along the centerline. When the approaching plane gets within 3000 feet of the runway, the lights begin to blink red and white; when the plane gets within 1000 feet, the lights become solid red. Precision instrument runways have runway end lights and edge lights. Runway end lights run the width of both ends of the runway: from the ground these lights appear red, while they appear green from above. Runway edge lights run the length of the runway on both sides. This lighting typically changes color as well when the plane gets within a certain distance of the front end of the runway. There are similar lights marking the boundaries of taxiways. An Approach Lighting System is a set of strobelights and/or lightbars that indicate the end of the runway from which descending aircraft should arrive. Runway end identification lights are synchronized lights that flash at the runway thresholds. At some airports these lights face in every direction, while at others they only face the direction from which planes approach. Runway end identification lights are useful when the runway doesn't stand out from the surrounding area, or when visibility is poor.

Some big airports also have **Visual Approach Slope Indicators**, which give the incoming pilot useful information. In a typical VASI system, white lights indicate the lower glide path limits and red lights indicate the upper. The VASI should be visible for twenty miles at night, and for three to five miles during the day under normal conditions. An effective VASI should keep the plane clear of obstructions so long as it remains within approximately ten degrees of the extended runway centerline and within four nautical miles of the runway threshold.

ASVAB Practice Test

Arithmetic Reasoning

The Arithmetic Reasoning Test on the ASVAB consists of a 36 minute section with 30 questions.

The Arithmetic Reasoning Test will test your ability to recognize problem types and apply those principles at a slightly higher level of understanding. You may still find extremely similar problems on both the Arithmetic Reasoning and Mathematics Knowledge tests.

A detailed knowledge of algebra and trigonometry is NOT necessary to answer to succeed on ASVAB Arithmetic Reasoning problems. Don't be intimidated by the questions presented on the Arithmetic Reasoning Test. They require nothing more than a basic knowledge of math, and the ability to recognize problems types and apply the right formulas to solve them.

1. A man buys two shirts. One is $7.50 and the other is $3.00. A 6% tax is added to his total. What is his total?
   A. $10.50
   B. $11.13
   C. $14.58
   D. $16.80

2. If a chef can make 25 pastries in a day, how many can he make in a week?
   A. 32
   B. 74
   C. 126
   D. 175

3. A woman must earn $250 in the next four days to pay a traffic ticket. How much will she have to earn each day?
   A. $45.50
   B. $62.50
   C. $75.50
   D. $100.50

4. A car lot has an inventory of 476 cars. If 36 people bought cars in the week after the inventory was taken, how many cars will remain in inventory at the end of that week?
   A. 440
   B. 476
   C. 484
   D. 512

5. A woman has $450 in a bank account. She earns 5% interest on her end-of-month balance. How much interest will she earn for the month?
   A. $5.00
   B. $22.50
   C. $427.50
   D. $472.50

6. Three children decide to buy a gift for their father. The gift costs $78.00. One child contributes $24.00. The second contributes $15.00 less than the first. How much will the third child have to contribute?
    A. $15.00
    B. $39.00
    C. $45.00
    D. $62.00

7. Two women have credit cards. One earns 3 points for every dollar she spends. The other earns 6 points for every dollar she spends. If they each spend $5.00, how many combined total points will they earn?
    A. 15
    B. 30
    C. 45
    D. 60

8. A company employing 540 individuals plans to increase its workforce by 13%. How many people will the company employ after the expansion?
    A. 527
    B. 547
    C. 553
    D. 610

9. A 13 story building has 65 apartments. If each floor has an equal number of apartments, how many apartments are on each floor?
    A. 2
    B. 3
    C. 4
    D. 5

10. If 5 people buy 3 pens each and 3 people buy 7 pencils each, what is the ratio of the total number of pens to the total number of pencils?
    A. 15:21
    B. 3:7
    C. 5:7
    D. 1:1

11. A man earns $15.23 per hour and gets a raise of $2.34 per hour. What is his new hourly rate of pay?
    A. $12.89
    B. $15.46
    C. $17.57
    D. $35.64

12. How many people can travel on 6 planes if each carries 300 passengers?
    A. 1800
    B. 1200
    C. 600
    D. 350

13. In a town, the ratio of men to women is 2:1. If the number of women in the town is doubled, what will be the new ratio of men to women?
    A. 1:2
    B. 1:1
    C. 2:1
    D. 3:1

14. A woman weighing 250 pounds goes on a diet. During the first week, she loses 3% of her body weight. During the second week, she loses 2%. At the end of the second week, how many pounds has she lost?
    A. 12.50
    B. 10
    C. 12.35
    D. 15

15. A woman is traveling to a destination 583 km away. If she drives 78 km every hour, how many hours will it take for her to reach her destination?
    A. 2.22
    B. 3.77
    C. 5.11
    D. 7.47

16. If one gallon of paint can paint 3 rooms, how many rooms can be painted with 28 gallons of paint?
    A. 10
    B. 25
    C. 56
    D. 84

17. Five workers earn $135/day. What is the total amount earned by the five workers?
    A. $675
    B. $700
    C. $725
    D. $750

18. A girl scores a 99 on her math test. On her second test, her score drops by 15. On the third test, she scores 5 points higher than she did on her second. What was the girl's score on the third test?
    A. 79
    B. 84
    C. 89
    D. 99

19. A man goes to the mall with $50.00. He spends $15.64 in one store and $7.12 in a second store. How much does he have left?
    A. $27.24
    B. $34.36
    C. $42.88
    D. $57.12

20. 600 students must share a school that has 20 classrooms. How many students will each classroom contain if there are an equal number of students in each class?
    A. 15
    B. 20
    C. 25
    D. 30

21. Four workers at a shelter agree to care for the dogs over a holiday. If there are 48 dogs, how many must each worker look after?
    A. 8
    B. 10
    C. 12
    D. 14

22. One worker has an office that is 20 feet long. Another has an office that is 6 feet longer. What is the combined length of both offices?
    A. 26 feet
    B. 36 feet
    C. 46 feet
    D. 56 feet

23. Four friends go shopping. They purchase items that cost $6.66 and $159.23. If they split the cost evenly, how much will each friend have to pay?
    A. $26.64
    B. $39.81
    C. $41.47
    D. $55.30

24. A 140 acre forest is cut in half to make way for development. What is the size of the new forest's acreage?
    A. 70
    B. 80
    C. 90
    D. 100

25. A farmer has 360 cows. He decides to sell 45. Shortly after, he purchases 85 more cows. How many cows does he have?
    A. 230
    B. 315
    C. 400
    D. 490

26. A couple plans to buy a car. They have $569 in a joint bank account. The man has $293 in additional cash and the woman has $189. What is the most expensive car they will be able to afford?
    A. $482
    B. $758
    C. $862
    D. $1051

27. The temperature of a cup of coffee is 98 degrees. If its temperature decreases by 2 degrees per minute, what will its temperature be after 4 minutes?
    A. 100 degrees
    B. 98 degrees
    C. 94 degrees
    D. 90 degrees

28. A man's lawn grass is 3 inches high. He mows the lawn and cuts off 30% of its height. How tall will the grass be after the lawn is mowed?
    A. 0.9 inches
    B. 2.1 inches
    C. 2.7 inches
    D. 2.9 inches

29. Three outlets are selling concert tickets. One ticket outlet sells 432; another outlet sells 238; the third outlet sells 123. How many concert tickets were sold in total?
    A. 361
    B. 555
    C. 670
    D. 793

30. A boy has a bag with 26 pieces of candy inside. He eats 8 pieces of candy, then divides the rest evenly between two friends. How many pieces of candy will each friend get?
    A. 7
    B. 9
    C. 11
    D. 13

Word Knowledge

The Word Knowledge Test on the ASVAB consists of an 11 minute section with 35 questions.

You will be given a sentence that has an underlined, boldfaced word. From the four answer choices provided, you must choose which answer choice most nearly means the same as the underlined word. In other words, you have to identify a synonym of the underlined word.

1. **Generous** most nearly means
    A. giving
    B. truthful
    C. selfish
    D. harsh

2. The math test was quite **challenging.**
    A. reasonable
    B. lengthy
    C. difficult
    D. simple

3. **Instructor** most nearly means
    A. pupil
    B. teacher
    C. survivor
    D. dictator

4. The audience applauded after the woman **concluded** her presentation.
    A. delivered
    B. prepared
    C. attended
    D. finished

5. **Residence** most nearly means
    A. home
    B. area
    C. plan
    D. resist

6. The company **instantly** agreed to the terms of the contract.
    A. reluctantly
    B. eventually
    C. immediately
    D. definitely

7. **Gigantic** most nearly means
    A. small
    B. great
    C. huge
    D. scary

8. The new car was very **costly**.
    A. expensive
    B. cheap
    C. attractive
    D. rare

9. **Opportunity** most nearly means
    A. event
    B. plan
    C. direction
    D. chance

10. The woman's **response** to the question was correct.
    A. hesitation
    B. answer
    C. decision
    D. concern

11. **Frequently** most nearly means
    A. difficulty
    B. freely
    C. often
    D. easy

12. He **observed** the eagles with binoculars.
    A. watched
    B. hunted
    C. scared
    D. sold

13. **Purchased** most nearly means
    A. sold
    B. bargained
    C. complained
    D. bought

14. She is having **difficulties** with her new computer.
    A. experiences
    B. solutions
    C. pleasures
    D. problems

15. **Entire** most nearly means
    A. whole
    B. divide
    C. tired
    D. basic

16. The woman's performance was **superior** to the man's.
    A. short
    B. similar
    C. better
    D. weak

17. **Remark** most nearly means
    A. rebuke
    B. comment
    C. lecture
    D. replace

18. **Selecting** the best person for the job was difficult.
    A. locating
    B. contacting
    C. choosing
    D. informing

19. **Commence** most nearly means
    A. begin
    B. progress
    C. finish
    D. comment

20. The fox ran **swiftly** after its prey.
    A. surely
    B. quickly
    C. slowly
    D. lightly

21. **Overdue** most nearly means
    A. overall
    B. early
    C. punctual
    D. late

22. She felt intense **anguish** when her parents divorced.
    A. loneliness
    B. confusion
    C. anger
    D. sorrow

23. **Solitary** most nearly means
    A. single
    B. solid
    C. sturdy
    D. stoic

24. The class **chuckled** when the professor dropped his notes.
    A. helped
    B. commented
    C. laughed
    D. chose

25. **Depart** most nearly means
    A. leave
    B. describe
    C. arrive
    D. portion

26. The child was unable to **locate** his toy.
    A. buy
    B. find
    C. enjoy
    D. share

27. **Soiled** most nearly means
    A. dirty
    B. sullen
    C. sultry
    D. dainty

28. The bear **slumbered** in its cave.
    A. hunted
    B. fed
    C. slept
    D. explored

29. **Puzzled** most nearly means
    A. admired
    B. retired
    C. confused
    D. understand

30. The woman **desired** a new car.
    A. purchased
    B. wanted
    C. described
    D. intended

31. **Cheap** most nearly means
    A. cheer
    B. doubtfully
    C. cheat
    D. inexpensive

32. She is a very **intelligent** lady.
    A. pretty
    B. nice
    C. smart
    D. mysterious

33. **Object** most nearly means
    A. disagree
    B. state
    C. concur
    D. relate

34. The workers **constructed** the home over a three month period.
    A. purchased
    B. explored
    C. improved
    D. built

35. **Required** most nearly means
    A. needed
    B. wished
    C. studied
    D. wanted

Paragraph Comprehension

The Paragraph Comprehension Test on the ASVAB consists of a 13 minute section with 15 questions.

You will be given one or more paragraphs of information to read followed by a question or incomplete statement. From the four answer choices provided, you must choose which answer choice best completes the statement or answers the question.

1. Mitosis refers to the process of cell division that occurs in most higher life forms. During interphase, all of the genetic material within the cell is replicated. Then, the strands that contain the genetic material, which are known as chromatin, become compacted and condensed. Centrosomes then move to opposite ends of the cells, after which the nucleus contained in the original single cell disintegrates.

Immediately before the chromatin becomes compacted,
    A. the genetic material in the cell is copied.
    B. the centrosomes move to opposite ends of the cell.
    C. the nucleus inside of the cell disintegrates.
    D. the process of mitosis takes place.

2. Obesity in the Western world has reached epidemic levels. While exercise is important to maintaining a healthy weight, healthy eating is even more important. Even the most active people cannot burn off hundreds of excess calories if they are consumed on a daily basis.

It can be concluded that
    A. most people need to exercise more.
    B. many people do not practice healthy eating.
    C. obesity would not be a problem if people were active.
    D. most people do not know how to choose healthy foods.

3. Racism is still a widespread problem in North America. For example, one man of Arab descent was held at an airport for hours for no apparent reason. The man said he was later told that officials thought he was carrying a bomb. An African-American woman was passed over for a promotion. She says the job went to a less qualified applicant.

The author is constructing her argument by
    A. relying on studies conducted by others.
    B. relying on the self-reported experiences of others.
    C. relying on events that she personally witnessed.
    D. relying on accepted facts and proven statistics.

4. *Beowulf* is an epic poem that is important because it is often viewed as the first significant work of English literature. Although it was first written in 700A.D., it is thought to be even hundreds of years older than that. It is believed that the story of *Beowulf* was told for centuries before it ever made its way on to paper. It is still taught today in various schools and universities.

Beowulf is a significant poem because
    A. it was first written down in 700A.D.
    B. it was told for centuries before it was written down.
    C. it is still taught today in academic settings.
    D. it is the first important work of English literature.

5. Cats are by far a superior pet compared to dogs. They are perfect for the working person, as they don't mind being left alone during the day. In addition, they are very easy to litter train. They require little care or effort, and are much cheaper to feed compared to dogs.

The author's purpose in writing this passage is
    A. to compare cats and dogs.
    B. to convince the reader that cats are better pets.
    C. to show why cats are cheaper to feed.
    D. to explain why cats are preferred to dogs.

6. Malaria is a dangerous disease that is still common in many countries. It is carried and transmitted to humans by female mosquitoes. When they bite humans or animals, the malaria parasite is introduced into the human's or animal's bloodstream. The parasites travel to the liver. There, they multiply, and soon they infect red blood cells. Malaria symptoms like fever will begin to be experienced at this point.

After the malaria parasites infect the red blood cells
    A. symptoms like fever occur.
    B. the parasites multiply.
    C. the parasites travel to the liver.
    D. malaria is transmitted by mosquitoes.

7. Many people feel that summer water sports are dangerous. While accidents do occur, the vast majority are preventable. For example, boating is a relatively safe activity, provided that life vests are worn and a reasonable speed is maintained. When diving, it's important to know the area well and be certain that water is deep enough. Finally, alcohol should never be consumed while engaging in water sports or activities. Following these simple tips would prevent many accidents.

The main idea expressed in the passage is
    A. summer sports are dangerous.
    B. many people who engage in water activities are careless.
    C. boating is a safe activity if precautions are followed.
    D. summer water activities aren't necessarily dangerous.

8. At one time, people who wanted to be writers had to write a query letter to a magazine and then wait weeks, sometimes months, for a response, which was usually a rejection. Today, with blogs, virtually anybody can put their work out there for others to view. It's as easy as setting up your blog, naming it, and posting anything you want: opinions, poems, short stories, news articles, etc. Of

course, while people who have blogs may choose to call themselves writers or journalists, it's unlikely they are making a living by putting their random thoughts out there into cyberspace.

It can be concluded that
    A. it is easier to make a living as a writer now.
    B. there are more people who want to be writers now.
    C. most people do not submit query letters to magazines anymore.
    D. there is no approval process for getting a blog.

9. Organic food has become quite popular in the last number of years, but the price of organic foods still prevents many people from eating them on a regular basis. People choose to buy organic foods because they are becoming more concerned about what types of chemicals, fertilizers, and pesticides are in their food. Organic food is usually relatively easily obtained by visiting a grocery store or a farmer's market, and the selection of foods is also quite good.

People don't eat organic food because
    A. it's hard to find.
    B. it contains chemicals.
    C. the selection is poor.
    D. it's too expensive.

10. Studies have shown again and again that birth order strongly influences the person one will eventually become. Oldest children have been shown to be more responsible and perform better in school. Younger children tend to do less well in school and be more free spirited. Cindy is a good example. She graduated on the Dean's list, and her parents report she always did her chores as a child.

It can be concluded that
    A. Cindy was an oldest child.
    B. Cindy was a youngest child.
    C. There have been no studies done on middle children.
    D. Birth order only matters if there are exactly two children.

11. Weddings put a substantial amount of stress on the bride, the groom, and their families. There is a large amount of planning that goes into a wedding, but the expenses are the major source of grief. The cake, the dress, and meals must all be purchased. The hall must be rented, the dj hired, and the invitations bought. By the time all is said and done, even a relatively simple wedding can cost in excess of $10,000.

The main idea expressed in the passage is
    A. weddings are stressful.
    B. weddings require a lot of planning.
    C. weddings are a source of grief.
    D. weddings are expensive.

12. According to the laws of supply and demand, consumers will demand less of a good if the price is higher and more if it is lower. Conversely, suppliers will produce more of a good when the price is higher and less when it is lower.

If a supplier wanted to sell more of a good, they would
　A.　reduce the supply.
　B.　reduce the price.
　C.　raise the price.
　D.　increase the supply.

13. North America is currently dealing with a crisis. People have accumulated substantial amounts of debt. While consumers are partially to blame, much of the blame lies with the companies who actually granted the credit to consumers. This crisis wouldn't have happened if loans were not granted to people who could not afford them.

Many people in North America are in debt because
　A.　they choose not to pay their debts.
　B.　they borrowed money they couldn't repay.
　C.　companies refused to grant credit.
　D.　they dealt with serious personal crises.

14. Knowing how to perform CPR properly can save another person's life. If you come across someone who is in trouble, call 911 right away. Then, lay them on their back. Next, open their airway by raising the chin. After this, but before beginning mouth to mouth, spend a few moments to determine whether they are breathing. If not, mouth to mouth and chest compressions will be necessary.

Before checking to see whether a person is breathing
　A.　open their airway.
　B.　begin mouth to mouth.
　C.　learn how to perform CPR.
　D.　begin chest compressions.

15. Using animals for fur is a barbaric practice that should be stopped. All across the world, animals are killed every day just to provide fashionable clothing. Fur is no longer necessary as it may once have been. There are now synthetic materials that can keep people just as warm during the winter months. In addition, the conditions on many fur farms are inhumane. Considering that fur-bearing animals are suffering to provide unnecessary luxury items, it is impossible to justify this sort of cruelty to animals.

The main idea expressed in this passage is
　A.　nobody needs fur.
　B.　fur animals are treated badly.
　C.　fur is a luxury item.
　D.　other materials are warmer than fur.

Mathematics Knowledge

The Mathematics Knowledge Test on the ASVAB consists of a 24 minute section with 25 questions.

The Mathematics Knowledge Test will test your general understanding of mathematic principles. You may find similar problems on both the Mathematics Knowledge and Arithmetic tests.

1. A rectangle has a width of 7cm and a length of 9cm. What is its perimeter?
    A. 16cm
    B. 32cm
    C. 48cm
    D. 62cm

2. In the following inequality, solve for q.
-3q + 12 ≥ 4q − 30
    A. q ≥ 6
    B. q = 6
    C. q ≠ 6
    D. q ≤ 6

3. If x − 6 = 0, then x is equal to
    A. 0
    B. 3
    C. 6
    D. 9

4. If x = -3, calculate the value of the following expression:
$3x^3 + (3x + 4) - 2x^2$
    A. -104
    B. -58
    C. 58
    D. 104

5. If 3x - 30 = 45 − 2x, what is the value of x?
    A. 5
    B. 10
    C. 15
    D. 20

6. Solve for x in the following inequality.
$\frac{1}{4}x - 25 \geq 75$
    A. x ≥ 400
    B. x ≤ 400
    C. x ≥ 25
    D. x ≤ 25

7. If $x^2 - 5 = 20$, what is the value of x?
    A. 5
    B. 10
    C. 12.5
    D. 15

8. What is the area of a square that has a perimeter of 8cm?
    A. $2cm^2$
    B. $4cm^2$
    C. $32cm^2$
    D. $64cm^2$

9. If x = 4 and y = 2, what is the value of the following expression:
$3xy - 12y + 5x$
    A. -4
    B. 10
    C. 12
    D. 20

10. If .65x + 10 = 15, what is the value of x?
    A. 4.92
    B. 5.78
    C. 6.45
    D. 7.69

11. Simplify the following:
(3x + 5) (4x – 6)
    A. $12x^2 - 38x - 30$
    B. $12x^2 + 2x - 30$
    C. $12x^2 - 2x - 1$
    D. $12x^2 + 2x + 30$

12. Simplify the following expression:
$$\frac{50x^{18}t^6w^3z^{20}}{5x^5t^2w^2z^{19}}$$
    A. $10x^{13}t^3wz$
    B. $10x^{13}t^4wz$
    C. $10x^{12}t^4wz$
    D. $10x^{13}t^4wz^2$

13. 4! =
    A. 4
    B. 12
    C. 16
    D. 24

14. If a cube is 5cm long, what is the volume of the cube?
    A. 15cm³
    B. 65cm³
    C. 105cm³
    D. 125cm³

15. Solve for x by factoring:
$x^2 -13x + 42 = 0$
    A. x = 6, 7
    B. x = -6, -7
    C. x = 6, -7
    D. x = -6, 7

16. A triangle has a base measuring 12cm and a height of 12cm. What is its area?
    A. 24cm²
    B. 56cm²
    C. 72cm²
    D. 144cm²

17. Simplify the following expression:
$(3x^2 * 7x^7)+ (2y^3 * 9y^{12})$
    A. $21x^{14} + 18y^{26}$
    B. $10x^9 + 11y^{15}$
    C. $21x^{14} + 18y^{15}$
    D. $21x^9 + 18y^{15}$

18. If x/3 + 27 = 30, what is the value of x?
    A. 3
    B. 6
    C. 9
    D. 12

19. What is the slope of a line with points A (4,1) and B (-13,8)?
    A. 7/17
    B. -7/17
    C. -17/7
    D. 17/7

20. If x is 20% of 200, what is the value of x?
    A. 40
    B. 80
    C. 100
    D. 150

21. If a bag of balloons consists of 47 white balloons, 5 yellow balloons, and 10 black balloons, what is the probability that a balloon chosen randomly from the bag will be black?
    A. 19%
    B. 16%
    C. 21%
    D. 33%

22. In a lottery game, there are 2 winners for every 100 tickets sold. If a man buys 10 tickets, what are the chances that he is a winner?
    A. 1 in 2
    B. 1 in 5
    C. 2 in 5
    D. 2 in 2

23. What is the volume of a rectangular prism with a height of 10cm, a length of 5cm, and a width of 6cm?
    A. 30cm³
    B. 60cm³
    C. 150cm³
    D. 300cm³

24. What is the midpoint of point A (6, 20) and point B (10, 40)?
    A. (30, 8)
    B. (16, 60)
    C. (8, 30)
    D. (7, 15)

25. If 5x + 60 = 75, what is the value of x?
    A. 3
    B. 4
    C. 5
    D. 6

General Science

1. What is the name for any substance that stimulates the production of antibodies?
    A. collagen
    B. hemoglobin
    C. lymph
    D. antigen

2. Which of the following correctly lists the cellular hierarchy from the simplest to the most complex structure?
    A. tissue, cell, organ, organ system, organism
    B. organism, organ system, organ, tissue, cell
    C. organ system, organism, organ, tissue, cell
    D. cell, tissue, organ, organ system, organism

3. If a cell is placed in a hypertonic solution, what will happen to the cell?
    A. It will swell.
    B. It will shrink.
    C. It will stay the same.
    D. It does not affect the cell.

4. Which group of major parts and organs make up the immune system?
    A. lymphatic system, spleen, tonsils, thymus, and bone marrow
    B. brain, spinal cord, and nerve cells
    C. heart, veins, arteries, and capillaries
    D. nose, trachea, bronchial tubes, lungs, alveolus, and diaphragm

5. The rate of a chemical reaction depends on all of the following except
    A. temperature.
    B. surface area.
    C. presence of catalysts.
    D. amount of mass lost.

6. Which of the answer choices provided best defines the following statement?
For a given mass and constant temperature, an inverse relationship exists between the volume and pressure of a gas.
    A. Ideal Gas Law
    B. Boyle's Law
    C. Charles' Law
    D. Stefan-Boltzmann Law

7. Which of the following statements correctly compares prokaryotic and eukaryotic cells?
    A. Prokaryotic cells have a true nucleus, eukaryotic cells do not.
    B. Both prokaryotic and eukaryotic cells have a membrane.
    C. Prokaryotic cells do not contain membrane-bound organelles, eukaryotic cells do.
    D. Prokaryotic cells are more complex than eukaryotic cells.

8. What is the role of ribosomes?
    A. make proteins
    B. waste removal
    C. transport
    D. storage

9. If an organism is *AaBb*, which of the following combinations in the gametes is impossible?
    A. AB
    B. aa
    C. aB
    D. Ab

10. What is the oxidation number of hydrogen in CaH_2?
    A. +1
    B. −1
    C. 0
    D. +2

11. Which hormone stimulates milk production in the breasts during lactation?
    A. norepinephrine
    B. antidiuretic hormone
    C. prolactin
    D. oxytocin

12. What is the typical result of mitosis in humans?
    A. two diploid cells
    B. two haploid cells
    C. four diploid cells
    D. four haploid cells

13. Which of the following does *not* exist as a diatomic molecule?
    A. boron
    B. fluorine
    C. oxygen
    D. nitrogen

14. Which of the following structures has the lowest blood pressure?
    A. arteries
    B. arteriole
    C. venule
    D. vein

15. How does water affect the temperature of a living thing?
    A. Water increases temperature.
    B. Water keeps temperature stable.
    C. Water decreases temperature.
    D. Water does not affect temperature.

16. What is another name for aqueous HI?
    A. hydroiodate acid
    B. hydrogen monoiodide
    C. hydrogen iodide
    D. hydriodic acid

17. Which of the heart chambers is the most muscular?
    A. left atrium
    B. right atrium
    C. left ventricle
    D. right ventricle

18. Which of the following is *not* a product of the Krebs cycle?
    A. carbon dioxide
    B. oxygen
    C. adenosine triphosphate (ATP)
    D. energy carriers

19. What is the name for the reactant that is entirely consumed by the reaction?
    A. limiting reactant
    B. reducing agent
    C. reaction intermediate
    D. reagent

20. Which part of the brain interprets sensory information?
    A. cerebrum
    B. hindbrain
    C. cerebellum
    D. medulla oblongata

21. What kind of bond connects sugar and phosphate in DNA?
    A. hydrogen
    B. ionic
    C. covalent
    D. overt

22. What is the mass (in grams) of 7.35 mol water?
    A. 10.7 g
    B. 18 g
    C. 132 g
    D. 180.6 g

23. Which of the following proteins is produced by cartilage?
    A. actin
    B. estrogen
    C. collagen
    D. myosin

24. How are lipids different than other organic molecules?
    A. They are indivisible.
    B. They are not water soluble.
    C. They contain zinc.
    D. They form long proteins.

25. Which of the following orbitals is the last to fill?
    A. 1s
    B. 3s
    C. 4p
    D. 6s

26. Which component of the nervous system is responsible for lowering the heart rate?
    A. central nervous system
    B. sympathetic nervous system
    C. parasympathetic nervous system
    D. distal nervous system

27. Which of the following is *not* a steroid?
    A. cholesterol
    B. estrogen
    C. testosterone
    D. hemoglobin

28. What is the name of the binary molecular compound NO_5?
    A. cnitro pentoxide
    B. ammonium pentoxide
    C. nitrogen pentoxide
    D. pentnitrogen oxide

29. In which of the following muscle types are the filaments arranged in a disorderly manner?
    A. cardiac
    B. smooth
    C. skeletal
    D. rough

30. Which hormone is produced by the pineal gland?
    A. insulin
    B. testosterone
    C. melatonin
    D. epinephrine

Assembling Objects

In this portion of the exam test takers need to figure out how an object would look after being properly assembled. The first picture in each problem displays all the parts that need to be assembled. The next five illustrations display five different methods of assembly, four of which are wrong, and one of which is right.

Every part has at least one letter marking it; some have more than one. Each letter represents a place on the assembled part. Some letters are also shown that correspond to unseen areas. These are displayed with a dotted line that points to the side underneath, or the unseen area.

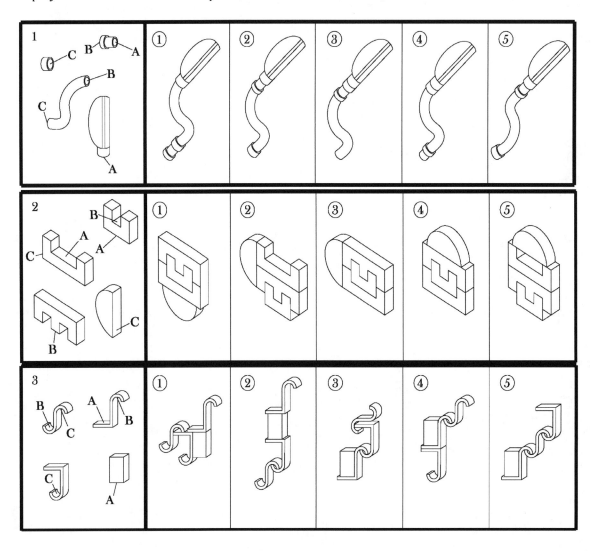

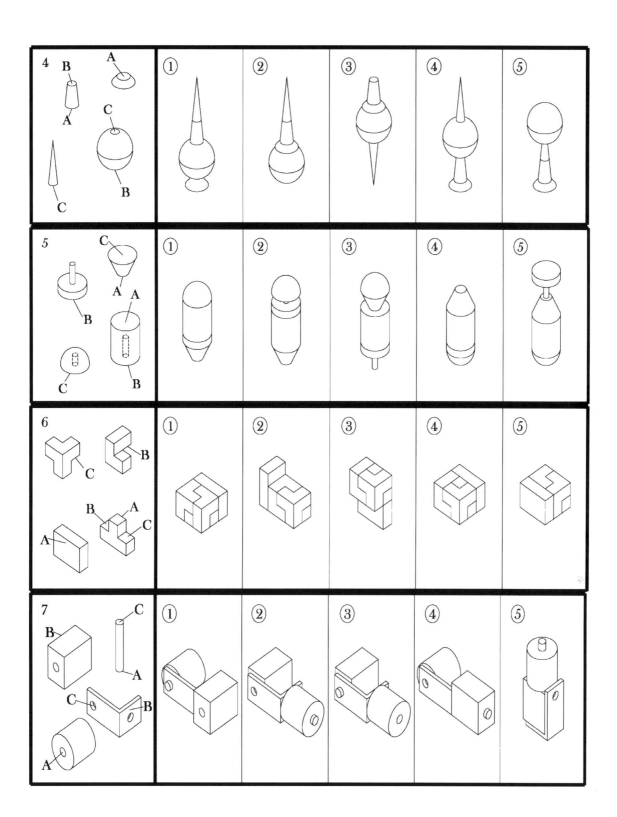

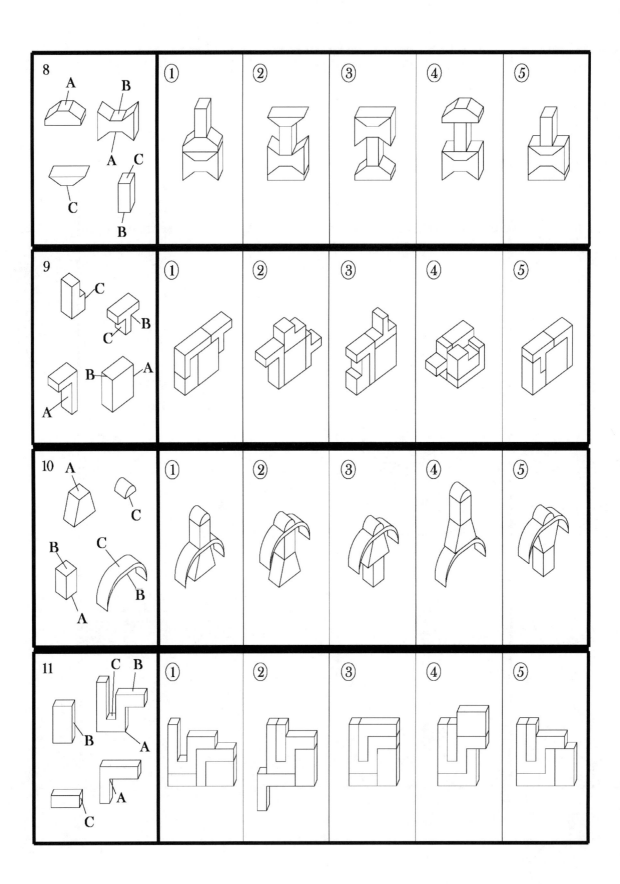

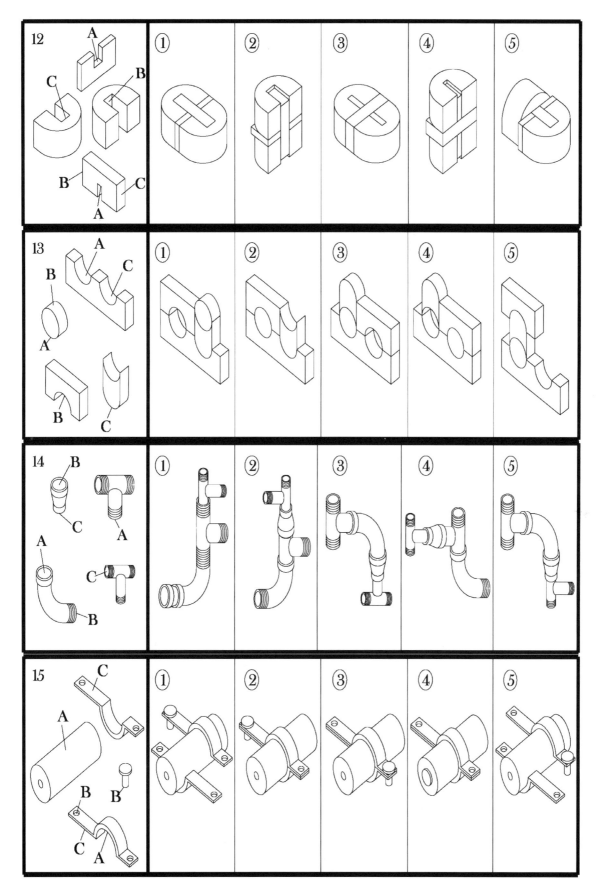

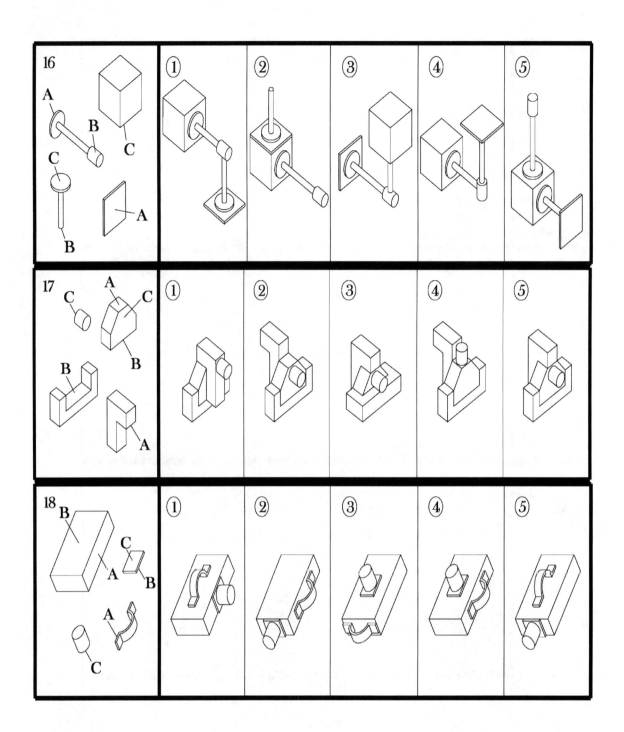

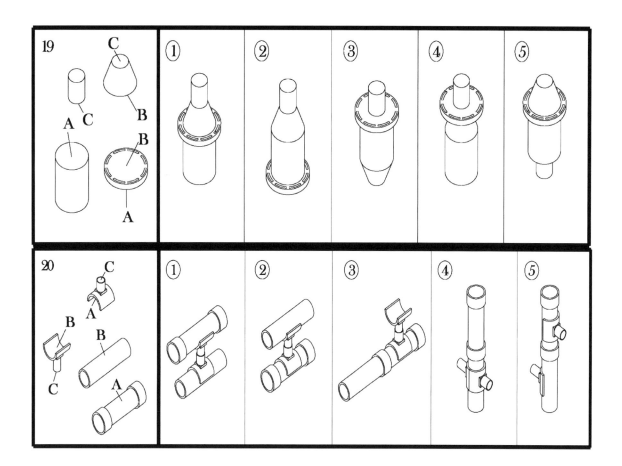

Mechanical Comprehension

This is a test of your ability to understand mechanical concepts. Each question has a picture, a question and three possible answers. Read each question carefully, study the picture, and decide which answer is correct.

1. Objects 1 and 2 are submerged in separate tanks, both filled with water. In which tank (A or B) will the water level be the highest? (If equal, mark C)

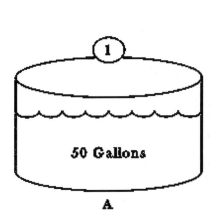
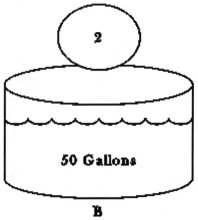

2. If ball 1 and ball 2 are of equal weight and moving at the same speed, in which direction (A, B or C) will ball 1 tend to go when it collides with ball 2 at point X?

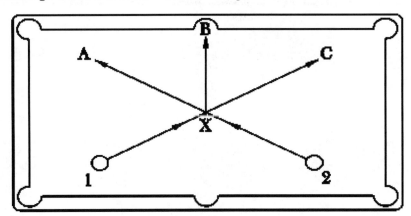

3. In which direction (A or B) will gear 5 spin if gear 1 is spinning counter-clockwise? (If both, mark C)

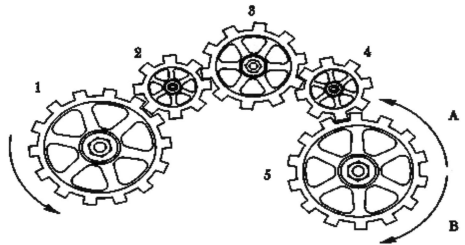

4. Which of the two identical objects (A or B) will launch a higher distance when the springs are released? (If equal, mark C)

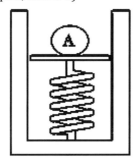 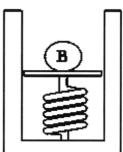

5. A watering can is filled with water. Which of the pictures (A or B) shows a more accurate representation of how the water will rest?

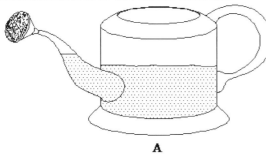 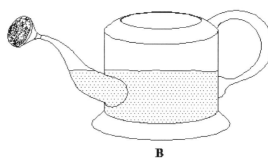

6. Among this arrangement of three pulleys, which pulley (A, B or C) turns fastest?

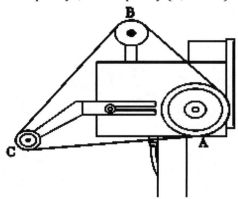

7. Which of the two scenarios (A or B) requires more effort to pull the weight up off the ground? (If equal, mark C)

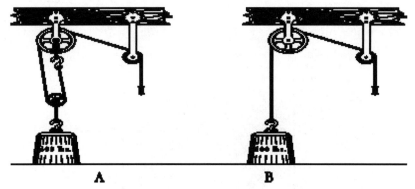

8. Which switch (A, B or C) should be closed in order to start the pump motor?

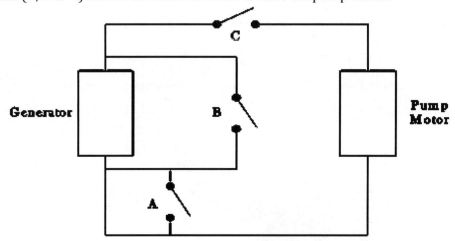

9. Which situation (A or B) requires more force to peddle the bicycle up the ramp? (If equal, mark C)

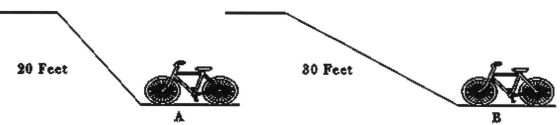

10. When the spring is released, the ball travels away from the spring to its highest point (A) and then begins to travel back towards its place of origin. At which point (A, B or C) will the ball travel to after it hits the spring a second time?

11. Which of the two boulders of equal weight (A or B) requires more force to push up the hill? (If equal, mark C)

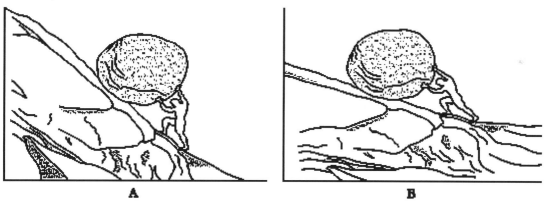

12. At which point (A, B or C) will the cannonball be traveling the slowest?

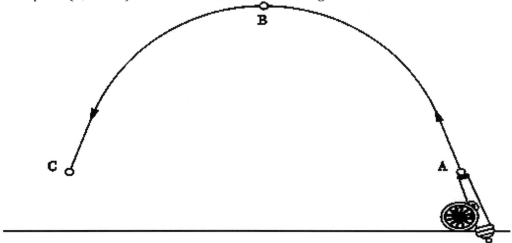

13. On which side of the pipe (A or B) would the water speed be slower? (If equal, mark C)

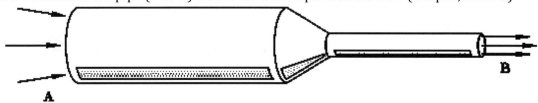

14. In which of the two figures (A or B) is the person bearing more weight? (If equal, mark C)

15. Which of the two lift trucks (A or B) carrying the same amount of weight is more likely to tip over? (If equal, mark C)

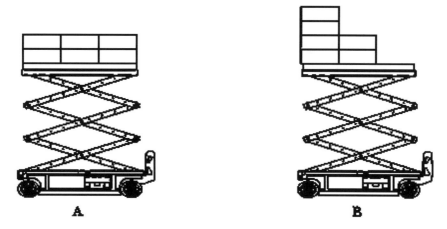

16. The weight of the boxes is being carried by the two men shown below. Which of the two men (A or B) is carrying more weight? (If equal, mark C)

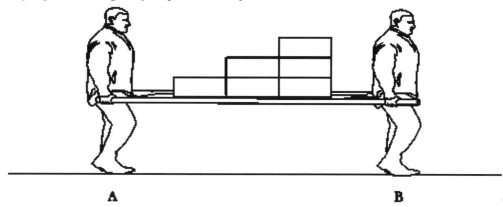

17. In the pictures below, which of the angles (A or B) is braced more solidly? (If equal, mark C)

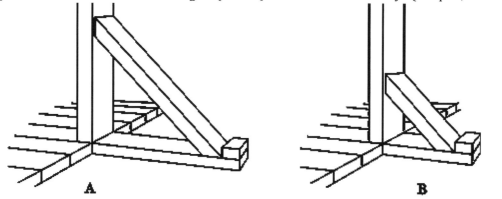

18. Given two birds sitting on branches of a tree at different elevations. Both drop objects of identical size and weight. Which object (A or B) will hit the ground with bigger force? (If equal, mark C)

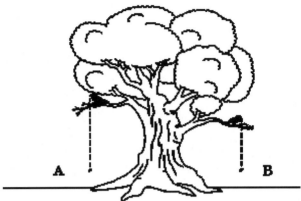

19. Which of the two wagons (A or B) of equal size and weight would be easier to drag up the hill? (If equal, mark C)

20. In which of the three positions (A, B or C) will it be easiest to accurately measure the amount of liquid in the graduated cylinder?

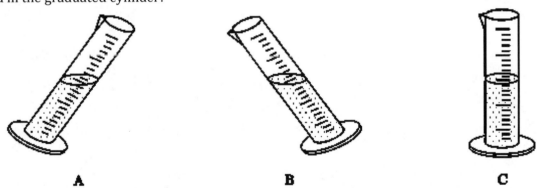

21. In which of the two figures (A or B) will the person require less force to lift a 100 pound weight? (If equal, mark C)

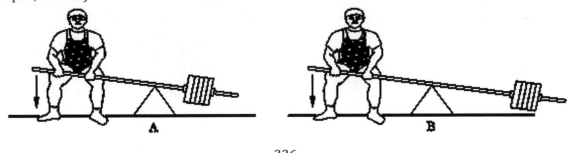

22. Which switch (A, B or C) should be closed to give power to the light?

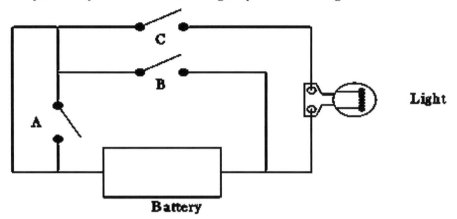

23. If the baseball and bowling ball are moving at the same speed, in which direction will the bowling ball tend to go when it collides with the baseball at point X?

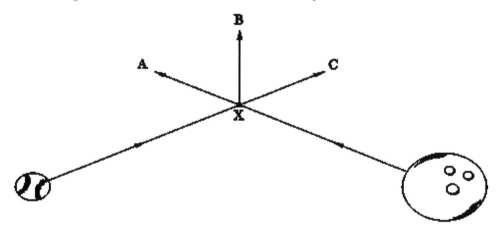

24. Which of the two rolls of paper towels (A or B) will undergo more revolutions if the ends of each roll were pulled downward with the same amount of force? (If equal, mark C)

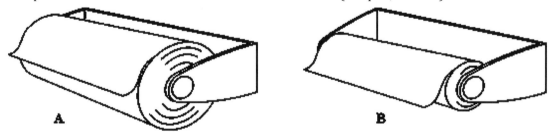

25. In which of the two containers (A or B) will water that is boiled to the same temperature cool more slowly? (If equal, mark C)

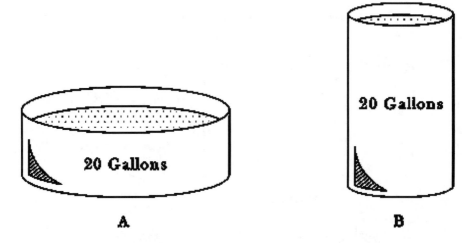

26. The water container A contains 50% salt. The water container B contains 25% salt. In which of the two containers (A or B) is an egg more likely to float? (If equal, mark C)

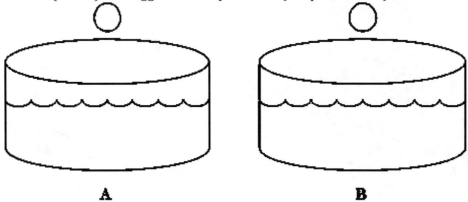

27. Which reflector (A or B) on the bicycle wheel is going to be traveling a greater distance when the wheel turns? (If equal, mark C)

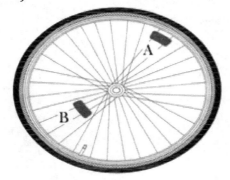

28. A javelin is thrown into the air. At which point (A, B or C) will the javelin be traveling the fastest?

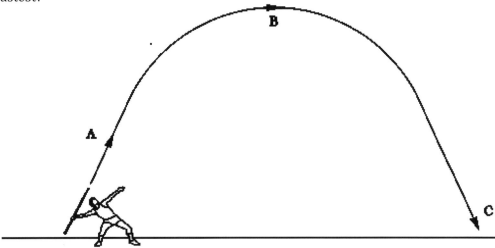

29. A child is released on the seat of a swing set at the position shown. To which point (A, B or C) will the child travel before he/she begins to return back to the point of origin?

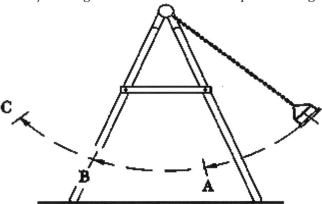

30. On which side of the pipe (A or B) would the water speed be slower? (If equal, mark C)

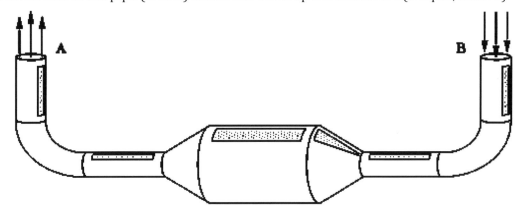

Answer Key

Arithmetic Reasoning

1. B: First, find the total before taxes: $7.50 + $3.00 = $10.50
Then, calculate 6% of the total: $10.50 * .06 = $0.63
Finally, add the tax to find the total cost: $10.50 + $0.63 = $11.13

2. D: There are 7 days in a week. Knowing that the chef can make 25 pastries in a day, the weekly number can be calculated:
25 * 7 = 175

3. B: The woman has four days to earn $250. To find the amount she must earn each day, divide the amount she must earn ($250) by 4:
$250 / 4 = $62.50

4. A: To find the number of cars remaining, subtract the number of cars that were sold from the original number: 476 – 36 = 440

5. B: Calculate 5% of $450: $450 * 0.05 = $22.50
This is the amount of interest she will earn.

6. C: First, figure out how much the second child contributed: $24.00 - $15.00 = $9.00
Then, calculate how much the first two children contributed in total: 24 + 9 = $33.00
Finally, figure out how much the third child will have to contribute:
$78.00 - $33.00 = $45.00

7. C: First, figure out how many points the first woman will earn: 3 * 5 = 15
Then, figure out how many points the second woman will earn: 6 * 5 = 30
Then, add these two values together: 30 + 15 = 45 points total.

8. D: First, calculate 13% of 540 = 70
Then, add this value onto the original number of workers: 540 + 70 = 610
610 is the number of people that the company will employ after the expansion.

9. D: To find the number of apartments on each floor, divide the total number of apartments by the number of floors:
65 / 13 = 5

10. A: First, find the total number of pens: 5 * 3 = 15
Then, find the total number of pencils: 3 * 7 = 21
Finally, express it as a ratio 15 : 21

11. C: To calculate his new salary, add his raise to his original salary:
$15.23 + $2.34 = $17.57

12. A: To find the total number of passengers, multiply the number of planes by the number of passengers each can hold: 6 * 300 = 1800

13. B: Currently, there are two men for every woman. If the number of women is doubled (1 * 2 = 2), then the new ratio is 2:2. This is equivalent to 1:1.

14. C: First, calculate 3% of 250 pounds: 250 * 0.03 = 7.5 pounds
Calculate how much she weighs at the end of the first week: 250 – 7.5 = 242.5 pounds
Calculate 2% of 242.5: 242.5 * 0.02 = 4.85 pounds
Add the two values together to get the total: 7.5 + 4.85 = 12.35

15. D: Divide the total distance she must travel (583km) by the number of kilometers she drives each hour (78km) to figure out how many hours it will take to reach her destination:
583 km / 78 km = 7.47 hours

16. D: One gallon of paint can paint three rooms, so to find out how many 28 gallons can do, that number must be multiplied by 3: 28 * 3 = 84 rooms

17. A: Each earns $135, so to find the total earned, that amount must be multiplied by the number of workers: $135 * 5 = $675

18. C: First, calculate her score on the second test: 99 – 15 = 84
Then, calculate her score on the third test: 84 + 5 = 89

19. A: To find out how much he has remaining, both numbers must be subtracted from the original amount ($50.00): $50.00 - $15.64 - $7.12 = $27.24

20. D: Divide the number of students (600) by the number of classrooms they will share (20):
600 / 20 = 30

21. C: To calculate this value, divide the number of dogs (48) by the number of workers that are available to care for them (4):
48 / 4 = 12

22. C: First, calculate the length of the second office: 20 + 6 = 26 feet
Then, add both values together to get a combined length: 26 + 20 = 46 feet

23. C: Find the total cost of the items: $6.66 + $159.23 = $165.89
Then, calculate how much each individual will owe: $165.89 / 4 = $41.47

24. A: To answer this question, simply calculate half of 140 acres: 140 / 2 = 70 acres

25. C: First, calculate how many he has after selling 45: 360 – 45 = 315
Then, calculate how many he has after buying 85: 315 + 85 = 400

26. D: Calculate the total amount of money the couple has available to spend, which is the amount in the joint bank account and the amount that each has:
$569 + $293 + $189 = $1051

27. D: First, find out what the total temperature decrease will be after 4 minutes:
2 * 4 = 8 degrees
Then, subtract that from the original temperature: 98 – 8 = 90 degrees

28. B: First, calculate 30% of 3 inches: 3 * 0.3 = 0.9 inches.
Then, subtract this value from the original length: 3 − 0.9 = 2.1

29. D: Add the number of tickets that were sold at each location to get the total number of tickets sold: 432 + 238 + 123 = 793

30. B: First, figure out how many pieces of candy are in the bag before they are divided: 26 − 8 = 18
Then, figure out how many pieces each friend will get by dividing by 2: 18 / 2 = 9

Word Knowledge

1. A: When it is said that someone is generous, it usually means they are giving and unselfish.

2. C: When something is described as challenging, it usually means that it is difficult or demanding.

3. B: A teacher provides instruction and information to an individual or group of individuals. An instructor functions in the same capacity, that is, in the practice of teaching.

4. D: When something is concluded, it means that it is finished or completed.

5. A: A residence is a place where a person lives; the term is often used to refer to someone's home.

6. C: To say something was done instantly means that it was done immediately and without hesitation.

7. C: Something that is described as gigantic is extremely large, or huge, in size.

8. A: Something that is costly is expensive. It costs a lot of money.

9. D: An opportunity is a chance to do something. For example, saying someone was given the *opportunity* to go to school or saying somebody was given the *chance* to go to school conveys the same meaning.

10. B: A response is also commonly known as an answer. A response to a question carries the same meaning as an answer to a question.

11. C: To say that something is done frequently implies that it is done regularly or often.

12. A: Something that is being observed is being watched. Binoculars are used to see things more clearly, so it makes sense that the man would be observing or watching eagles with binoculars.

13. D: Saying that somebody purchased something and saying they bought something conveys the same meaning.

14. D: The word difficulty implies hardship. When it is said that somebody is having difficulties with another person or thing, it usually means they are experiencing problems.

15. A: When it is said that something is entire, it usually means that it is still whole. For example, if someone says they ate an entire apple, it is the same as saying they ate a whole apple.

16. C: To say something is superior to something else usually implies that it is better.

17. B: A remark is a spoken statement, also commonly known as a comment. To say somebody made a remark conveys the same meaning as saying they made a comment.

18. C: When a selection is being made, it involves making a choice between several options. Selecting something is the same as choosing something.

19. A: To commence something is to begin or start something.

20. B: When something is done swiftly, it means that it is done fast or quickly.

21. D: When something is overdue, it means that it is late. For example, when a bill is overdue, it means that it has not been paid on time.

22. D: Somebody who is experiencing or feeling anguish is experiencing sorrow or sadness.

23. A: Solitary can mean a number of different things, but one meaning is single. For example, if you said there was a solitary tree in a yard, you would mean that there was a single tree.

24. C: Chuckled is a synonym for laughed. To say somebody chuckled or to say that somebody laughed conveys the same meaning.

25. A: If somebody or something is departing from somewhere, it means they are leaving. For example, to say the train departed from the station is the same as saying the train left the station.

26. B: To locate something that is lost or misplaced is to find it. For example, the child could not locate (find) his toy.

27. A: Something that is soiled is stained or dirty. When somebody says their clothing is soiled, it is the same as saying their clothing is dirty.

28. C: Slumber is another word for sleep. Saying someone slumbered is the same as saying they slept.

29. C: If somebody is puzzled about something, it implies confusion or bewilderment. For example, to say that a man was puzzled by the woman's reaction means that the man was confused by her reaction.

30. B: To desire something is to want something. Saying that a woman desired a new car has the same meaning as saying the woman wanted a new car.

31. D: Something that is cheap does not cost a substantial amount of money. Saying something is cheap and saying that something is inexpensive conveys the same meaning.

32. C: When somebody is described as intelligent, it usually means that they are smart.

33. A: When somebody objects to something, it means that they disagree with it. A person who objects to the expression of a specific political opinion may be said to disagree with it.

34. D: Constructed most nearly means built. If you say a home is constructed out of wood, for example, it conveys the same meaning as saying the home is built out of wood.

35. A: Something that is required is needed. For example, if you say you require ten dollars to buy lunch, it implies that you need ten dollars for lunch.

Paragraph Comprehension

1. A: The passage states: "During interphase, all of the genetic material within the cell is replicated. Then, the strands that contain the genetic material, which are known as chromatin, become compacted and condensed." According to the passage, the material in the cell is copied (replicated) before the chromatin becomes compacted into strands.

2. B: The conclusion that many people do not practice healthy eating can be made based on two points in the passage. First, we are told that many people are obese. Second, it is stated that healthy eating is more important to maintaining a healthy weight than exercise activity. Therefore, if a large number of people are overweight, and eating unhealthy foods is the major contributor to obesity, it can be concluded that many people do not practice healthy eating.

3. B: The author is relying on experiences that were reported by others. This is evident through the use of such phrases as "he said that he was later told" and "she says the job went to a less qualified applicant."

4. D: The work is important because "it is often viewed as the first significant work of English literature."

5. B: The purpose of the passage is to convince the reader that cats are better pets than dogs. This is stated in the opening sentence when the author states "Cats are by far a superior pet compared to dogs." Also, the remainder of the passage describes the advantages of cats as pets. Any time cats are compared to dogs, cats are described as superior pets.

6. A: After the malaria parasites infect the red blood cells, symptoms like fever occur. The passage states that "There, they multiply, and soon they infect red blood cells. Malaria symptoms like fever will begin to be experienced at this point."

7. D: The main idea in this passage is that water sports don't have to be dangerous. The passage states "Many people feel that summer time water sports are dangerous. While accidents do occur, the vast majority are preventable." Also, the rest of the passage focuses on outlining what precautions can be taken to ensure that water sports are less risky.

8. D: This can be concluded based on the section of the passage that states "Today, with blogs, virtually anybody can put their work out there for others to view. It's as easy as setting up your blog, naming it, and posting anything you want."

9. D: Many people still do not buy organic food because it is too expensive. The passage states "the price of organic foods still prevents many people from eating them on a regular basis."

10. A: It is reasonable to conclude that the author mentioned Cindy because it would strengthen the argument that birth order affects personality traits. The passage states that oldest children are more responsible and perform better in school. Since Cindy was on the Dean's list (good academic performance) and did her chores as a child (a sign of responsibility), it is reasonable to assume that Cindy was an oldest child.

11. D: The main idea discussed in the passage is that weddings are expensive. The author states that the expense of a wedding is a major source of stress. Then, the author discusses many of the expenses involved, finally providing an example of the high cost of a simple wedding.

12. B: A supplier should reduce the price of a good if they want to sell more of it because, according to the passage, "consumers will demand less of a good if the price is higher *and more if it is lower.*"

13. B: According to the passage, many people are in debt in North America because they borrowed money they could not afford to repay. The passage states that "much of the blame lies with the companies who actually granted the credit to consumers. This debt crisis wouldn't have happened if loans were not granted to people who could not afford them."

14. A: The person's airway should be opened before checking whether they are breathing. The passage advises the rescuer to "open their airway by raising the chin. After this, but before beginning mouth to mouth, spend a few moments to determine whether they are breathing." Mouth to mouth and chest compressions are only done if the person is not breathing.

15. A: The main idea of the passage is that fur is an unnecessary material good. The author provides two statements to support the main theme. First, the author notes that there are synthetic materials which can keep people warm, so fur is no longer necessary. Second, the author states that fur is used to produce luxury items which are not really needed by anyone.

Mathematics Knowledge

1. B: The perimeter of a figure is the sum of all of its sides. Since a rectangle's width and length will be the same on opposite sides, the perimeter of a rectangle can be calculated by using the following formula: perimeter = 2(width) + 2(length)
Using the numbers given in the question:
perimeter = 2(7cm) + 2(9cm)
perimeter = 14cm + 18cm
perimeter = 32cm

2. D: First, gather the like terms on opposite sides of the equation to make it easier to solve:
$-3q - 4q \geq -30 - 12$
$-7q \geq -42$
Then, divide both sides by -7 to solve for q:
$-7q/-7 \geq -42/-7$
$q \geq 6$
Finally, when both sides are divided by a negative number, the direction of the sign must be reversed:
$q \leq 6$

3. C: To solve for x, it is necessary to add 6 to both sides to isolate the variable:
x − 6 + 6 = 0 + 6
x = 6

4. A: To calculate the value of this expression, substitute -3 for x each time it appears in the expression: $3(-3)^3 + (3(-3)+ 4) − 2(-3)^2$
According to the order of operations, any operations inside of brackets must be done first:
$3(-3)^3 + (-9+ 4) − 2(-3)^2$
$3(-3)^3 + -5 − 2(-3)^2$
Then, the value of the expression can be calculated:
3(-27) + -5 − 2(9)
-81 + -5 − 18
-104

5. C: First, combine like terms to make the equation easier to solve:
3x + 2x = 45 + 30
5x = 75
Then, divide both sides by 5 to solve for x:
5x/5 = 75/5
x = 15

6. A: First, add 25 to both sides to isolate x:
1/4x − 25 + 25 ≥ 75 + 25
1/4x ≥ 100
Then, multiply both sides by 4 to solve for x:
1/4x * 4 ≥ 100 * 4
x ≥ 400

7. A: First, add 5 to both sides to isolate x:
x^2 − 5 + 5 = 20 + 5
x^2 = 25
Then, take the square root of both sides to solve for x
$\sqrt{x^2} = \sqrt{25}$
x = 5

8. B: First, we must calculate the length of one side of the square. Since we know the perimeter is 8cm, and that a square has 4 equal sides, the length of each side can be calculated by dividing the perimeter (8cm) by 4: 8cm / 4 = 2cm
The formula for the area of a square is length^2
Therefore, to calculate the area of this square: 2cm^2 or 2cm * 2cm
Area = 4cm^2

9. D: To find the value of this expression, substitute the given values for x and y into the expression:
3(4)(2) − 12(2) + 5(4)
Then, calculate the value of the expression:
3*8 − 12*2 + 5*4
24 − 24 + 20
20

10. D: First, subtract 10 from both sides to isolate x:
$0.65x + 10 - 10 = 15 - 10$
$0.65x = 5$
Then, divide both sides by 0.65 to solve for x:
$0.65x/0.65 = 5/0.65$
$x = 7.69$

11. B: Use the FOIL method (first, outside, inside, and last) to get rid of the brackets:
$12x^2 - 18x + 20x - 30$
Then, combine like terms to simplify the expression:
$12x^2 - 18x + 20x - 30$
$12x^2 + 2x - 30$

12. B: To simplify this expression, it is necessary to follow the law of exponents that states: $x^n/x^m = x^{n-m}$
First, the 50 can be divided by 5: $50/5 = 10$
Then, it is simply a matter of using the law of exponents described above to simplify the expression:
$10x^{18-5}t^{6-2}w^{3-2}z^{20-19}$
$10x^{13}t^4wz$

13. D: To calculate the value of this permutation, it is necessary to multiply each number between one and 4: $1 * 2 * 3 * 4 = 24$

14. D: Because it is a cube, it is known that the width and the height of the cube is also 5cm. Therefore, to find the volume of the cube, we must cube 5cm: $5cm^3$
This is the same as: $5 * 5 * 5 = 125$
The volume of the cube is $125cm^3$.

15. A: First, factor this equation to make solving for x easier:
$(x - 6)(x - 7) = 0$
Then, solve for both values of x:
1) $x - 6 = 0$
$x = 6$
2) $x - 7 = 0$
$x = 7$

16. C: The area of a triangle can be calculated by using the following formula: $A = 1/2b*h$
Therefore, by using the values given in the question:
$A = 1/2(12cm) * 12cm$
$A = 6cm * 12cm$
$A = 72cm^2$

17. D: To simplify this expression, it is necessary to observe the law of exponents that states:
$x^n * x^m = x^{n+m}$
Therefore: $3*7x^{7+2} + 2*9y^{12+3}$
$21x^9 + 18y^{15}$

18. C: First, subtract 27 from both sides to isolate x:
x/3 + 27 − 27 = 30 − 27
x/3 = 3
Then, both sides must be multiplied by 3 to solve for x:
3(x/3) = 3 * 3
x = 9

19. B: To find the slope of a line, it is necessary to calculate the change in y and the change in x:
Change in y: 1 − 8 = -7
Change in x: 4 − (-13) = 17
The slope of a line is expressed as change in y over change in x: -7/17

20. A: To solve for x, it is necessary to calculate the value of 20% of 200:
200 * 0.20 = 40
Therefore, x = 40

21. B: First, calculate the total number of balloons in the bag: 47 + 5 + 10 = 62
Ten of these are black, so divide this number by 62, then multiply by 100 to express the probability as a percentage:
10 / 62 = 0.16
0.16 * 100 = 16%

22. B: First, it is easier to find out how many tickets are sold for one winner.
If there are 2 winners for every 100 tickets, there is 1 winner for every 50 tickets.
If ten tickets are bought, the chances of winning are 10 in 50.
This can also be expressed as 1 in 5.

23. D: To find the volume of a rectangular prism, the formula is length * width * height.
Therefore, for this rectangular prism, volume = 10cm * 5cm * 6cm
The volume of this rectangular prism is 300cm^3

24. C: To calculate the midpoint of a line, find the sum of the points and divide by two.
For x, the midpoint can be calculated as follows: 6 + 10 = 16   16/2 = 8
For y, the midpoint can be calculated as follows: 40 + 20 = 60   60/2 = 30
Therefore, the midpoint is (8, 30)

25. A: First, subtract 60 from both sides to isolate x:
5x + 60 − 60 = 75 − 60
5x = 15
Then, divide both sides by 5 to solve for x:
5x/5 = 15/5
x = 3

General Science

1. D: The name for a substance that stimulates the production of antibodies is an *antigen*. An antigen is any substance perceived by the immune system as dangerous. When the body senses an antigen, it produces an antibody. *Collagen* is one of the components of bone, tendon, and cartilage. It is a spongy protein that can be turned into gelatin by boiling. *Hemoglobin* is the part of red blood

cells that carries oxygen. In order for the blood to carry enough oxygen to the cells of the body, there has to be a sufficient amount of hemoglobin. *Lymph* is a near-transparent fluid that performs a number of functions in the body: It removes bacteria from tissues, replaces lymphocytes in the blood, and moves fat away from the small intestine. Lymph contains white blood cells. As you can see, some of the questions in the vocabulary section will require technical knowledge.

2. D: The cellular hierarchy starts with the cell, the simplest structure, and progresses to organisms, the most complex structures.

3. B: A hypertonic solution is a solution with a higher particle concentration than in the cell, and consequently lower water content than in the cell. Water moves from the cell to the solution, causing the cell to experience water loss and shrink.

4. A: The immune system consists of the lymphatic system, spleen, tonsils, thymus and bone marrow.

5. D: The rate at which a chemical reaction occurs does not depend on the amount of mass lost, since the law of conservation of mass (or matter) states that in a chemical reaction there is no loss of mass.

6. B: Boyle's law states that for a constant mass and temperature, pressure and volume are related inversely to one another: PV = c, where c = constant.

7. C: Prokaryotic cells are simpler cells that do not have membrane-bound organelles, whereas eukaryotic cells have several membrane-bound organelles.

8. A: A ribosome is a structure of eukaryotic cells that makes proteins.

9. B: It is impossible for an *AaBb* organism to have the *aa* combination in the gametes. It is impossible for each letter to be used more than one time, so it would be impossible for the lowercase *a* to appear twice in the gametes. It would be possible, however, for *Aa* to appear in the gametes, since there is one uppercase *A* and one lowercase *a*. Gametes are the cells involved in sexual reproduction. They are germ cells.

10. B: The oxidation number of the hydrogen in CaH_2 is –1. The oxidation number is the positive or negative charge of a monoatomic ion. In other words, the oxidation number is the numerical charge on an ion. An ion is a charged version of an element. Oxidation number is often referred to as oxidation state. Oxidation number is sometimes used to describe the number of electrons that must be added or removed from an atom in order to convert the atom to its elemental form.

11. C: *Prolactin* stimulates the production of breast milk during lactation. *Norepinephrine* is a hormone and neurotransmitter secreted by the adrenal gland that regulates heart rate, blood pressure, and blood sugar. *Antidiuretic hormone* is produced by the hypothalamus and secreted by the pituitary gland. It regulates the concentration of urine and triggers the contractions of the arteries and capillaries. *Oxytocin* is a hormone secreted by the pituitary gland that makes it easier to eject milk from the breast and manages the contractions of the uterus during labor.

12. A: The typical result of mitosis in humans is two diploid cells. *Mitosis* is the division of a body cell into two daughter cells. Each of the two produced cells has the same set of chromosomes as the

parent. A diploid cell contains both sets of homologous chromosomes. A haploid cell contains only one set of chromosomes, which means that it only has a single set of genes.

13. A: Boron does not exist as a diatomic molecule. The other possible answer choices, fluorine, oxygen, and nitrogen, all exist as diatomic molecules. A diatomic molecule always appears in nature as a pair: The word *diatomic* means "having two atoms." With the exception of astatine, all of the halogens are diatomic. Chemistry students often use the mnemonic BrINClHOF (pronounced "brinkelhoff") to remember all of the diatomic elements: bromine, iodine, nitrogen, chlorine, hydrogen, oxygen, and fluorine. Note that not all of these diatomic elements are halogens.

14. D: Of the given structures, veins have the lowest blood pressure. *Veins* carry oxygen-poor blood from the outlying parts of the body to the heart. An *artery* carries oxygen-rich blood from the heart to the peripheral parts of the body. An *arteriole* extends from an artery to a capillary. A *venule* is a tiny vein that extends from a capillary to a larger vein.

15. B: Water stabilizes the temperature of living things. The ability of warm-blooded animals, including human beings, to maintain a constant internal temperature is known as *homeostasis*. Homeostasis depends on the presence of water in the body. Water tends to minimize changes in temperature because it takes a while to heat up or cool down. When the human body gets warm, the blood vessels dilate and blood moves away from the torso and toward the extremities. When the body gets cold, blood concentrates in the torso. This is the reason why hands and feet tend to get especially cold in cold weather.

16. D: Hydriodic acid is another name for aqueous HI. In an aqueous solution, the solvent is water. Hydriodic acid is a polyatomic ion, meaning that it is composed of two or more elements. When this solution has an increased amount of oxygen, the *-ate* suffix on the first word is converted to *-ic*.

17. C: Of the four heart chambers, the left ventricle is the most muscular. When it contracts, it pushes blood out to the organs and extremities of the body. The right ventricle pushes blood into the lungs. The atria, on the other hand, receive blood from the outlying parts of the body and transport it into the ventricles. The basic process works as follows: Oxygen-poor blood fills the right atrium and is pumped into the right ventricle, from which it is pumped into the pulmonary artery and on to the lungs. In the lungs, this blood is oxygenated. The blood then reenters the heart at the left atrium, which when full pumps into the left ventricle. When the left ventricle is full, blood is pushed into the aorta and on to the organs and extremities of the body.

18. B: Oxygen is not one of the products of the Krebs cycle. The *Krebs cycle* is the second stage of cellular respiration. In this stage, a sequence of reactions converts pyruvic acid into carbon dioxide. This stage of cellular respiration produces the phosphate compounds that provide most of the energy for the cell. The Krebs cycle is also known as the citric acid cycle or the tricarboxylic acid cycle.

19. A: A limiting reactant is entirely used up by the chemical reaction. Limiting reactants control the extent of the reaction and determine the quantity of the product. A reducing agent is a substance that reduces the amount of another substance by losing electrons. A reagent is any substance used in a chemical reaction. Some of the most common reagents in the laboratory are sodium hydroxide and hydrochloric acid. The behavior and properties of these substances are known, so they can be effectively used to produce predictable reactions in an experiment.

20. A: The *cerebrum* is the part of the brain that interprets sensory information. It is the largest part of the brain. The cerebrum is divided into two hemispheres, connected by a thin band of tissue called the corpus callosum. The *cerebellum* is positioned at the back of the head, between the brain stem and the cerebrum. It controls both voluntary and involuntary movements. The *medulla oblongata* forms the base of the brain. This part of the brain is responsible for blood flow and breathing, among other things.

21. C: The sugar and phosphate in DNA are connected by covalent bonds. A *covalent bond* is formed when atoms share electrons. It is very common for atoms to share pairs of electrons. An *ionic bond* is created when one or more electrons are transferred between atoms. *Ionic bonds*, also known as *electrovalent bonds*, are formed between ions with opposite charges. There is no such thing as an *overt bond* in chemistry.

22. C: The mass of 7.35 mol water is 132 grams. You should be able to find the mass of various chemical compounds when you are given the number of mols. The information required to perform this function is included on the periodic table. To solve this problem, find the molecular mass of water by finding the respective weights of hydrogen and oxygen. Remember that water contains two hydrogen molecules and one oxygen molecule. The molecular mass of hydrogen is roughly 1, and the molecular mass of oxygen is roughly 16. A molecule of water, then, has approximately 18 grams of mass. Multiply this by 7.35 mol, and you will obtain the answer 132.3, which is closest to answer choice C.

23. C: *Collagen* is the protein produced by cartilage. Bone, tendon, and cartilage are all mainly composed of collagen. *Actin* and *myosin* are the proteins responsible for muscle contractions. Actin makes up the thinner fibers in muscle tissue, while myosin makes up the thicker fibers. Myosin is the most numerous cell protein in human muscle. *Estrogen* is one of the steroid hormones produced mainly by the ovaries. Estrogen motivates the menstrual cycle and the development of female sex characteristics.

24. B: Unlike other organic molecules, lipids are not water soluble. Lipids are typically composed of carbon and hydrogen. Three common types of lipid are fats, waxes, and oils. Indeed, lipids usually feel oily when you touch them. All living cells are primarily composed of lipids, carbohydrates, and proteins. Some examples of fats are lard, corn oil, and butter. Some examples of waxes are beeswax and carnauba wax. Some examples of steroids are cholesterol and ergosterol.

25. D: Of these orbitals, the last to fill is 6s. Orbitals fill in the following order: 1s, 2s, 2p, 3s, 3p, 4s, 3d, 4p, 5s, 4d, 5p, 6s, 4f, 5d, 6p, 7s, 5f, 6d, and 7p. The number is the orbital number, and the letter is the sublevel identification. Sublevel s has one orbital and can hold a maximum of two electrons. Sublevel p has three orbitals and can hold a maximum of six electrons. Sublevel d has five orbitals and can hold a maximum of 10 electrons. Sublevel f has seven orbitals and can hold a maximum of 14 electrons.

26. C: The parasympathetic nervous system is responsible for lowering the heart rate. It slows down the heart rate, dilates the blood vessels, and increases the secretions of the digestive system. The central nervous system is composed of the brain and the spinal cord. The sympathetic nervous system is a part of the autonomic nervous system; its role is to oppose the actions taken by the parasympathetic nervous system. So, the sympathetic nervous system accelerates the heart, contracts the blood vessels, and decreases the secretions of the digestive system.

27. D: *Hemoglobin* is not a steroid. It is a protein that helps to move oxygen from the lungs to the various body tissues. Steroids can be either synthetic chemicals used to reduce swelling and inflammation or sex hormones produced by the body. *Cholesterol* is the most abundant steroid in the human body. It is necessary for the creation of bile, though it can be dangerous if the levels in the body become too high. *Estrogen* is a female steroid produced by the ovaries (in females), testes (in males), placenta, and adrenal cortex. It contributes to adolescent sexual development, menstruation, mood, lactation, and aging. *Testosterone* is the main hormone produced by the testes; it is responsible for the development of adult male sex characteristics.

28. C: Nitrogen pentoxide is the name of the binary molecular compound NO_5. The format given in answer choice C is appropriate when dealing with two nonmetals. A prefix is used to denote the number of atoms of each element. Note that when there are seven atoms of a given element, the prefix *hepta-* is used instead of the usual *septa-*. Also, when the first atom in this kind of binary molecular compound is single, it does not need to be given the prefix *mono-*.

29. B: Smooth muscle tissue is said to be arranged in a disorderly fashion because it is not striated like the other two types of muscle: cardiac and skeletal. Striations are lines that can only be seen with a microscope. *Smooth* muscle is typically found in the supporting tissues of hollow organs and blood vessels. *Cardiac* muscle is found exclusively in the heart; it is responsible for the contractions that pump blood throughout the body. *Skeletal* muscle, by far the most preponderant in the body, controls the movements of the skeleton. The contractions of skeletal muscle are responsible for all voluntary motion. There is no such thing as *rough* muscle.

30. C: *Melatonin* is produced by the pineal gland. One of the primary functions of melatonin is regulation of the circadian cycle, which is the rhythm of sleep and wakefulness. *Insulin* helps regulate the amount of glucose in the blood. Without insulin, the body is unable to convert blood sugar into energy. *Testosterone* is the main hormone produced by the testes; it is responsible for the development of adult male sex characteristics. *Epinephrine*, also known as adrenaline, performs a number of functions: It quickens and strengthens the heartbeat and dilates the bronchioles. Epinephrine is one of the hormones secreted when the body senses danger.

Assembling Objects and Mechanical Comprehension

1. B: Object 2 is larger than object 1, so it will displace more water and cause the water level in tank B to be higher than that of tank A.

2. A: Since momentum is conserved in all collisions, and there is no indication that the balls merge into one upon colliding, ball 1 must rebound off ball 2 toward the upper left pocket.

3. A: Consecutive gears alternate rotation direction, which means all odd numbered gears turn the same direction. Since 1 and 5 are both odd, both are spinning counter-clockwise in this problem.

4. B: The spring being compressed under object B is being compressed further than the spring under object A, and therefore has more potential energy stored up to launch the ball higher into the air.

5. B: Water (along with nearly every other substance) seeks the lowest energy state in which to rest. Functionally, this means that the water level will be equally high in all parts of the watering can.

6. C: Every point on the belt, and consequently every point on the outside of each pulley, is moving at the same linear speed. Therefore, the pulley with the smallest circumference will rotate the fastest.

Assembling Objects	
Question	Answer
1.	2
2.	3
3.	5
4.	4
5.	3
6.	4
7.	1
8.	2
9.	5
10.	2
11.	4
12.	1
13.	2
14.	5
15.	2
16.	3
17.	5
18.	4
19.	1
20.	2

7. B: A pulley only reduces the amount of force required to lift an object if the weight is distributed across multiple segments of the rope, as is done in A.

8. C: Only switch C creates a closed loop between the generator and the motor. Closing B creates a short circuit, and closing A does nothing.

9. A: More force is required to propel a bicycle up a steeper slope.

10. B: Because of friction losses within the spring and between the ball and the surface, the ball will not travel as far the second time.

11. A: More force is required to push a boulder up a steeper slope.

12. B: In ballistic flight, the horizontal component of velocity is essentially constant. At point B, the vertical component of the cannonball's velocity is zero, making the peak of its arc the slowest point.

13. A: Since all the water must leave at the same rate at which it enters, the water must travel significantly faster at point B, since the opening is much smaller.

14. A: In figure A, the load is centered much closer to the man and much farther from the wheel than in figure B. This means that the man will have to bear a larger percentage of the weight of the load.

15. B: On truck A, the load is evenly distributed, while on truck B it is concentrated on one end, making it more likely to tip over.

16. B: The load on the stretcher is concentrated more closely to man B than man A, so man B is bearing more of the load.

17. A: The bracing in A is more solid because it extends higher up on the post.

18. A: Though the force of gravity is the same on both objects, object A will have had more time to build up speed, so it will hit the ground with more force than object B.

19. B: The wagon will roll more easily up the smoother slope.

20. C: The amount of liquid will be easiest to measure when the angle of the water line matches the lines drawn on the cylinder.

21. A: In figure A, the weight is much closer to the fulcrum, so it will require less force to raise.

22. C: Only switch C creates a closed loop between the battery and the light. Closing B creates a short circuit, and closing A does nothing.

23. A: Since a bowling ball weighs nearly 50 times as much as a baseball, the bowling ball's path will not be significantly affected by its collision with the baseball.

24. B: Roll B will turn faster, both because it is lighter, thus having a lower moment of inertia, and because it requires less paper to be pulled to undergo a revolution.

25. B: Water in container B will cool more slowly because less of the surface of the water is exposed to the air.

26. A: Adding salt to water increases the density of the fluid, making objects more likely to float in it.

27. A: Reflector A is farther from the center of the wheel. Therefore, it will travel more distance when the wheel turns.

28. C: Since air resistance on a javelin is negligible, and its horizontal velocity is effectively constant throughout its flight, the fastest point will be the point that has the greatest vertical velocity. Since point C is the lowest point, it is the point at which the maximum potential energy will have been converted to kinetic energy.

29. C: The child will travel to the approximately equivalent height on the other side of the swing set before returning to the initial side.

30. C: Since all the water must leave at the same rate at which it enters, and since both the entry and exit pipes have the same size, the water must be traveling at the same speed in both locations.

OAR Practice Test

Math Skills Test

1. If 16x + 4 = 100, what is the value of x?
    a. 6
    b. 7
    c. 8
    d. 9

2. Simplify the following expression:
(2x − 20) (5x + 10)
    a. $10x^2 - 80x - 200$
    b. $70x - 200$
    c. $10x^2 - 80x + 200$
    d. $10x^2 - 120x - 200$

3. Which of the following are complementary angles?
    a. 71° and 19°
    b. 90° and 90°
    c. 90° and 45°
    d. 15° and 30°

4. Simply the following expression:
$(2x^4)^3 + 2(y^5)^5$
    a. $8x^{64} + 2y^{3125}$
    b. $6x^7 + 2y^{10}$
    c. $6x^{12} + 2y^{25}$
    d. $8x^{12} + 2y^{25}$

5. If the measures of the three angles in a triangle are 2 : 6 : 10, what is the measure of the smallest angle?
    a. 20 degrees
    b. 40 degrees
    c. 60 degrees
    d. 80 degrees

6. If a circle has a diameter of 12cm, what is its area?
    a. 38cm²
    b. 113cm²
    c. 276cm²
    d. 452cm²

7. The length of a square is 15cm. What is its area?
    a. 30cm²
    b. 60cm²
    c. 150cm²
    d. 225cm²

8. A rectangular solid measures 12cm by 3cm by 9cm. What is its volume?
    a. 36cm³
    b. 108cm³
    c. 324cm³
    d. 407cm³

9. If $2x^2 = -4x^2 + 216$, what is the value of x?
    a. 4
    b. 5
    c. 6
    d. 7

10. The perimeter of a square is 24 m. What is its area?
    a. 30m²
    b. 36m²
    c. 42m²
    d. 24m²

11. If a rectangle has a length of 5cm and a width of 7cm, what is its area?
    a. 24cm²
    b. 35cm²
    c. 42cm²
    d. 56cm²

12. On a six-sided die, each side has a number between 1 and 6. What is the probability of throwing a 3 or a 4?
    a. 1 in 6
    b. 1 in 3
    c. 1 in 2
    d. 1 in 4

13. Solve for y in the following inequality:
$-2y \geq 24 + 6$
    a. $y \leq 15$
    b. $y \geq 15$
    c. $y \leq -15$
    d. $y \geq -15$

14. If $2x = 5x - 30$, what is the value of x?
    a. 10
    b. -10
    c. 4.3
    d. -4.3

15. Given the functions, $f(x) = 3x + 6$ and $g(x) = 2x - 8$, what is the solution of the equation, $f(x) = g(x)$?
    a. $x = -12$
    b. $x = -8$
    c. $x = -14$
    d. $x = -10$

16. Suppose the area of the square in the diagram to the right is 64 in². (The square is not shown actual size.) What is the area of the circle?

    a. 16π in²
    b. 64π in²
    c. $\frac{64}{\pi}$ in²
    d. $(64 + \pi)$ in²

17. Solve for x in the following inequality:
4x + 23 > -3x – 6
    a. x > -4.14
    b. x < -4.14
    c. x > 4.14
    d. x < 4.14

18. If 2x + 5x = 3x + x + 30, what is the value of x?
    a. 2.72
    b. 4.29
    c. 6
    d. 10

19. 3x²y + y/2 – 6x
If x=4 and y=10, what is the value of the expression
    a. 221
    b. 461
    c. 872
    d. 1916

20. At a school carnival, three students spend an average of $10. Six other students spend an average of $4. What is the average amount of money spent by all nine students?
    a. $5
    b. $6
    c. $7
    d. $8

21. If w=7, calculate the value of the following expression:
$8w^2 - 12w + (4w - 5) + 6$
   a. 279
   b. 285
   c. 337
   d. 505

22. If x/3 + 7 = 35, what is the value of x?
   a. 9.33
   b. 14
   c. 84
   d. 126

23. In the following equation, solve for x by factoring:
$2x^2 - 7x = x^2 - 12$
   a. x = -3, -4
   b. x = 3, 4
   c. x = 3, -4
   d. x = -3, 4

24. If x is 25% of 250, what is the value of x?
   a. 62.5
   b. 100
   c. 1000
   d. 6250

25. Which of the following inequalities is correct?
   a. $\frac{1}{3} < \frac{2}{7} < \frac{5}{12}$
   b. $\frac{2}{7} < \frac{1}{3} < \frac{5}{12}$
   c. $\frac{5}{12} < \frac{2}{7} < \frac{1}{3}$
   d. $\frac{5}{12} < \frac{1}{3} < \frac{2}{7}$

26. If the volume of a cube is 8cm³, what is the length of the cube?
   a. 1cm
   b. 2cm
   c. 3cm
   d. 4cm

27. Simply the following expression:
$(2x^2 + 3)(2x - 1)$
   a. $4x^3 - 2x^2 + 6x - 3$
   b. $2x^2 + 6x - 3$
   c. $4x^3 - 2x^2 + 6x + 3$
   d. $4x^3 - 2x^2 - 6x - 3$

28. Simply the following expression
$(2x^4y^7m^2z) * (5x^2y^3m^8)$
    a. $10x^6y^9m^{10}z$
    b. $7x^6y^{10}m^{10}z$
    c. $10x^5y^{10}m^{10}z$
    d. $10x^6y^{10}m^{10}z$

29. A classroom contains 13 boys and 18 girls. If a student's name is chosen randomly, what is the probability it will be a girl's name?
    a. 36%
    b. 42%
    c. 58 %
    d. 72%

30. What is 10% of 40%?
    a. 4%
    b. 30%
    c. 50%
    d. 400%

Reading Comprehension Test

Use the following passage to answer questions 1-2:

Inside the cockpit, three key items to be checked are: (1) battery and ignition switches—off, (2) control column locks—removed, (3) landing gear control— down and locked.

The fuel selectors should be checked for proper operation in all positions—including the OFF position. Stiff selectors, or ones where the tank position is hard to find, are unacceptable. The primer should also be exercised. The pilot should feel resistance when the primer is both pulled out and pushed in. The primer should also lock securely. Faulty primers can interfere with proper engine operation. The engine controls should also be manipulated by slowly moving each through its full range to check for binding or stiffness.

The airspeed indicator should be properly marked, and the indicator needle should read zero. If it does not, the instrument may not be calibrated correctly. Similarly, the vertical speed indicator (VSI) should also read zero when the airplane is on the ground. If it does not, a small screwdriver can be used to zero the instrument. The VSI is the only flight instrument that a pilot has the prerogative to adjust. All others must be adjusted by an FAA certificated repairman or mechanic.

1. According to the passage, which of the following setting configurations is correct?
a.
Battery and ignition switches: On
Control column locks: Removed
Landing gear control: Down and locked
Airspeed Indicator needle: 0

b.
Battery and ignition switches: Off
Control column locks: Removed
Landing gear control: Up and locked
Airspeed Indicator needle: 0

c.
Battery and ignition switches: On
Control column locks: In place
Landing gear control: Up and locked
Airspeed Indicator needle: 0

d.
Battery and ignition switches: Off
Control column locks: Removed
Landing gear control: Down and locked
Airspeed Indicator needle: 0

2. Which of the following is closest in meaning to *ineluctable*?
   a. Inevitable
   b. Extremely sad
   c. Indivisible
   d. Impossible to comprehend

Use the following passage to answer questions 3-4:
Radial engines were widely used during World War II and many are still in service today. With these engines, a row or rows of cylinders are arranged in a circular pattern around the crankcase. The main advantage of a radial engine is the favorable power-to-weight ratio. In-line engines have a comparatively small frontal area, but their power-to-weight ratios are relatively low. In addition, the rearmost cylinders of an air-cooled, in-line engine receive very little cooling air, so these engines are normally limited to four or six cylinders. V-type engines provide more horsepower than in-line engines and still retain a small frontal area.

3. According to this passage, how much use did in-line engines get during World War II?
    a. None at all
    b. A limited amount of use
    c. Quite a bit of use
    d. The article doesn't say

4 It takes only one harrowing experience to clarify the distinction between minimum practical knowledge and a thorough understanding of how to apply the procedures and techniques used in instrument flight. Your instrument training is never complete; it is adequate when you have absorbed every foreseeable detail of knowledge and skill to ensure a solution will be available if and when you need it.
In this passage, *harrowing* means:
    a. Rapidly descending
    b. Extremely disturbing
    c. Extremely unprepared
    d. Last minute

Use the following passage to answer questions 5-6:
One disadvantage of the float-type carburetor is its icing tendency. Carburetor ice occurs due to the effect of fuel vaporization and the decrease in air pressure in the venturi, which causes a sharp temperature drop in the carburetor. If water vapor in the air condenses when the carburetor temperature is at or below freezing, ice may form on internal surfaces of the carburetor, including the throttle valve. The reduced air pressure, as well as the vaporization of fuel, contributes to the temperature decrease in the carburetor. Ice generally forms in the vicinity of the throttle valve and in the venturi throat. This restricts the flow of the fuel/air mixture and reduces power. If enough ice builds up, the engine may cease to operate. Carburetor ice is most likely to occur when temperatures are below 70 degrees Fahrenheit (°F) or 21 degrees Celsius (°C) and the relative humidity is above 80 percent. Due to the sudden cooling that takes place in the carburetor, icing can occur even with temperatures as high as 100 °F (38 °C) and humidity as low as 50 percent. This temperature drop can be as much as 60 to 70 °F (15 to 21 °C). Therefore, at an outside air temperature of 100 °F (37 °C), a temperature drop of 70 °F (21 °C) results in an air temperature in the carburetor of 30 °F (-1 °C).

5. Which set of conditions would allow carburetor icing to occur?
    a. 50 degrees Fahrenheit and 50% relative humidity
    b. 90 degrees Fahrenheit and 40% relative humidity
    c. 75 degrees Fahrenheit and 60% relative humidity
    d. 80 degrees Fahrenheit and 30% relative humidity

6 In which set of conditions would carburetor icing be most likely to occur?
   a. 50 degrees Fahrenheit and 75% relative humidity
   b. 60 degrees Fahrenheit and 85% relative humidity
   c. 75 degrees Fahrenheit and 60% relative humidity
   d. 850 degrees Fahrenheit and 30% relative humidity

7. It is impossible to emphasize too strongly the necessity for forming correct habits in flying straight and level. All other flight maneuvers are in essence a deviation from this fundamental flight maneuver. Many flight instructors and students are prone to believe that perfection in straight-and-level flight will come of itself, but such is not the case. It is not uncommon to find a pilot whose basic flying ability consistently falls just short of minimum expected standards, and upon analyzing the reasons for the shortcomings to discover that the cause is the inability to fly straight and level properly.

Which of these human activities would most closely correlate to straight-and-level flying?
   a. Crawling
   b. Walking
   c. Jogging
   d. Running

8. The National Airspace System (NAS) is the network of United States airspace: air navigation facilities, equipment, services, airports or landing areas, aeronautical charts, information/services, rules, regulations, procedures, technical information, manpower, and material. Included are system components shared jointly with the military. The system's present configuration is a reflection of the technological advances concerning the speed and altitude capability of jet aircraft, as well as the complexity of microchip and satellite based navigation equipment. To conform to international aviation standards, the United States adopted the primary elements of the classification system developed by the International Civil Aviation Organization (ICAO).

Which of these would not be considered part of the National Airspace System?
   a. Airports
   b. Private Flight schools
   c. FAA
   d. Air traffic controllers

9. The effect of free stream density and velocity is a necessary consideration when studying the development of the various aerodynamic forces. Suppose that a particular shape of airfoil is fixed at a particular angle to the airstream. The relative velocity and pressure distribution will be determined by the shape of the airfoil and the angle to the airstream. If the same airfoil shape is placed at the same angle to an airstream with twice as great a dynamic pressure the magnitude of the pressure distribution will be twice as great but the relative shape of the pressure distribution will be the same.

Which of these is closest to the meaning of *velocity* in this passage?
   a. Attitude
   b. Thrust
   c. Speed
   d. Altitude

10. Using a supercharger, at 8,000 feet a typical engine may be able to produce 75 percent of the power it could produce at mean sea level (MSL) because the air is less dense at the higher altitude. The supercharger compresses the air to a higher density allowing a supercharged engine to produce the same manifold pressure at higher altitudes as it could produce at sea level. Thus, an engine at 8,000 feet MSL could still produce 25 "Hg of manifold pressure whereas without a supercharger it could produce only 22 "Hg. Superchargers are especially valuable at high altitudes (such as 18,000 feet) where the air density is 50 percent that of sea level. The use of a supercharger in many cases will supply air to the engine at the same density it did at sea level. With a normally aspirated engine, it is not possible to have manifold pressure higher than the existing atmospheric pressure. A supercharger is capable of boosting manifold pressure above 30 "Hg.

According to this passage, if an engine using a supercharger can produce 22 "Hg of manifold pressure at 8,000 feet, approximately how much manifold pressure can it produce at MSL?
  a. 29
  b. 32
  c. 16
  d. 27

11. Learning to manage the many information and automation resources now available to you in the cockpit is a big challenge. Specifically, you must learn how to choose which advanced cockpit systems to use, and when. There are no definitive rules. In fact, you will learn how different features of advanced cockpit avionics systems fall in and out of usefulness depending on the situation. Becoming proficient with advanced avionics means learning to use the right tool for the right job at the right time. In many systems, there are multiple methods of accomplishing the same function. The competent pilot learns all of these methods and chooses the method that works best for the specific situation, environment, and equipment. This handbook will help you get started in learning this important skill.

With which of these statements would the author agree?
  a. Choosing an advanced aviation system method is a personal decision best left to each pilot.
  b. Any decent advanced aviation system method will do just fine in most situations.
  c. There is no advanced aviation system method that is a one size fits all solution.
  d. Advanced aviation systems are somewhat overrated.

12. Keeping track of primary flight information is critical at all times, and pilots must become very familiar with their PFD. Flight instrument presentations on a PFD differ from conventional instrumentation not only in format, but sometimes in location as well. Airspeed and altitude indications are presented on vertical tape displays that appear on the left and right sides of the display. The vertical speed indicator is depicted using conventional analog presentation. Turn coordination is shown using a segmented triangle near the top of the attitude indicator. The rate-of-turn indicator appears as a curved line display at the top of the heading/navigation instrument in the lower half of the PFD.

In this passage, PFD most likely stands for:
  a. Pilot Flight Details
  b. Power and Features Display
  c. Pilot Flight Display
  d. Primary Flight Display

13. Turbochargers increase the pressure of the engine's induction air, which allows the engine to develop sea level or greater horsepower at higher altitudes. A turbocharger is comprised of two main elements: a compressor and turbine. The compressor section houses an impeller that turns at a high rate of speed. As induction air is drawn across the impeller blades, the impeller accelerates the air, allowing a large volume of air to be drawn into the compressor housing. The impeller's action subsequently produces high-pressure, high-density air, which is delivered to the engine. To turn the impeller, the engine's exhaust gases are used to drive a turbine wheel that is mounted on the opposite end of the impeller's drive shaft. By directing different amounts of exhaust gases to flow over the turbine, more energy can be extracted, causing the impeller to deliver more compressed air to the engine. The waste gate, essentially an adjustable butterfly valve installed in the exhaust system, is used to vary the mass of exhaust gas flowing into the turbine. When closed, most of the exhaust gases from the engine are forced to flow through the turbine. When open, the exhaust gases are allowed to bypass the turbine by flowing directly out through the engine's exhaust pipe.

High-density air is delivered to the engine after being produced by the:
    a. Impeller
    b. Inductor
    c. Turbocharger
    d. Butterfly valve

Use the following passage to answer questions 14-15:
One of the biggest safety concerns in aviation is the surface movement accident. As a direct result, the Federal Aviation Administration (FAA) has rapidly expanded the information available to pilots, including the addition of taxiway and runway information in FAA publications, particularly the IFR U.S. Terminal Procedures Publication (TPP) booklets and Airport/Facility Directory (A/FD) volumes. The FAA has also implemented new procedures and created educational and awareness programs for pilots, ATC, and ground operators. By focusing resources to attack this problem head on, the FAA hopes to reduce and eventually eliminate surface movement accidents.
Airport sketches and diagrams provide pilots of all levels with graphical depictions of the airport layout. Aeronautical Information Systems (AIS), formerly known as Aeronautical Products (AeroNav), provide an airport sketch on the lower left or right portion of every instrument approach chart. This sketch depicts the runways, their length, width and slope, the touchdown zone elevation, the lighting system installed on the end of the runway, and taxiways. Graphical depictions of NOTAMS are also available for selected airports as well as for temporary flight restriction (TFRs) areas on the defense internet NOTAM service (DINS) website.
For select airports, typically those with heavy traffic or complex runway layouts, AIS also prints an airport diagram. The diagram is located in the IFR TPP booklet following the instrument approach chart for a particular airport. It is a full page depiction of the airport that includes the same features of the airport sketch plus additional details, such as taxiway identifiers, airport latitude and longitude, and building identification. The airport diagrams are also available in the A/FD and on the AIS website, located at www.aeronav.faa.gov.

14. Which of the following acronyms is not defined by this passage?
    a. TPP
    b. A/FD
    c. NOTAMS
    d. AeroNav

15. Which of the following provides the airport sketches of the airport layout?
    a. AeroNav
    b. AIS
    c. TFR
    d. A/FD

16. Although the regulations specify minimum requirements, the amount of instructional time needed to earn an instrument rating is determined not by the regulation, but by the individual's ability to achieve a satisfactory level of proficiency. A professional pilot with diversified flying experience may easily attain a satisfactory level of proficiency in the minimum time required by regulation. Your own time requirements will depend upon a variety of factors, including previous flying experience, rate of learning, basic ability, frequency of flight training, type of aircraft flown, quality of ground school training, and quality of flight instruction, to name a few. The total instructional time you will need, the scheduling of such time, is up to the individual most qualified to judge your proficiency—the instructor who supervises your progress and endorses your record of flight training.

Which of these is not a factor in determining how long it takes a pilot to earn an instrument rating?
    a. Make and model of plane used
    b. Age of pilot
    c. Whether someone is a fast or slow learner
    d. How good the instructor is

Use the following passage to answer questions 17-18:
Flight deck crews almost seamlessly launch, recover and move aircraft and people all over the flight deck? Did you ever wonder how we do this? Simply put, because we have very dedicated professionals doing their job. However, whether we realize it or not, the success of those individuals who are just "doing their job" hinges upon thorough application of Operational Risk Management on three levels—in depth, when our leadership and acquisition folks provide the equipment, training and guidance for flight deck operations; deliberate, when we plan and brief for the events or operations of the day; and time critical, when we actually apply the risk controls or use the resources provided to us for getting the job done.

17. Which of the following best expresses the meaning of *seamlessly* in this passage?
    a. Without a hitch
    b. Around the clock
    c. Non-stop
    d. Automatically

18 Which of the following is closest in meaning to *deploy*?
    a. Retire from active service
    b. Disguise by using camouflage
    c. Move into position
    d. Recruit or enlist someone into military service

19. Displacing the cyclic forward causes the nose to pitch down initially, with a resultant increase in airspeed and loss of altitude. Aft cyclic causes the nose to pitch up initially, slowing the helicopter and causing it to climb; however, as the helicopter reaches a state of equilibrium, the horizontal stabilizer levels the helicopter airframe to minimize drag, unlike an airplane. Therefore, the helicopter has very little pitch deflection up or down when the helicopter is stable in a flight mode. The variation from absolutely level depends on the particular helicopter and the horizontal stabilizer function. Increasing collective (power) while maintaining a constant airspeed induces a climb while decreasing collective causes a descent. Coordinating these two inputs, down collective plus aft cyclic or up collective plus forward cyclic, results in airspeed changes while maintaining a constant altitude. The pedals serve the same function in both a helicopter and a fixed-wing aircraft, to maintain balanced flight. This is done by applying a pedal input in whichever direction is necessary to center the ball in the turn and bank indicator.

According to this article, what is one of the main factors when it comes to variation from absolute level?
    a. The horizontal stabilizer
    b. The direction of the wind
    c. The speed of the wind
    d. Whether the weight of passengers and cargo is properly balanced

20. Making good choices sounds easy enough. However, there are a multitude of factors that come into play when these choices, and subsequent decisions, are made in the aeronautical world. Many tools are available for pilots to become more self-aware and assess the options available, along with the impact of their decision. Yet, with all the available resources, accident rates are not being reduced. Poor decisions continue to be made, frequently resulting in lives being lost and/or aircraft damaged or destroyed. The Risk Management Handbook discusses aeronautical decision-making (ADM) and single-pilot resource management (SRM) in detail and should be thoroughly read and understood. While progress is continually being made in the advancement of pilot training methods, aircraft equipment and systems, and services for pilots, accidents still occur. Historically, the term "pilot error" has been used to describe the causes of these accidents. Pilot error means an action or decision made by the pilot was the cause of, or a contributing factor that led to, the accident. This definition also includes the pilot's failure to make a decision or take action. From a broader perspective, the phrase "human factors related" more aptly describes these accidents since it is usually not a single decision that leads to an accident, but a chain of events triggered by a number of factors. The poor judgment chain, sometimes referred to as the "error chain," is a term used to describe this concept of contributing factors in a human factors related accident. Breaking one link in the chain is often the only event necessary to change the outcome of the sequence of events.

Which of the following statements would the author be most likely to agree with?
    a. The problem of aircraft accidents has never been worse.
    b. Realistically, the number of aircraft accidents is probably as low as it's ever going to be.
    c. There is still much room for improvement when it comes to reducing aircraft accidents.
    d. Great strides have been made in recent years in reducing the number of aircraft accidents.

21. When an operator requests a Minimum Equipment List (MEL), and a Letter of Authorization (LOA) is issued by the FAA, then the use of the MEL becomes mandatory for that helicopter. All maintenance deferrals must be accomplished in accordance with the terms and conditions of the MEL and the operator-generated procedures document. Exercise extreme caution when hovering near buildings or other aircraft. The use of an MEL for rotorcraft operated under part 91 also allows for the deferral of inoperative items or equipment. The primary guidance becomes the FAA-approved MEL issued to that specific operator and N-numbered helicopter. The FAA has developed master minimum equipment lists (MMELs) for rotorcraft in current use. Upon written request by a rotorcraft operator, the local FAA Flight Standards District Office (FSDO) may issue the appropriate make and model MMEL, along with an LOA, and the preamble. The operator then develops operations and maintenance (O&M) procedures from the MMEL. This MMEL with O&M procedures now becomes the operator's MEL. The MEL, LOA, preamble, and procedures document developed by the operator must be on board the helicopter when it is operated. The FAA considers an approved MEL to be a supplemental type certificate (STC) issued to an aircraft by serial number and registration number. It therefore becomes the authority to operate that aircraft in a condition other than originally type certificated. With an approved MEL, if the position lights were discovered inoperative prior to a daytime flight, the pilot would make an entry in the maintenance record or discrepancy record provided for that purpose. The item is then either repaired or deferred in accordance with the MEL. Upon confirming that daytime flight with inoperative position lights is acceptable in accordance with the provisions of the MEL, the pilot would leave the position lights switch OFF, open the circuit breaker (or whatever action is called for in the procedures document), and placard the position light switch as INOPERATIVE.

What resource would a pilot use to determine if an inoperative part or system rendered daytime flight unacceptable?
    a. LOA
    b. O&M
    c. FSDO
    d. MEL

22. It is important to understand the three levels of Operational Risk Management, because each level plays a role in improving our chance of completing the mission successfully. In particular, the controls developed at each level are resources we can tap into to accomplish our job or mission during its execution. These resources make it easier to do our job and help catch errors that might be detrimental to task or mission success. Beyond the equipment itself and our fellow shipmates, there are other resources we can tap to help mitigate the risks associated with the hazards of the flight deck.

Which of the following is closest in meaning to *mitigate*?
    a. Eliminate
    b. Control
    c. Reduce
    d. Understand

23. Today, helicopters are quite reliable. However, emergencies do occur, whether as a result of mechanical failure or pilot error, and should be anticipated. Regardless of the cause, the recovery needs to be quick and precise. By having a thorough knowledge of the helicopter and its systems, a pilot is able to handle the situation more readily. Helicopter emergencies and the proper recovery procedures should be discussed and, when possible, practiced in flight. In addition, by knowing the conditions that can lead to an emergency, many potential accidents can be avoided. Emergencies should always be anticipated. Knowledge of the helicopter, possible malfunctions and failures, and methods of recovery can help the pilot avoid accidents and be a safer pilot. Helicopter pilots should always expect the worst hazards and possible aerodynamic effects and plan for a safe exit path or procedure to compensate for the hazard.

Which of the following best sums up this passage?
    a. Planning ahead is the best way to prepare for emergencies.
    b. Many helicopter emergencies are due to faulty equipment.
    c. In helicopter emergencies, there is little margin for error.
    d. Helicopter emergencies can arise at any time and for a variety of reasons.

24. Medium-frequency vibrations (1,000–2,000 cycles per minute) range between the low frequencies of the main rotor (100–500 cycles per minute) and the high frequencies (2,100 cycles per minute or higher) of the engine and tail rotor. Depending on the helicopter, medium-frequency vibration sources may be engine and transmission cooling fans, and accessories such as air conditioner compressors, or driveline components. Medium-frequency vibrations are felt through the entire airframe, and prolonged exposure to the vibrations will result in greater pilot fatigue. Most tail rotor vibrations fall into the high-frequency range (2,100 cycles per minute or higher) and can be felt through the tail rotor pedals as long as there are no hydraulic actuators to dampen out the vibration. This vibration is felt by the pilot through his or her feet, which are usually "put to sleep" by the vibration. The tail rotor operates at approximately a 6:1 ratio with the main rotor, meaning for every one rotation of the main rotor the tail rotor rotates 6 times. A main rotor operating rpm of 350 means the tail rotor rpm would be 2,100 rpm. Any imbalance in the tail rotor system is very harmful as it can cause cracks to develop and rivets to work loose. Piston engines usually produce a normal amount of high-frequency vibration, which is aggravated by engine malfunctions, such as spark plug fouling, incorrect magneto timing, carburetor icing and/or incorrect fuel/air mixture. Vibrations in turbine engines are often difficult to detect as these engines operate at a very high rpm. Turbine engine vibration can be at 30,000 rpm internally, but common gearbox speeds are in the 1,000 to 3,000 rpm range for the output shaft. The vibrations in turbine engines may be short lived as the engine disintegrates rapidly when damaged due to high rpm and the forces present.

Which frequencies can result in the pilot experiencing numbness in the pilot's leg?
    a. Low frequency vibrations
    b. Medium frequency vibrations
    c. High frequency vibrations
    d. Both medium and high frequency vibrations

25. Rigid rotor systems tend to behave like fully articulated systems through aerodynamics, but lack flapping or lead/ lag hinges. Instead, the blades accommodate these motions by bending. They cannot flap or lead/lag but they can be feathered. As advancements in helicopter aerodynamics and materials continue to improve, rigid rotor systems may become more common because the system is fundamentally easier to design and offers the best properties of both semirigid and fully articulated systems. The rigid rotor system is very responsive and is usually not susceptible to mast bumping like the semirigid or articulated systems because the rotor hubs are mounted solid to the main rotor mast. This allows the rotor and fuselage to move together as one entity and eliminates much of the oscillation usually present in the other rotor systems. Other advantages of the rigid rotor include a reduction in the weight and drag of the rotor hub and a larger flapping arm, which significantly reduces control inputs. Without the complex hinges, the rotor system becomes much more reliable and easier to maintain than the other rotor configurations. A disadvantage of this system is the quality of ride in turbulent or gusty air. Because there are no hinges to help absorb the larger loads, vibrations are felt in the cabin much more than with other rotor head designs. There are several variations of the basic three rotor head designs. The bearingless rotor system is closely related to the articulated rotor system, but has no bearings or hinges. This design relies on the structure of blades and hub to absorb stresses. The main difference between the rigid rotor system and the bearingless system is that the bearingless system has no feathering bearing—the material inside the cuff is twisted by the action of the pitch change arm. Nearly all bearingless rotor hubs are made of fiber-composite materials.

The author of this passage would probably agree most strongly with which statement about rigid rotor systems?
    a. They are less popular than they used to be.
    b. They result in fewer vibrations than semirigid and fully articulated systems.
    c. Overall, they're superior to semirigid and fully articulated systems.
    d. They are not as safe as semirigid and fully articulated systems.

26. The roots of aviation are firmly based on curiosity. Where would we be today had it not been for the dreams of Leonardo da Vinci, the Wright Brothers, and Igor Sikorsky? They all were infatuated with flight, a curiosity that led to the origins of aviation. The tale of aviation is full of firsts: first flight, first helicopter, first trans-Atlantic flight, and so on. But, along the way there were many setbacks, fatalities, and lessons learned. Today, we continue to learn and investigate the limits of aviation. We've been to the moon, and soon beyond. Our curiosity will continue to drive us to search for the next challenge. However, curiosity can also have catastrophic consequences. Despite over 100 years of aviation practice, we still see accidents that are caused by impaired judgment formed from curious behavior. Pilots commonly seek to determine the limits of their ability as well as the limits of the aircraft. Unfortunately, too often this leads to mishaps with deadly results. Inquisitive behavior must be harnessed and displayed within personal and material limits. Deadly curiosity may not seem as obvious to some as it is to others. Simple thoughts such as, "Is visibility really as bad as what the ATIS is reporting?" or "Does the 20 minute fuel light really indicate only 20 minutes worth of fuel?" can lead to poor decisions and disastrous outcomes. Some aviators blatantly violate rules and aircraft limitations without thinking through the consequences. "What indications and change in flight characteristics will I see if I fly this helicopter above its maximum gross weight?" or "I've heard this helicopter can do aerobatic flight. Why is it prohibited?" are examples of extremely harmful curiosity. Even more astounding is their ignoring the fact that the damage potentially done to the aircraft will probably manifest later in the aircraft's life, affecting other crews. Spontaneous excursions in aviation can be deadly. Curiosity is natural, and promotes learning. Airmen should abide by established procedures until proper and complete hazard assessment and risk management can be completed.

Which of these statements most closely matches the theme of this passage?
    a. Fortune favors the bold.
    b. An ounce of prevention is worth a pound of cure.
    c. Curiosity killed the cat.
    d. Fools rush in where angels fear to tread.

27. Although helicopters were developed and built during the first half-century of flight, some even reaching limited production; it was not until 1942 that a helicopter designed by Igor Sikorsky reached full-scale production, with 131 aircraft built. Even though most previous designs used more than one main rotor, it was the single main rotor with an antitorque tail rotor configuration design that would come to be recognized worldwide as the helicopter. In 1951, at the urging of his contacts at the Department of the Navy, Charles H. Kaman modified his K-225 helicopter with a new kind of engine, the turbo-shaft engine. This adaptation of the turbine engine provided a large amount of horsepower to the helicopter with a lower weight penalty than piston engines, heavy engine blocks, and auxiliary components. On December 11, 1951, the K-225 became the first turbine-powered helicopter in the world. Two years later, on March 26, 1954, a modified Navy HTK-1, another Kaman helicopter, became the first twin-turbine helicopter to fly. However, it was the Sud Aviation Alouette II that would become the first helicopter to be produced with a turbine engine. Reliable helicopters capable of stable hover flight were developed decades after fixed-wing aircraft. This is largely due to higher engine power density requirements than fixed-wing aircraft. Improvements in fuels and engines during the first half of the 20th century were a critical factor in helicopter development. The availability of lightweight turbo-shaft engines in the second half of the 20th century led to the development of larger, faster, and higher-performance helicopters. The turbine engine has the following advantages over a reciprocating engine: less vibration, increased aircraft performance, reliability, and ease of operation. While smaller and less expensive helicopters still use piston engines, turboshaft engines are the preferred powerplant for helicopters today.

27. Which of these is most responsible for a huge increase in the number of helicopters in use?
    a. The military demand for helicopters in World War II
    b. The development of an antitorque tail rotor configuration design
    c. The development of turbine engine powered helicopters
    d. The development of piston engine helicopters

Mechanical Comprehension Test

1. A cannon fires off a ship up towards a mountain range. Neglecting air resistance, where will the velocity of the projectile be greatest?

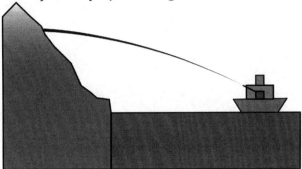

   a. Exiting the muzzle
   b. Halfway to the mountains
   c. As it impacts the mountains

2. These pulleys are connected by belts. Which pulley travels the fastest?

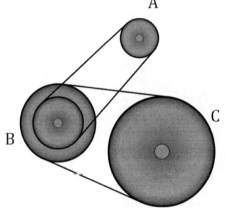

   a. Pulley A
   b. Pulley B
   c. Pulley C

3. If Gear A is traveling at 10 rpm, how many times will Gear C rotate in 3 minutes?

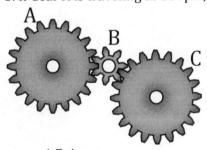

   a. 1.7 times
   b. 3 times
   c. 30 times

4. Where should the fulcrum be located to balance this beam?

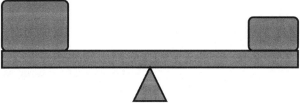

   a. closer to the large mass
   b. closer to the small mass
   c. exactly between the two masses

5. Which orientation will require more force to pull?

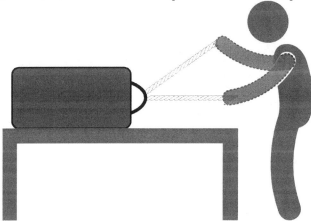

   a. with the rope at an angle to the table
   b. with the rope parallel to the table
   c. both orientations are equal

6. The larger piston has four times as much horizontal area as the smaller piston. If the small piston is compressed 8 inches, how far will the larger piston move?

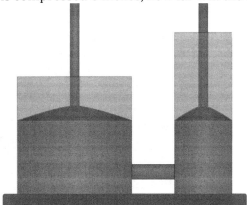

   a. 8 inches
   b. 2 inches
   c. 32 inches

7. A wing in flight has a set of pressures causing the overall forces on the top and bottom of the wing. Where will the total force on the wing point?

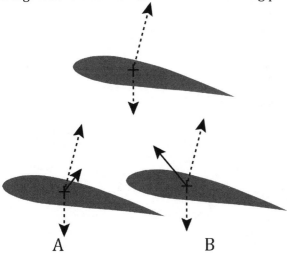

   a. up and to the right
   b. up and to the left
   c. neither A or B

8. River water enters a section where the water spreads out into a wide, deep area. Is the water likely to speed up, slow down, or remain at a constant speed?

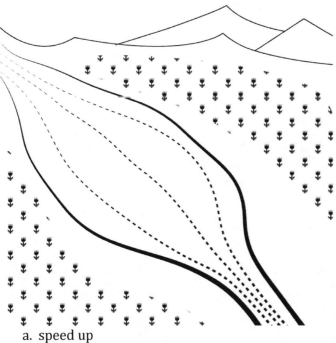

   a. speed up
   b. slow down
   c. remain at a constant speed

9. A magnet is placed in the middle of two identical, anchored magnets. Which direction will the magnet go?

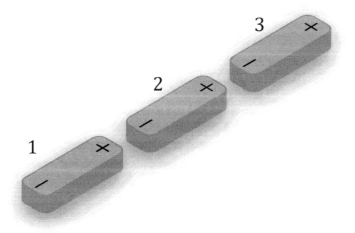

   a. towards magnet 1
   b. towards magnet 2
   c. the magnet won't move

10. A solid substance melts at -21°C. If the object is known to change phase at 81°C, will the object be a solid, liquid, or gas at 90°C?

   a. solid
   b. liquid
   c. gas

11. If the resistors in the circuits are identical, which circuit will have the greatest overall resistance?

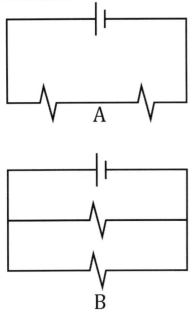

   a. circuit A
   b. circuit B
   c. circuit A and B have the same overall resistance

12. A pendulum swings back and forth once per second. The pendulum is shortened by removing half of the string. The new frequency is 1.4 Hz (Hz=1/sec). How often will the pendulum swing back and forth in a minute?

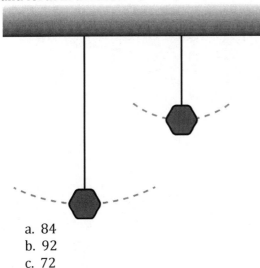

   a. 84
   b. 92
   c. 72

13. Two identical pistons are connected by a pipe. What is the mechanical advantage of the piston system?

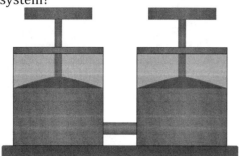

   a. 0.5
   b. 1
   c. 2

14. A ball is thrown horizontally off a cliff at the same time that an identical ball is dropped off a cliff. How long after the dropped ball hits the ground will the thrown ball hit?

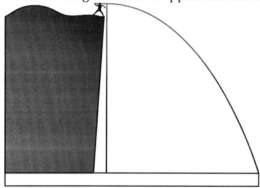

   a. approximately 1 second after
   b. approximately 2 seconds after
   c. they will hit at the same time

15. The cam rotates at 5 rpm. How many times will the follower (needle) move up and down in a minute?

   a. 20
   b. 72
   c. 140

16. Which of the following are not ways to increase the torque applied to a wrench?

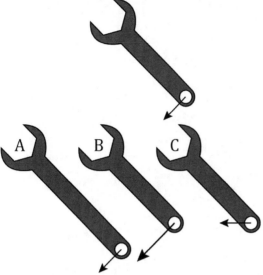

    a. increase the length from the center to the applied force
    b. increase the force
    c. angle the force toward the center

17. A ball is pushed down into a vertical spring. The ball is released and flies upward. Which best describes the states of energy the ball underwent?

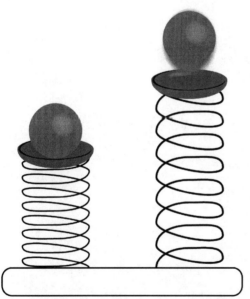

    a. convective energy to potential energy
    b. potential energy to kinetic energy
    c. kinetic energy to potential energy

18. A vacuum tank is held by weights at a depth of 50 feet underwater. If the tank is raised to a depth of 25 feet, will the pressure on the walls of the tank increase, decrease, or stay the same?

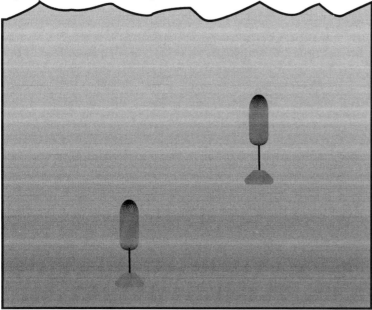

   a. increase
   b. decrease
   c. stay the same

19. Adding salt to water raises its density. Will salt water have a lower, higher, or the same specific gravity than 1?

   a. lower
   b. higher
   c. the same

20. Which of the following is an example of convective heat transfer?

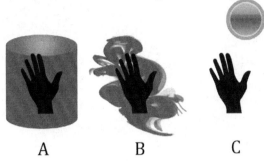

   a. a man burns his hand on a hot pot
   b. a man burns his hand in steam
   c. a man gets a sun burn

21. Which of the following is the electrical component which holds a voltage across a gap between two conductive materials?
   a. resistor
   b. inductor
   c. capacitor

22. Which of these wrenches is likely to provide the greatest torque with a set force?

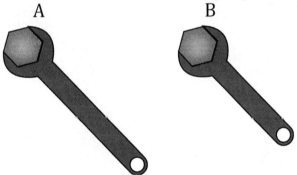

   a. wrench A
   b. wrench B
   c. both wrenches will provide the same torque

23. Which of the following is true of this circuit?

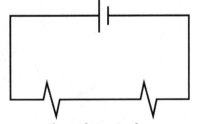

   a. the voltage is the same everywhere
   b. the current is the same everywhere
   c. there is no current

24. Which device does not measure current in an electrical system?
    a. ammeter
    b. multimeter
    c. voltmeter

25. A ball is thrown straight into the air with an initial kinetic energy of 100 ft-lb. What is the potential energy of the ball at half of the height of the flight path?

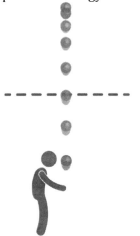

    a. 33 ft-lb
    b. 50 ft-lb
    c. 66 ft-lb

26. An increase in mechanical advantage with a set motion for the load and a set applied force necessitates an increase in _____ the applied force?
    a. the distance traveled by
    b. the angle of action of
    c. the potential energy behind

27. A windlass drum has two sections with different circumferences. When winding the drum, one side of the rope winds around the large section and the other unwinds from the small section. If the large section of the drum has a circumference of 3.5 ft and the other section has a circumference of 1 ft, how far will the weight lift with two full turns of the drum?

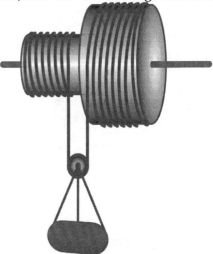

a. 2.5 feet
b. 9 feet
c. 4.5 feet

28. Which type of situation will lead to condensation on the outside of pipes?
    a. hot liquid in the pipe and cold air outside the pipe
    b. cold liquid in the pipe and colder air outside the pipe
    c. cold liquid in the pipe and hot air outside the pipe

29. Which color will absorb the most radiation?
    a. black
    b. green
    c. dark yellow

30. Two gears (30 and 18 teeth) mesh. If the smaller gear rotates 3 times, how many times will the larger gear rotate?

a. 1.6 times
b. 1.8 times
c. 3 times

Answer Key

Math Skills Test

1. A: First, subtract 4 from both sides to isolate x:
16x + 4 – 4 = 100 – 4
16x = 96
Then, divide both sides by 16 to solve for x:
16x/16 = 96/16
x = 6

2. A: Use the FOIL method (first, outside, inside, and last) to get rid of the brackets:
$10x^2 + 20x - 100x - 200$
Then, combine like terms to simplify the expression:
$10x^2 - 80x - 200$

3. A: Complementary angles are two angles that equal 90° when added together.

4. D: To simplify this expression, the law of exponents that states that $(x^m)^n = x^{m*n}$ must be observed:
$2^3 x^{4*3} + 2\ (y^{5*5})$
$8x^{12} + 2y^{25}$

5. A: The sum of the measures of the three angles of any triangle is 180. The equation of the angles of this triangle can be written as 2x + 6x + 10x = 180, or 18x = 180. Therefore, x = 10. Therefore, the measure of the smallest angle is 20.

6. B: The formula for the area of a circle is πr^2. The diameter of a circle is equal to twice its radius. Therefore, to find the radius of this circle, it is necessary to divide the diameter by 2: 12 / 2 = 6cm
Then, use the formula to find the area of the circle: $\pi 6^2$
$\pi * 36 = 113 cm^2$

7. D: The general equation to find the area of a quadrilateral is length * width.
Since the length and width of a square are equal, we can calculate the area of the square described in the question:
A = l * w
A = 15cm * 15cm
$A = 225 cm^2$

8. C: To find the volume of a rectangular solid, the formula is length * width * height.
Therefore, this solid's volume = 12cm * 3cm * 9cm = $324 cm^3$

9. C: First, add $4x^2$ to both sides to isolate x:
$2x^2 + 4x^2 = -4x^2 + 4x^2 + 216$
$6x^2 = 216$
Then, divide both sides by 6:
$6x^2/6 = 216/6$
$x^2 = 36$

Finally, take the square root of both sides to solve for x:
$\sqrt{x^2} = \sqrt{36}$
x = 6

10. B: The general equation to find the area of a quadrilateral is length * width.
Since the length and width of a square are equal, we can calculate the area of the square described in the question. We can divide the perimeter by 4 since all sides are equal length. Once we know each side is 6 m we can multiply 6*6 to get an area of 36 m^2.

11. B: The formula for the area of a rectangle is length * width. Using the measurements given in the question, the area of the rectangle can be calculated:
A = length * width
A = 5cm * 7cm
A = 35cm²

12. B: On a six-sided die, the probability of throwing any number is 1 in 6. The probability of throwing a 3 or a 4 is double that, or 2 in 6. This can be simplified by dividing both 2 and 6 by 2. Therefore, the probability of throwing either a 3 or 4 is 1 in 3.

13. C: First, add the 24 and the 6:
-2y ≥ 30
Then, divide both sides by -2 to solve for y:
-2y/-2 ≥ 30/-2
y ≥ -15
Finally, when both sides are divided by a negative number, the direction of the sign must be reversed:
y ≤ -15

14. A: First, bring the 5x to the left side of the equation to make it easier to solve:
2x – 5x = -30
-3x = -30
Then, divide both sides by -3 to solve for x:
-3x/-3 = -30/-3
x = 10

15. C: The solution of $f(x) = g(x)$ can be determined by setting the two functions equal to one another. Thus, the following may be written $3x + 6 = 2x - 8$. Solving for x gives $x = -14$.

16. A: The area of a square is equal to the square of the length of one side. If the area is 64 in², the side length must therefore be $\sqrt{64in^2} = 8$ in. The circle is inscribed in the square, so the side length of the square is the same as the circle's diameter. If the circle's diameter is 8 in, then the circle's radius must be half of that, or 4 in. The area of a circle is equal to $A = \pi r^2 = \pi(4 \text{ in})^2 = 16\pi \text{ in}^2$.

17. A: First, bring the -3x to the left side of the equation and the 23 to the right side of the equation to make it easier to solve:
4x + 3x > -6 – 23
7x > -29
Then, divide both side by 7 to solve for x:
7x/7 > -29/7

x > -4.14

18. D: First, bring all of the terms containing x to the left side of the equation to make it easier to solve: 2x + 5x – 3x – x = 30
7x – 4x = 30
3x = 30
Then, divide both sides by 3 to solve for x:
3x/3 = 30/3
x=10

19. B: First, substitute the given values for x and y into the expression:
$3(4)^2$ 10 + 10/2 – 6(4)
Then, calculate the value of the expression:
According to the order of operations, any multiplying and dividing must be done first:
3*16*10 + 5 – 24
480 + 5 – 24
Then, any addition or subtraction should be completed:
485 – 24
461

20. B: The average is the total amount spent divided by the number of students. The first three students spend an average of $10, so the total amount they spend is 3 × $10 = $30. The other six students spend an average of $4, so the total amount they spend is 6 × $4 = $24. The total amount spent by all nine students is $30 + $24 = $54, and the average amount they spend is $54 ÷ 9 = $6.

21. C: First, substitute the given value of w (7) into the expression each time it appears.
$8*7^2$ – 12(7) + (4*7 – 5) + 6
According to the order of operations, any calculations inside of the brackets must be done first:
$8*7^2$ – 12(7) + (23) + 6
Finally, calculate the value of the expression:
8*49 – 84 + 23 + 6
392 – 84 + 23 + 6
337

22. C: First, subtract 7 from both sides to isolate x:
x/3 + 7 – 7 = 35 – 7
x/3 = 28
Then, multiply both sides by three to solve for x:
x/3 * 3 = 28 * 3
x = 84

23. B: First, bring all terms to the left side of the equation to make it easier to solve:
2x² -7x - x² +12 = 0
Combine like terms:
x² -7x + 12 = 0
Then, factor the equation:
(x – 4) (x – 3) =0
Finally, solve for x in both instances:
x – 4 = 0
x = 4
x – 3 = 0
x = 3
x = 3, 4

24. A: Another way of expressing the fact that x is 25% of 250 is:
x = (0.25)250
Then, it is simply a matter of multiplying out the right side of the equation to calculate the value of x: x = 62.5

25. B: The volume of a cube is calculated by cubing the length, width, or height of the cube (the value for all three of these is the same.
Therefore, the volume of a cube equals = length³
In this case 8cm³= x * x * x, where x can represent the length of the cube.
To find the length, we must figure out which number cubed equals 8.
The answer is 2cm: 2cm * 2cm * 2cm = 8cm³

26. B: One way to compare fractions is to convert them to equivalent fractions which have common denominators. In this case the lowest common denominator of the three fractions is $7 \times 12 = 84$.
Converting each of the fractions to this denominator, $\frac{1}{3} = \frac{1 \times 28}{3 \times 28} = \frac{28}{84}, \frac{2}{7} = \frac{2 \times 12}{7 \times 12} = \frac{24}{84}$, and $\frac{5}{12} = \frac{5 \times 7}{12 \times 7} = \frac{35}{84}$. Since $24 < 28 < 35$, it must be the case that $\frac{2}{7} < \frac{1}{3} < \frac{5}{12}$.

27. A: Use the FOIL (first, outside, inside, last) to expand the expression:
4x³ -2x² +6x – 3
There are no like terms, so the expression cannot be simplified any further.

28. D: To simplify this expression, the law of exponents that states that $x^m * x^n = x^{m+n}$ must be observed.
$10x^{4+2}y^{7+3}m^{2+8}z$
Therefore, $10x^6y^{10}m^{10}z$ is the simplified expression.

29. C: First, find the total number of students in the classroom: 13 + 18 = 31
There is an 18 in 31 chance that a name chosen randomly will be a girl's name.
To express this as a percentage, divide 18 by 31, then multiply that number by 100:
18/31 * 100% = 58%

30. A: x percent is the same thing as $\frac{x}{100}$, and finding x percent of a number is the same as multiplying that number by x percent. This is true even when the number is itself a percent. So, 10% of 40% is $40\% \times 10\% = 40\% \times \frac{10}{100} = 40\% \times \frac{1}{10} = 4\%$.

Reading Comprehension Test

1 D: Battery and ignition switches: Off
Control column locks: Removed
Landing gear control: Down and locked
Airspeed Indicator needle: 0

2. A: Of the four answer choices, *inevitable* is closest in meaning to *ineluctable*

3. D: While the article does state that radial engines were widely used during WW II, it doesn't say how much use in-line engines got.

4. B: *Harrowing* means *extremely disturbing*

5. C: 75 degrees Fahrenheit and 60% relative humidity

6. B: 60 degrees Fahrenheit and 85% relative humidity

7. B: Walking is the fundamental way of moving for a mature human being, and jogging and running are deviations of walking, just as straight-and-level flying is the fundamental flight maneuver, and all other flight maneuvers are considered to be deviations of it.

8. B: Flight schools not affiliated with the government or the military would not be considered part of the National Airspace System. They might conduct some of their training at airports, which are part of the NAS, but the school itself would not be considered an NAS component.

9. C: Of the four answer choices, *Speed* is the closest in meaning to *velocity*.

10. A: 29. Using a supercharger, at 8,000 feet a typical engine may be able to produce 75 percent of the power it could produce at mean sea level (MSL) because the air is less dense at the higher altitude. So, if it can produce 22 "Hg at 8,000 feet, that represents 75% of its potential production at MSL. Dividing 22 by .75 gives us 29.333.

11. C: The author makes it clear that there is no advanced aviation system method that works best in every situation, and pilots need to choose the right method based on the circumstances.

12. D: Primary Flight Display. From the context, we know that PFD most likely refers to both *primary* and *display*, and D is the only answer choice that contains both words.

13. A: Impeller. The passage states that the impeller's action produces high-pressure, high-density air, which is then delivered to the engine.

14. C: NOTAMS is used in the passage, but isn't defined. It stands for Notices to Airmen. The singular is NOTAM, for Notice to Airmen. A NOTAM is a notification to pilots of factors along a flight path or at a destination that could be hazardous.

15. B: AIS stands for Aeronautical Information Systems, which is a division of the Federal Aviation Administration.

16. B: Although the passage states that how much experience a pilot has is important, the age of the pilot is not listed as a factor in how quickly one can earn an instrument rating. All of the other answer choices are listed in the passage as factors, although not in the same exact words.

17. A: *Seamlessly* in this context means *easily and without interruptions*, and *without a hitch* is another way of expressing that.

18. C: Of the four answer choices, *move into position* best expresses the meaning of *deploy*.

19. A: The passages states: "The variation from absolutely level depends on the particular helicopter and the horizontal stabilizer function."

20. C: This statement is the only one backed up by the contents of the passage.

21. D: The Minimum Equipment List (MEL) is the correct answer, based on this part of the passage: The FAA considers an approved MEL to be a supplemental type certificate (STC) issued to an aircraft by serial number and registration number. It therefore becomes the authority to operate that aircraft in a condition other than originally type certificated.

22. C: *Reduce* and *mitigate* have almost the same meaning.

23. A: The author's main point in this passage is that pilots can best prepare for emergencies by planning ahead – taking the time to master the helicopter and its systems, and anticipating emergencies and how to respond to them before they happen.

24. C: The passage states that high frequency vibrations can cause numbness in the pilot's leg.

25. C: The author says that rigid rotor systems offer the best properties of semirigid and fully articulated systems, and are easier to design, too.

26. D: *Fools rush in where angels fear to tread* expresses the idea that people who are new or inexperienced at something will often take dangerous chances that wiser or more experienced people would steer clear of, which most closely matches the author's theme that unbridled curiosity in a pilot can lead to disaster. He is not saying curiosity is bad in and of itself; only that it needs to have limits. That's why *Curiosity killed the cat* is incorrect.

27. C: Taking the entire passage as a whole, it shows that the development of turbine engine powered helicopters is the factor that was most responsible for a huge increase in the number of helicopters in use.

Mechanical Comprehension Test

1. A, The velocity is made up of two components, the x and y components. The x component is not changing during flight, but the weight of the projectile decreases the positive y component of the velocity. Thus, the total velocity will be greatest before the y component has decreased.

2. A, Because the linear speed of two connected pulleys is the same, the pulley with the smaller radius spins faster. The largest pulley will spin slower than the middle pulley. The smallest pulley will spin faster than the middle pulley, making it the fastest pulley.

3. C, Gear A and gear C have the same number of teeth. Thus, gears A and C will have the same speed. Since gear C is rotating at 10 rpm, the total number of rotations is calculated by multiplying the rpm by the number of minutes.

4. A, Because the large mass will produce a greater torque at the same distance from the fulcrum as the small mass, the distance from the large mass to the fulcrum should be shortened. Then, the torque produced by the large mass will decrease and the torque produced by the small mass will increase.

5. A, When the rope is not parallel to the intended path of motion, the force is divided into useful force (x direction) and not useful force (y direction). If only some of the force is useful, then the man will need to apply more force to achieve the same pulling force as if the rope were parallel to the table.

6. B, Because the volume of the liquid doesn't change, when the small piston is compressed, the volume decrease in one piston is the volume increase in the other piston. As the volume is the area times the height, the height of the larger piston only needs to raise one fourth the height that the small piston moved.

7. A, The downward force decreases part of the y component of the top force, but does not affect the x component of the force. Thus, the resultant force is up and to the right.

8. B, because the same volume of water has to flow through all parts of the river, the water will slow down to fill the wide section.

9. C, The negative side of magnet 2 will be attracted to the positive side of magnet 1. The positive side of magnet 2 will be attracted to the negative side of magnet 3 with the same force. Because the magnitudes of the forces are equal and the directions are opposite, the sum of the forces will be zero.

10. C, The first phase change is from solid to liquid at -21°C. The next phase change is from liquid to gas at 81°C. 90°C is only slightly higher than 81°C, making it safe to say that the substance is still a gas.

11. A, Substitute token values for the resistors to solve. Using 1 Ω resistors, the resistance of circuit A, having resistors in series, is the simple sum of the two resistors: 2 Ω. Because the resistors in circuit B are in parallel, the resistance of circuit B is the reciprocal of the sum of the reciprocals of the resistances, or $\frac{1}{\frac{1}{1}+\frac{1}{1}}$. the result is 1/2 Ω.

12. A, To find the number of swings in a time period from the frequency, multiply the frequency times the time period, after converting the time period into seconds to match the frequency. The final calculation is $\frac{1.4\ swings}{second} * 1\ minute * \frac{60\ seconds}{1\ minute} = 84\ swings$

13. B, The mechanical advantage is calculated by the output force divided by the input force. Because both pistons are the same size, the output force will equal the input force, resulting in a mechanical advantage of one.

14. C, Because the horizontal component of the thrown ball's velocity will not affect the vertical component, the vertical component of the thrown ball's velocity will be identical to the dropped ball's velocity. The balls will hit at the same time.

15. A, The cam has four bumps on it. The needle will move up and down for each bump. The cam will rotate five times in the time period of one minute. The total times the needle will move up and down will be five times four.

16. C, Torque is the product of a force perpendicular to the arm and the length of the arm. Options A and B each increase one part of the torque calculation. However, angling the force towards the center would decrease the part of the force that is acting perpendicular to the arm, as some of the force will be acting inward.

17. B, When the ball is compressed into the spring, the ball has potential stored in the spring. When the ball is flying upwards, the ball has kinetic energy associated with the motion.

18. B, Pressure increases with depth in water. When the tank was lower it experienced more pressure. Thus, when the tank is higher it experiences less pressure.

19. B, Specific gravity can be calculated as the ratio of the density of the liquid in question to the density of water. Because salt water has a higher density than water, the ratio will be greater than one.

20. B, Convective heat transfer deals with the transfer of heat by fluids (including gas). Steam is a fluid which transfers heat to objects, like a hand, with lower temperatures than it.

21. C, A capacitor stores voltage across a gap between two conductive materials.

22. A, Torque is the product of a force perpendicular to the arm and the length of the arm. Wrench A, with the longer arm, will be able to achieve greater amounts of torque with a set force.

23. B, Because the circuit only has one path and the two resistors are in series, the current is the same everywhere in the circuit. The voltage will drop over both resistors. Also, because the circuit is complete, there is current in the circuit.

24. C, Ammeters measure current (think amps). Multimeters measure current and voltage. Voltmeters only measure voltage.

25. B, When the ball is flying upwards, the kinetic energy is being converted into potential energy. Potential energy increases linearly with height, meaning that an object at 2 feet over the ground has twice the potential energy of the object at 1 foot over the ground. Thus, if all of the energy of the ball

will be converted from kinetic energy, and half of the energy will be converted at half the height, the potential energy of the ball will be 50 ft-lb.

26. A, Because mechanical advantage is the ratio of output force to input force, an increase in mechanical advantage means, in this case, that the output force will be increasing. However, energy in simple machines is conserved. This means that the work, or force times distance, done to the input will need to increase, while keeping the force the same. Increasing the distance of the applied force will increase the work, allowing for an increased force for the output.

27. A, As the drum spins one full turn, the hanging rope increases length by 1 foot and decreases length by 3.5 feet. Thus, every spin decreases the rope length by 2.5 feet. In two turns, the rope will decrease length by 5 feet. The pulley makes the weight lift half the distance that the rope decreased. Thus, the weight raises 2.5 feet.

28. C, Condensation from the air occurs when the water vapor in the air cools down enough to change phase from vapor to liquid water. If a pipe is cold and the air is warm, the water vapor will condense on the pipe.

29. A, The color, black, will absorb the most heat from radiation.

30. B, The gear ratio between the small and large gears is 18/30 or 3/5. Multiply the number of rotations of the small gear times the gear ratio to get (3 rotations)*(3/5) = 1.8 rotations.

AFOQT Practice Test

Verbal Analogies

1. CHASTISE is to REPRIMAND as IMPETUOUS is to
    a. punish
    b. rash
    c. considered
    d. poor
    e. calm

2. ARM is to HUMERUS as LEG is to
    a. ulna
    b. clavicle
    c. femur
    d. mandible
    e. metacarpal

3. MULTIPLICATION is to DIVISION as PRODUCT is to
    a. quotient
    b. divisor
    c. integer
    d. dividend
    e. multiplier

4. WEAR is to SWEATER as EAT is to
    a. shirt
    b. top hat
    c. asparagus
    d. looks
    e. mouth

5. MONEY is to IMPECUNIOUS as FOOD is to
    a. famished
    b. nauseated
    c. distracted
    d. antagonistic
    e. impoverished

6. DENIGRATE is to MALIGN as DEMUR is to
    a. protest
    b. defer
    c. slander
    d. benumb
    e. belittle

7. OBEISANCE is to DEFERENCE as MUNIFICENT is to
    a. benevolent
    b. magnificent
    c. squalid
    d. generous
    e. avarice

8. GOAT is to NANNY as PIG is to
    a. shoat
    b. ewe
    c. cub
    d. sow
    e. calf

9. CACHE is to RESERVE as DEARTH is to
    a. stockpile
    b. paucity
    c. cudgel
    d. dirge
    e. somber

10. ARABLE is to FARMABLE as ASYLUM is to
    a. famine
    b. danger
    c. arid
    d. fertile
    e. refuge

11. MYRIAD is to FEW as STATIONARY is to
    a. peripatetic
    b. many
    c. several
    d. halted
    e. parked

12. HOUSE is to MANSION as BOTTLE is to
    a. flagon
    b. container
    c. vessel
    d. pot
    e. flask

13. DICTIONARY is to DEFINITIONS as THESAURUS is to
    a. pronunciations
    b. synonyms
    c. explanations
    d. pronouns
    e. definitions

14. ABSTRUSE is to ESOTERIC as ADAMANT is to
    a. yielding
    b. stubborn
    c. keen
    d. forthright
    e. flexible

15. BEES is to HIVE as CATTLE is to
    a. swarm
    b. pod
    c. herd
    d. flock
    e. pack

16. LATITUDE is to LONGITUDE as PARALLEL is to
    a. strait
    b. line
    c. equator
    d. aquifer
    e. meridian

17. PREVENTION is to DETERRENCE as INCITEMENT is to
    a. excitement
    b. provocation
    c. request
    d. disregard
    e. disgust

18. VALUE is to WORTH as MEASURE is to
    a. gauge
    b. allowance
    c. demerit
    d. insignificance
    e. large

19. ENERVATE is to ENERGIZE as ESPOUSE is to
    a. oppose
    b. wed
    c. equine
    d. epistolary
    e. marry

20. HAMMER is to CARPENTER as STETHOSCOPE is to
    a. patient
    b. hearing
    c. heartbeat
    d. doctor
    e. pedometer

21. ARMOIRE is to BEDROOM as DESK is to
    a. chair
    b. office
    c. computer
    d. work
    e. building

22. PLANKTON is to WHALES as BAMBOO is to
    a. predators
    b. grasses
    c. pandas
    d. fast-growing
    e. China

23. TUNDRA is to ARCTIC as SAVANNA is to
    a. prairie
    b. lush
    c. tropic
    d. Georgia
    e. jungle

24. THIRSTY is to PARCHED as HUNGRY is to
    a. famished
    b. fed
    c. satiated
    d. satisfied
    e. full

25. FELICITY is to SADNESS as IGNOMINY is to
    a. shame
    b. slander
    c. crime
    d. indict
    e. honor

Arithmetic Reasoning

1. A man buys two shirts. One is $7.50 and the other is $3.00. A 6% tax is added to his total. What is his total?
   a. $10.50
   b. $11.13
   c. $14.58
   d. $16.80
   e. $18.21

2. If a chef can make 25 pastries in a day, how many can he make in a week?
   a. 32
   b. 74
   c. 126
   d. 175
   e. 250

3. A woman must earn $250 in the next four days to pay a traffic ticket. How much will she have to earn each day?
   a. $45.50
   b. $62.50
   c. $75.50
   d. $100.50
   e. $125.00

4. A car lot has an inventory of 476 cars. If 36 people bought cars in the week after the inventory was taken, how many cars will remain in inventory at the end of that week?
   a. 440
   b. 476
   c. 484
   d. 512
   e. 536

5. A woman has $450 in a bank account. She earns 0.5% interest on her end-of-month balance. How much interest will she earn for the month?
   a. $0.50
   b. $2.25
   c. $4.28
   d. $4.73
   e. $6.34

6. Three children decide to buy a gift for their father. The gift costs $78. One child contributes $24. The second contributes $15 less than the first. How much will the third child have to contribute?
   a. $15
   b. $39
   c. $45
   d. $62
   e. $69

7. Two women have credit cards. One earns 3 points for every dollar she spends. The other earns 6 points for every dollar she spends. If they each spend $5.00, how many combined total points will they earn?
   a. 15
   b. 30
   c. 45
   d. 60
   e. 75

8. A company employing 540 individuals plans to increase its workforce by 13%. How many people will the company employ after the expansion?
   a. 527
   b. 547
   c. 553
   d. 570
   e. 610

9. A 13 story building has 65 apartments. If each floor has an equal number of apartments, how many apartments are on each floor?
   a. 2
   b. 3
   c. 4
   d. 5
   e. 6

10. If 5 people buy 3 pens each and 3 people buy 7 pencils each, what is the ratio of the total number of pens sold to the total number of pencils sold?
    a. 15:21
    b. 3:7
    c. 5:7
    d. 1:1
    e. 5:3

11. A man earns $15.23 per hour and gets a raise of $2.34 per hour. What is his new hourly rate of pay?
    a. $12.89
    b. $15.46
    c. $17.57
    d. $23.40
    e. $35.64

12. How many people can travel on 6 planes if each carries 300 passengers?
    a. 1800
    b. 1200
    c. 600
    d. 350
    e. 300

13. In a town, the ratio of men to women is 2:1. If the number of women in the town is doubled, what will be the new ratio of men to women?
   a. 1:2
   b. 1:1
   c. 2:1
   d. 3:1
   e. 4:1

14. A woman weighing 250 pounds goes on a diet. During the first week, she loses 3% of her body weight. During the second week, she loses 2%. At the end of the second week, how many pounds has she lost?
   a. 12.5
   b. 10
   c. 12.35
   d. 15
   e. 17.5

15. A woman is traveling to a destination 583 km away. If she drives 78 km every hour, how many hours will it take for her to reach her destination?
   a. 2.22
   b. 3.77
   c. 5.11
   d. 7.47
   e. 8.32

16. If one gallon of paint can paint 3 rooms, how many rooms can be painted with 28 gallons of paint?
   a. 10
   b. 25
   c. 56
   d. 84
   e. 92

17. Five workers earn $135/day. What is the total amount earned by the five workers?
   a. $675
   b. $700
   c. $725
   d. $750
   e. $775

18. A girl scores a 99 on her math test. On her second test, her score drops by 15. On the third test, she scores 5 points higher than she did on her second. What was the girl's score on the third test?
   a. 79
   b. 84
   c. 89
   d. 99
   e. 104

19. A man goes to the mall with $50.00. He spends $15.64 in one store and $7.12 in a second store. How much does he have left?
    a. $27.24
    b. $32.76
    c. $34.36
    d. $42.8
    e. $57.12

20. 600 students must share a school that has 20 classrooms. How many students will each classroom contain if there are an equal number of students in each class?
    a. 10
    b. 15
    c. 20
    d. 25
    e. 30

21. Four workers at a shelter agree to care for the dogs over a holiday. If there are 48 dogs, how many must each worker look after?
    a. 8
    b. 10
    c. 12
    d. 14
    e. 16

22. One worker has an office that is 20 feet long. Another has an office that is 6 feet longer. What is the combined length of both offices?
    a. 26 feet
    b. 36 feet
    c. 46 feet
    d. 56 feet
    e. 66 feet

23. Four friends go shopping. They purchase items that cost $6.66 and $159.23. If they split the cost evenly, how much will each friend have to pay?
    a. $26.64
    b. $39.81
    c. $41.47
    d. $55.30
    e. $82.95

24. A 140 acre forest is cut in half to make way for development. What is the size of the new forest's acreage?
    a. 70
    b. 80
    c. 90
    d. 100
    e. 120

25. A farmer has 360 cows. He decides to sell 45. Shortly after, he purchases 85 more cows. How many cows does he have?
    a. 230
    b. 315
    c. 400
    d. 490
    e. 530

Word Knowledge

1. The word **spoiled** most nearly means
    a. ruined
    b. splendid
    c. told
    d. believed
    e. hated

2. He made an **oath** to his king.
    a. delivery
    b. promise
    c. statement
    d. criticism
    e. threat

3. **Inquire** most nearly means
    a. invest
    b. ask
    c. tell
    d. release
    e. inquest

4. Spanish is a difficult language to **comprehend**.
    a. learn
    b. speak
    c. understand
    d. appreciate
    e. commemorate

5. **Apparent** most nearly means
    a. clear
    b. occasional
    c. angry
    d. applied
    e. father

6. They enjoyed the **silence** of the night.
    a. darkness
    b. excitement
    c. quiet
    d. mood
    e. quaint

7. **Absolutely** most nearly means
    a. assuredly
    b. rapidly
    c. never
    d. weakly
    e. completely

8. He **modified** his schedule so he could attend the staff lunch.
    a. checked
    b. shortened
    c. considered
    d. changed
    e. lengthened

9. **Delicate** most nearly means
    a. fragile
    b. sturdy
    c. loud
    d. soft
    e. lovely

10. She attended the New Year's **festivities**.
    a. commitments
    b. celebrations
    c. crowds
    d. dates
    e. funeral

11. **Exhausted** most nearly means
    a. excited
    b. tired
    c. worried
    d. energized
    e. animated

12. She **cleansed** her face in the morning.
    a. examined
    b. washed
    c. touched
    d. dried
    e. motivated

13. **Battled** most nearly means
    a. fought
    b. attempt
    c. bold
    d. saw
    e. excited

14. He **wandered** around the mall.
    a. looked
    b. shopped
    c. roamed
    d. searched
    e. lived

15. **Abruptly** most nearly means
    a. homely
    b. commonly
    c. wisely
    d. ugly
    e. suddenly

16. He was **tricked** into giving her money.
    a. conned
    b. begged
    c. convinced
    d. nagged
    e. criticized

17. **Extremely** most nearly means
    a. almost
    b. slightly
    c. very
    d. clearly
    e. happily

18. She was **doubtful** whether the plan would work.
    a. uncertain
    b. panicked
    c. pondering
    d. indifferent
    e. confused

19. **Peculiar** most nearly means
    a. original
    b. novel
    c. dull
    d. strange
    e. awesome

20. He is a very **courteous** young man.
    a. handsome
    b. polite
    c. inconsiderate
    d. odd
    e. unrelenting

21. **Troubled** most nearly means
    a. relieved
    b. satisfied
    c. bothered
    d. relaxed
    e. persistent

22. **Perspiration** most nearly means
    a. sweat
    b. work
    c. help
    d. advice
    e. job

23. The child **trembled** with fear.
    a. spoke
    b. shook
    c. wept
    d. ducked
    e. cowered

24. **Adhered** most nearly means
    a. stuck
    b. went
    c. spoke
    d. altered
    e. stunk

25. She kept her house **tidy**.
    a. furnished
    b. warm
    c. locked
    d. neat
    e. inviting

Math Knowledge

1. A rectangle has a width of 7 cm and a length of 9 cm. What is its perimeter?
    a. 16 cm
    b. 32 cm
    c. 48 cm
    d. 62 cm
    e. 63 cm

2. In the following inequality, solve for q.
$-3q + 12 \geq 4q - 30$
    a. $q \geq 6$
    b. $q = 6$
    c. $q \neq 6$
    d. $q \leq 6$
    e. q does not exist

3. If $x - 6 = 0$, then x is equal to
    a. 0
    b. 3
    c. 6
    d. 9
    e. 12

4. If $x = -3$, calculate the value of the following expression:
$3x^3 + (3x + 4) - 2x^2$
    a. -104
    b. -58
    c. 58
    d. 104
    e. 0

5. If $3x - 30 = 45 - 2x$, what is the value of x?
    a. 5
    b. 10
    c. 15
    d. 20
    e. 25

6. Solve for x in the following inequality.
$\frac{1}{4}x - 25 \geq 75$
    a. $x \geq 400$
    b. $x \leq 400$
    c. $x \geq 25$
    d. $x \leq 25$
    e. $x \geq 50$

7. If $x^2 - 5 = 20$, what is possible value of x?
   a. 5
   b. 10
   c. 12.5
   d. 15
   e. 25

8. What is the area of a square that has a perimeter of 8 cm?
   a. 2 cm²
   b. 4 cm²
   c. 32 cm²
   d. 64 cm²
   e. 160 cm²

9. If x = 4 and y = 2, what is the value of the following expression:
$3xy - 12y + 5x$
   a. -4
   b. 10
   c. 12
   d. 20
   e. 24

10. If $0.65x + 10 = 15$, what is the value of x?
    a. 4.92
    b. 5.78
    c. 6.45
    d. 7.69
    e. 8.12

11. Simplify the following:
$(3x + 5)(4x - 6)$
    a. $12x^2 - 30x - 30$
    b. $12x^2 + 2x - 30$
    c. $12x^2 - 2x - 1$
    d. $12x^2 + 2x + 30$
    e. $12x^2 + 7x - 30$

12. Simplify the following expression:
$$\frac{50x^{18}t^6w^3z^{20}}{5x^5t^2w^2z^{19}}$$
    a. $10x^{13}t^3wz$
    b. $10x^{13}t^4wz$
    c. $10x^{12}t^4wz$
    d. $10x^{13}t^4wz^2$
    e. $10x^{12}t^3w^2z^2$

13. 4! =
    a. 4
    b. 12
    c. 16
    d. 20
    e. 24

14. If a cube is 5 cm long, what is the volume of the cube?
    a. 15 cm³
    b. 65 cm³
    c. 105 cm³
    d. 125 cm³
    e. 225 cm³

15. Solve for x by factoring:
$x^2 - 13x + 42 = 0$
    a. x = 6, 7
    b. x = -6, -7
    c. x = 6, -7
    d. x = -6, 7
    e. x = 7 only

16. A triangle has a base measuring 12 cm and a height of 12 cm. What is its area?
    a. 24 cm²
    b. 56 cm²
    c. 72 cm²
    d. 144 cm²
    e. 288 cm²

17. Simplify the following expression:
$(3x^2 * 7x^7) + (2y^3 * 9y^{12})$
    a. $21x^{14} + 18y^{26}$
    b. $10x^9 + 11y^{15}$
    c. $21x^{14} + 18y^{15}$
    d. $21x^9 + 18y^{15}$
    e. $10x^{14} + 11y^{26}$

18. If x/3 + 27 = 30, what is the value of x?
    a. 3
    b. 6
    c. 9
    d. 12
    e. 27

19. What is the slope of a line with points A (4,1) and B (-13,8)?
    a. 7/17
    b. -7/17
    c. -17/7
    d. 17/7
    e. none of the above

20. If x is 20% of 200, what is the value of x?
    a. 40
    b. 80
    c. 100
    d. 150
    e. 180

21. If a bag of balloons consists of 47 white balloons, 5 yellow balloons, and 10 black balloons, what is the probability that a balloon chosen randomly from the bag will be black?
    a. 19%
    b. 16%
    c. 21%
    d. 33%
    e. 10%

22. In a lottery game, there are 2 winners for every 100 tickets sold. If a man buys 10 tickets, what are the chances that he is a winner?
    a. 1 in 2
    b. 1 in 5
    c. 2 in 5
    d. 2 in 2
    e. 1 in 50

23. What is the volume of a rectangular prism with a height of 10cm, a length of 5cm, and a width of 6cm?
    a. 30 cm³
    b. 60 cm³
    c. 150 cm³
    d. 240 cm³
    e. 300 cm³

24. What is the midpoint of point A (6, 20) and point B (10, 40)?
    a. (30, 8)
    b. (16, 60)
    c. (8, 30)
    d. (7, 15)
    e. (15, 8)

25. If 5x + 60 = 75, what is the value of x?
    a. 3
    b. 4
    c. 5
    d. 6
    e. 7

Reading Comprehension

Passage 1

Since air is a gas, it can be compressed or expanded. When air is compressed, a greater amount of air can occupy a given volume. Conversely, when pressure on a given volume of air is decreased, the air expands and occupies a greater space. That is, the original column of air at a lower pressure contains a smaller mass of air. In other words, the density is decreased. In fact, density is directly proportional to pressure. If the pressure is doubled, the density is doubled, and if the pressure is lowered, so is the density. This statement is true only at a constant temperature.

Increasing the temperature of a substance decreases its density. Conversely, decreasing the temperature increases the density. Thus, the density of air varies inversely with temperature. This statement is true only at a constant pressure. In the atmosphere, both temperature and pressure decrease with altitude, and have conflicting effects upon density. However, the fairly rapid drop in pressure as altitude is increased usually has the dominant effect. Hence, pilots can expect the density to decrease with altitude.

The preceding paragraphs are based on the presupposition of perfectly dry air. In reality, it is never completely dry. The small amount of water vapor suspended in the atmosphere may be negligible under certain conditions, but in other conditions humidity may become an important factor in the performance of an aircraft. Water vapor is lighter than air; consequently, moist air is lighter than dry air. Therefore, as the water content of the air increases, the air becomes less dense, increasing density altitude and decreasing performance. It is lightest or least dense when, in a given set of conditions, it contains the maximum amount of water vapor.

Humidity, also called relative humidity, refers to the amount of water vapor contained in the atmosphere, and is expressed as a percentage of the maximum amount of water vapor the air can hold. This amount varies with the temperature; warm air can hold more water vapor, while colder air can hold less. Perfectly dry air that contains no water vapor has a relative humidity of zero percent, while saturated air that cannot hold any more water vapor has a relative humidity of 100 percent. Humidity alone is usually not considered an essential factor in calculating density altitude and aircraft performance; however, it does contribute.

The higher the temperature, the greater amount of water vapor that the air can hold. When comparing two separate air masses, the first warm and moist (both qualities making air lighter) and the second cold and dry (both qualities making it heavier), the first must be less dense than the second. Pressure, temperature, and humidity have a great influence on aircraft performance because of their effect upon density. There is no rule-of-thumb or chart used to compute the effects of humidity on density altitude, but it must be taken into consideration. Expect a decrease in overall performance in high humidity conditions.

1. The primary purpose of the passage is to
    a. explain the qualities of air that may affect flight.
    b. explain g-force and how it works.
    c. describe the constituent elements of air.
    d. explain humidity.
    e. describe the ideal air conditions for flight.

2. In the second paragraph, *inversely* most nearly means
    a. severely
    b. incredibly
    c. in the opposite direction
    d. in an unrelated fashion
    e. concurrently

3. If the air temperature drops while a plane is gaining altitude, the pilot can expect
    a. the density of the air to increase.
    b. the humidity of the air to increase.
    c. the air pressure to increase.
    d. the density of the air to decrease.
    e. aircraft performance to decrease.

4. With which one of the following claims about air quality would the author most likely agree?
    a. Pilots never need to pay attention to relative humidity.
    b. For a pilot, the density of air is more important than the relative humidity.
    c. Completely dry air is very rare.
    d. Aircraft performance is unrelated to humidity.
    e. The best conditions for flying are very hot and humid.

5. What is the most likely reason why there is no chart for assessing the effects of humidity on density altitude?
    a. Humidity does not affect density altitude.
    b. It is impossible to measure humidity.
    c. Humidity does not affect flight performance very much.
    d. Humidity varies a great deal in relatively small areas.
    e. Density altitude never varies.

Passage 2

The climb performance of an aircraft is affected by certain variables. The conditions of the aircraft's maximum climb angle or maximum climb rate occur at specific speeds, and variations in speed will produce variations in climb performance. There is sufficient latitude in most aircraft that small variations in speed from the optimum do not produce large changes in climb performance, and certain operational considerations may require speeds slightly different from the optimum. Of course, climb performance would be most critical with high gross weight, at high altitude, in obstructed takeoff areas, or during malfunction of a powerplant. Then, optimum climb speeds are necessary.

Weight has a very pronounced effect on aircraft performance. If weight is added to an aircraft, it must fly at a higher angle of attack (AOA) to maintain a given altitude and speed. This increases the induced drag of the wings, as well as the parasite drag of the aircraft. Increased drag means that additional thrust is needed to overcome it, which in turn means that less reserve thrust is available for climbing. Aircraft designers go to great effort to minimize the weight since it has such a marked effect on the factors pertaining to performance.

A change in an aircraft's weight produces a twofold effect on climb performance. First, a change in weight will change the drag and the power required. This alters the reserve power available, which in turn, affects both the climb angle and the climb rate. Secondly, an increase in weight will reduce the maximum rate of climb, but the aircraft must be operated at a higher climb speed to achieve the smaller peak climb rate.

An increase in altitude also will increase the power required and decrease the power available. Therefore, the climb performance of an aircraft diminishes with altitude. The speeds for maximum rate of climb, maximum angle of climb, and maximum and minimum level flight airspeeds vary with altitude. As altitude is increased, these various speeds finally converge at the absolute ceiling of the aircraft. At the absolute ceiling, there is no excess of power and only one speed will allow steady, level flight. Consequently, the absolute ceiling of an aircraft produces zero rate of climb. The service ceiling is the altitude at which the aircraft is unable to climb at a rate greater than 100 feet per minute (fpm). Usually, these specific performance reference points are provided for the aircraft at a specific design configuration.

In discussing performance, it frequently is convenient to use the terms power loading, wing loading, blade loading, and disk loading. Power loading is expressed in pounds per horsepower and is obtained by dividing the total weight of the aircraft by the rated horsepower of the engine. It is a significant factor in an aircraft's takeoff and climb capabilities. Wing loading is expressed in pounds per square foot and is obtained by dividing the total weight of an airplane in pounds by the wing area (including ailerons) in square feet. It is the airplane's wing loading that determines the landing speed. Blade loading is expressed in pounds per square foot and is obtained by dividing the total weight of a helicopter by the area of the rotor blades. Blade loading is not to be confused with disk loading, which is the total weight of a helicopter divided by the area of the disk swept by the rotor blades.

6. Which of the following would be the best title for this passage?
    a. The Importance of Weight
    b. Climb Performance and You
    c. Power Loading, Wing Loading, and Disk Loading
    d. Influences on Climb Performance
    e. Achieving Maximum Climb Angle

7. In the second paragraph, *pronounced* most nearly means
    a. selective
    b. intoned
    c. detrimental
    d. spoken
    e. noticeable

8. Which of the following is NOT one of the effects of increased weight on flight performance?
    a. diminished reserve power
    b. decreased climb rate
    c. lower angle of attack required to maintain altitude
    d. diminished maximum rate of climb
    e. increased drag

9. With which one of the following claims about climb performance would the author most likely agree?
    a. Optimal climb performance can be achieved even with heavy cargo.
    b. At the end of a long journey, a plane will have a higher maximum rate of climb.
    c. A plane can handle any amount of weight, though climb performance will be affected.
    d. Pilots have no influence over climb performance.
    e. The climb performance of a two-engine plane will remain the same even if one engine fails.

10. If a helicopter weighs two tons and its rotor blades cover an area of five hundred square feet, what is its disc loading measure?
    a. 4 pounds per square foot
    b. 250 square foot-pounds
    c. 0.25 tons per square foot
    d. 4 metric tons
    e. 3.65 ton-feet

Passage 3

The aerodynamic properties of an aircraft generally determine the power requirements at various conditions of flight, while the powerplant capabilities generally determine the power available at various conditions of flight. When an aircraft is in steady, level flight, a condition of equilibrium must prevail. An unaccelerated condition of flight is achieved when lift equals weight, and the powerplant is set for thrust equal to drag. The power required to achieve equilibrium in constant-altitude flight at various airspeeds is depicted on a power required curve. The power required curve illustrates the fact that at low airspeeds near the stall or minimum controllable airspeed, the power setting required for steady, level flight is quite high.

Flight in the region of normal command means that while holding a constant altitude, a higher airspeed requires a higher power setting and a lower airspeed requires a lower power setting. The majority of aircraft flying (climb, cruise, and maneuvers) is conducted in the region of normal command.

Flight in the region of reversed command means flight in which a higher airspeed requires a lower power setting and a lower airspeed requires a higher power setting to hold altitude. It does not imply that a decrease in power will produce lower airspeed. The region of reversed command is encountered in the low speed phases of flight. Flight speeds below the speed for maximum endurance (lowest point on the power curve) require higher power settings with a decrease in airspeed. Since the need to increase the required power setting with decreased speed is contrary to the normal command of flight, the regime of flight speeds between the speed for minimum required power setting and the stall speed (or minimum control speed) is termed the region of reversed command. In the region of reversed command, a decrease in airspeed must be accompanied by an increased power setting in order to maintain steady flight.

An airplane performing a low airspeed, high pitch attitude power approach for a short-field landing is an example of operating in the region of reversed command. If an unacceptably high sink rate should develop, it may be possible for the pilot to reduce or stop the descent by applying power. But without further use of power, the airplane would probably stall or be incapable of flaring for the landing. Merely lowering the nose of the airplane to regain flying speed in this situation, without the use of power, would result in a rapid sink rate and corresponding loss of altitude.

If during a soft-field takeoff and climb, for example, the pilot attempts to climb out of ground effect without first attaining normal climb pitch attitude and airspeed, the airplane may inadvertently enter the region of reversed command at a dangerously low altitude. Even with full power, the airplane may be incapable of climbing or even maintaining altitude. The pilot's only recourse in this situation is to lower the pitch attitude in order to increase airspeed, which will inevitably result in a loss of altitude. Airplane pilots must give particular attention to precise control of airspeed when operating in the low flight speeds of the region of reversed command.

11. The primary purpose of the passage is to
    a. instruct pilots on proper airspeed.
    b. discuss the interrelationships of airspeed, power, and pitch attitude.
    c. explain reversed command.
    d. discuss the physics of flight at low airspeeds.
    e. persuade the reader to fly faster aircraft.

12. In the fifth paragraph, *inadvertently* most nearly means
    a. unintentionally
    b. indirectly
    c. sequentially
    d. primarily
    e. eventually

13. In which region of command does most flight occur?
    a. inverse command
    b. normal command
    c. direct command
    d. reverse command
    e. decreased command

14. With which one of the following statements about flight would the author most likely agree?
    a. As speed increases, the power required to descend decreases.
    b. As speed increases, the power required to descend remains constant.
    c. As speed decreases, the power required to climb remains constant.
    d. As speed increases, the power required to maintain altitude increases.
    e. As speed decreases, the power required to maintain altitude increases.

15. Which of the following would be the best title for this passage?
    a. How to Avoid Reversed Command
    b. Normal Command Flight
    c. Learning to Fly
    d. Power Requirements During Flight
    e. Reversed Command and the Modern Pilot

Passage 4

In many cases, the landing distance of an aircraft will define the runway requirements for flight operations. The minimum landing distance is obtained by landing at some minimum safe speed, which allows sufficient margin above stall and provides satisfactory control and capability for a go-around. Generally, the landing speed is some fixed percentage of the stall speed or minimum control speed for the aircraft in the landing configuration. As such, the landing will be accomplished at some particular value of lift coefficient and AOA. The exact values will depend on the aircraft characteristics but, once defined, the values are independent of weight, altitude, and wind.

To obtain minimum landing distance at the specified landing speed, the forces that act on the aircraft must provide maximum deceleration during the landing roll. The forces acting on the aircraft during the landing roll may require various procedures to maintain landing deceleration at the peak value.

A distinction should be made between the procedures for minimum landing distance and an ordinary landing roll with considerable excess runway available. Minimum landing distance will be obtained by creating a continuous peak deceleration of the aircraft; that is, extensive use of the brakes for maximum deceleration. On the other hand, an ordinary landing roll with considerable excess runway may allow extensive use of aerodynamic drag to minimize wear and tear on the tires and brakes. If aerodynamic drag is sufficient to cause deceleration, it can be used in deference to the brakes in the early stages of the landing roll; i.e., brakes and tires suffer from continuous hard use, but aircraft aerodynamic drag is free and does not wear out with use.

The use of aerodynamic drag is applicable only for deceleration to 60 or 70 percent of the touchdown speed. At speeds less than 60 to 70 percent of the touchdown speed, aerodynamic drag is so slight as to be of little use, and braking must be utilized to produce continued deceleration. Since the objective during the landing roll is to decelerate, the powerplant thrust should be the smallest possible positive value (or largest possible negative value in the case of thrust reversers). In addition to the important factors of proper procedures, many other variables affect the landing performance. Any item that alters the landing speed or deceleration rate during the landing roll will affect the landing distance.

The effect of gross weight on landing distance is one of the principal items determining the landing distance. One effect of an increased gross weight is that a greater speed will be required to support the aircraft at the landing AOA and lift coefficient. For an example of the effect of a change in gross weight, a 21 percent increase in landing weight will require a ten percent increase in landing speed to support the greater weight.

When minimum landing distances are considered, braking friction forces predominate during the landing roll and, for the majority of aircraft configurations, braking friction is the main source of deceleration.

The minimum landing distance will vary in direct proportion to the gross weight. For example, a ten percent increase in gross weight at landing would cause a:
- Five percent increase in landing velocity
- Ten percent increase in landing distance

A contingency of this is the relationship between weight and braking friction force.

The effect of wind on landing distance is large and deserves proper consideration when predicting landing distance. Since the aircraft will land at a particular airspeed independent of the wind, the principal effect of wind on landing distance is the change in the groundspeed at which the aircraft touches down. The effect of wind on deceleration during the landing is identical to the effect on acceleration during the takeoff.

The effect of pressure altitude and ambient temperature is to define density altitude and its effect on landing performance. An increase in density altitude increases the landing speed but does not alter the net retarding force. Thus, the aircraft at altitude lands at the same IAS as at sea level but, because of the reduced density, the TAS is greater. Since the aircraft lands at altitude with the same weight and dynamic pressure, the drag and braking friction throughout the landing roll have the same values as at sea level. As long as the condition is within the capability of the brakes, the net retarding force is unchanged, and the deceleration is the same as with the landing at sea level. Since an increase in altitude does not alter deceleration, the effect of density altitude on landing distance is due to the greater TAS.

16. The main purpose of the passage is to
    a. give some examples of near accidents during landing.
    b. improve landing skills.
    c. explain the effects of varying pressure altitudes.
    d. advocate safer protocols for landing.
    e. describe the factors that influence landing distance.

17. Why will a pilot rely on aerodynamic drag when making a normal landing?
    a. To increase the rate of deceleration
    b. To avoid wearing down the brakes and tires
    c. To avoid unnecessary turbulence
    d. To mitigate a large gross weight
    e. To simplify landing procedures

18. In the fourth paragraph, *principal* most nearly means
    a. most important
    b. easiest
    c. moral
    d. first
    e. value

19. Why must a heavier plane land at a higher speed?
    a. To diminish fuel supplies and thereby decrease gross weight
    b. To encourage a stall just before landing
    c. To avoid hitting the runway with too much force
    d. To improve handling on the runway
    e. To allow for the longest landing distance

20. With which one of the following claims about landing performance would the author most likely agree?
    a. Many landings occur without any use of the brakes.
    b. Ambient temperature has no effect on minimum landing distance.
    c. Gross weight and minimum landing distance are positively correlated.
    d. Runway length is less important than gross weight in determining the appropriate airspeed during landing.
    e. Minimum landing distance is generally consistent for aircraft of the same size.

Passage 5
For over 25 years, the importance of good pilot judgment, or aeronautical decision-making (ADM), has been recognized as critical to the safe operation of aircraft, as well as accident avoidance. The airline industry, motivated by the need to reduce accidents caused by human factors, developed the first training programs based on improving ADM. Crew resource management (CRM) training for flight crews is focused on the effective use of all available resources: human resources, hardware, and information supporting ADM to facilitate crew cooperation and improve decision-making. The goal of all flight crews is good ADM and the use of CRM is one way to make good decisions.

Research in this area prompted the Federal Aviation Administration (FAA) to produce training directed at improving the decision-making of pilots and led to current FAA regulations that require that decision-making be taught as part of the pilot training curriculum. ADM research, development, and testing culminated in 1987 with the publication of six manuals oriented to the decision-making needs of variously rated pilots.

These manuals provided multifaceted materials designed to reduce the number of decision related accidents. The effectiveness of these materials was validated in independent studies where student pilots received such training in conjunction with the standard flying curriculum. When tested, the pilots who had received ADM training made fewer inflight errors than those who had not received ADM training. The differences were statistically significant and ranged from about 10 to 50 percent fewer judgment errors. In the operational environment, an operator flying about 400,000 hours annually demonstrated a 54 percent reduction in accident rate after using these materials for recurrency training.

Contrary to popular opinion, good judgment can be taught. Tradition held that good judgment was a natural by-product of experience, but as pilots continued to log accident-free flight hours, a corresponding increase of good judgment was assumed. Building upon the foundation of conventional decision-making, ADM enhances the process to decrease the probability of human error and increase the probability of a safe flight. ADM provides a structured, systematic approach to analyzing changes that occur during a flight and how these changes might affect a flight's safe outcome. The ADM process addresses all aspects of decision-making in the flight deck and identifies the steps involved in good decision-making.

Steps for good decision-making are:
1. Identifying personal attitudes hazardous to safe flight
2. Learning behavior modification techniques
3. Learning how to recognize and cope with stress
4. Developing risk assessment skills
5. Using all resources
6. Evaluating the effectiveness of one's ADM skills

Risk management is an important component of ADM. When a pilot follows good decision-making practices, the inherent risk in a flight is reduced or even eliminated. The ability to make good decisions is based upon direct or indirect experience and education.

Consider automotive seat belt use. In just two decades, seat belt use has become the norm, placing those who do not wear seat belts outside the norm, but this group may learn to wear a seat belt by either direct or indirect experience.

For example, a driver learns through direct experience about the value of wearing a seat belt when he or she is involved in a car accident that leads to a personal injury. An indirect learning experience occurs when a loved one is injured during a car accident because he or she failed to wear a seat belt.

While poor decision-making in everyday life does not always lead to tragedy, the margin for error in aviation is thin. Since ADM enhances management of an aeronautical environment, all pilots should become familiar with and employ ADM.

21. The primary purpose of the passage is to
    a. list the steps in good decision-making.
    b. improve the decision-making abilities of the reader.
    c. outline the relationship between aeronautical decision-making and crew resource management.
    d. discuss aeronautical decision-making.
    e. inspire the reader to make better decisions.

22. According to the passage, how is aviation safety distinguished from other forms of safety?
    a. Aviation safety is much simpler than most other areas of safety.
    b. Aviation safety is no different than most other areas of safety.
    c. Aviation safety is only important to a small percentage of the population.
    d. There is a smaller margin for error in aviation.
    e. Aviation safety can be systematized.

23. In the second paragraph, *conjunction* most nearly means
    a. combination
    b. linking word
    c. opposition
    d. collection
    e. organization

24. With which one of the following claims about aeronautical decision-making would the author most likely agree?
    a. The body of knowledge about ADM is increasing, and this will have a positive effect on flight safety.
    b. Aeronautical decision-making is the responsibility of the pilot alone.
    c. Eventually, researchers will establish a perfect set of decision-making tools for pilots.
    d. Aeronautical decision-making is the only tool required for flight safety.
    e. Aeronautical decision-making has no applications in areas other than flight.

25. When a pilot reads the account of a recent aviation accident, this is an opportunity for a(n)
    a. recertification.
    b. direct learning experience.
    c. implicit learning experience.
    d. reorientation of learning.
    e. indirect learning experience.

Situational Judgment

Situation 1:
You are approached by a senior officer, who requests a private meeting with you. He asks you for your candid opinion on your immediate supervisor, who is a subordinate to the senior officer. You have a generally favorable opinion of the supervisor, but you do have a few complaints about her performance. Specifically, you feel that she does a poor job of running staff meetings.
Possible actions:
    a. Decline to meet with the senior officer.
    b. Meet with the senior officer, and focus on the ways your supervisor could improve staff meetings.
    c. Give your candid opinion of your supervisor, including your criticisms, but emphasizing your overall positive opinion.
    d. Write a letter to the senior officer, explaining your opinions about your supervisor.
    e. Give the senior officer a glowing report of the supervisor, without mentioning your complaints.

1. Select the MOST EFFECTIVE action in response to the situation.
2. Select the LEAST EFFECTIVE action in response to the situation.

Situation 2:
While performing administrative work with another officer, you notice that he is manipulating the numbers on some reports. Specifically, the officer is inflating the amounts of time spent on certain training exercises. These reports are sent on to senior officers, who use them to assess the readiness of the soldiers for more sophisticated and complicated missions. If men and women are unprepared for these more difficult missions, there is a greater risk of accident or injury, to themselves or others.
Possible actions:
    a. Immediately report this infraction to your superior.
    b. Wait until you can acquire clear proof of the data manipulation.
    c. Say nothing at present, but keep an eye on this other officer.
    d. Write an anonymous letter to the other officer, encouraging him to stop manipulating the data.
    e. Go back over the other officer's work, correcting the data as necessary.

3. Select the MOST EFFECTIVE action in response to the situation.
4. Select the LEAST EFFECTIVE action in response to the situation.

Situation 3:
You and two of your fellow officers have been selected to interview candidates for a promotion. One of the three candidates (Candidate A) is known to be an old family friend of your commanding officer. During the interviews, Candidate A is, in your opinion, the most impressive. However, Candidate B also seems like an excellent choice for the position. Candidate C performs poorly. One of the other judges votes for Candidate A, and the other votes for Candidate B.
Possible actions:
    a. Vote for Candidate B to avoid the appearance of favoritism.
    b. Vote for Candidate C so that the decision will be passed on to another round of deliberation.
    c. Abstain from voting.
    d. Request a new set of candidates.
    e. Vote for Candidate A because you feel he is the most qualified.

5. Select the MOST EFFECTIVE action in response to the situation.
6. Select the LEAST EFFECTIVE action in response to the situation.

Situation 4:
You attend a meeting with two other officers. During the meeting, these officers get into a heated conflict over a possible change in policy. You have heard that they have a personal antipathy, though you are not aware of its origins. They ask you to settle their dispute.
Possible actions:
    a. Ignore the request, and instead lecture the two officers on the need for cooperation and goodwill in the military.
    b. Take the side of the officer who you believe will be the most help to you in the future.
    c. Refuse to take a side, citing the obvious personal differences between the two officers.
    d. Select the option that you think is best, leaving aside everything you know about their personal conflict.
    e. Take the side of the officer you like better.

7. Select the MOST EFFECTIVE action in response to the situation.
8. Select the LEAST EFFECTIVE action in response to the situation.

Situation 5:
You are asked to collaborate on an important project with an officer from a different unit. This officer is very close to retirement, and appears to have lost interest in his work. Consequently, he puts very little effort into the work you two are supposed to be sharing.
Possible actions:
    a. Request a different partner for the project.
    b. Inform your senior officer of the situation.
    c. Do your best on your work, but accept that the project probably will fail because of the poor attitude of your collaborator.
    d. Do your work and the work that was supposed to be done by your partner, since the most important thing is the successful completion of the project.
    e. Express your frustrations directly to your collaborator, emphasizing that the success of this project is very important to you, and agree upon a fair division of labor.

9. Select the MOST EFFECTIVE action in response to the situation.
10. Select the LEAST EFFECTIVE action in response to the situation.

Situation 6:
While working in a field office, you observe that one of your colleagues is being severely overworked. Despite doing an excellent job and working more than the required number of hours, she is unable to keep up with the amount of paperwork being sent to her by other offices. Her commanding officers do not seem to be aware that she is being asked to do an unfair amount of work.
Possible actions:
    a. Assign one of your subordinates to assist your overworked colleague.
    b. Request a meeting with your overworked colleague's commanding officer, and describe the problem to him.
    c. Ignore the problem, since it does not relate to your work.
    d. Express your sympathy with your overworked colleague.
    e. Help your overworked colleague whenever you finish your own work.

11. Select the MOST EFFECTIVE action in response to the situation.
12. Select the LEAST EFFECTIVE action in response to the situation.

Situation 7:
You have become suspicious that one of your junior officers is trying to undermine your work. This junior officer is very competent and very ambitious. You have even heard from other officers that this junior officer wants to take over your job. You have not yet mentioned your suspicion directly to the junior officer.
Possible actions:
    a. Meet with the junior officer and explain that you expect his full cooperation and support.
    b. Publicly reprimand and humiliate the junior officer.
    c. Ignore the situation, in the hopes that it will resolve itself.
    d. Report this insubordination to your commanding officer.
    e. Ask one of your fellow officers to talk with the junior officer and try to improve the situation.

13. Select the MOST EFFECTIVE action in response to the situation.
14. Select the LEAST EFFECTIVE action in response to the situation.

Situation 8:
You are extremely busy with paperwork, but a senior officer asks you to complete a set of special reports in addition to your normal work. You do not think it will be possible for you to complete this extra assignment without sacrificing the quality of your work. However, you would like to impress your senior officer by fulfilling his request.
Possible actions:
    a. Tell the senior officer that you are too busy to complete extra projects.
    b. Accept the extra work and resolve to complete it quickly no matter what.
    c. Ask the senior officer if you can have a few days to decide whether you will complete the extra work.
    d. Accept the extra work, but have a junior staff member complete it for you.
    e. Decline the extra work, but offer to pass it along to another qualified member of your team.

15. Select the MOST EFFECTIVE action in response to the situation.
16. Select the LEAST EFFECTIVE action in response to the situation.

Situation 9:
Over the past few months, you have noticed that office supplies are being used at a much greater pace than is usual. There is no clear reason for this, but you have begun to suspect that one of your fellow officers is taking office supplies home with her at night. You have no specific evidence to support your claims, but the office supplies seem to be disappearing after shifts in which this officer is alone with access to the supply closet.
Possible actions:
    a. Ignore the situation, since you know that you are not personally responsible for the thefts.
    b. Tell a senior officer about your suspicions.
    c. Confront the officer with your suspicions and ask for an explanation.
    d. Set up a hidden surveillance camera so that you can catch the officer stealing supplies.
    e. Ask some of your fellow officers if they have noticed any suspicious behavior.

17. Select the MOST EFFECTIVE action in response to the situation.
18. Select the LEAST EFFECTIVE action in response to the situation.

Situation 10:
You have been working in the same unit for the past two years. Although you have been successful, you are beginning to get burnt out, and you are thinking about requesting a transfer. You contact some friends and colleagues in other units, trying to determine whether you would like it there. After a few weeks, you realize that rumors about your interest in a transfer have begun to spread throughout your unit.
Possible actions:
    a. In a letter to your senior officer, acknowledge the truth of the rumors and request a transfer.
    b. Deny the rumors publicly, but continue to pursue a transfer.
    c. Ignore the rumors, but refocus on your work with your current unit.
    d. Acknowledge the rumors and refocus on your work with your current unit.
    e. Request a leave of absence so that you can decide how to handle the situation.

19. Select the MOST EFFECTIVE action in response to the situation.
20 Select the LEAST EFFECTIVE action in response to the situation.

Situation 11:
You have been asked to attend a meeting between the senior officers in your unit and a group of local community leaders. The meeting has been called because of rumors that your military base will be shut down due to budgetary constraints. The community leaders are willing to lobby on behalf of the base, so long as they receive assurances that the military leadership will cooperate with them on some local initiatives. Specifically, the community leaders would like the soldiers to help coordinate disaster relief efforts when necessary. At one point, a community leader turns to you and asks for your opinion on the subject.

Possible actions:
    a. Answer the question, but mention only the potential positive consequences of the proposal.
    b. Try to answer the question as honestly and completely as possible, but remind the community leader that you are only a junior officer.
    c. Answer the question thoroughly, even though you have little direct knowledge of the situation.
    d. Remind the community leader that you are a junior officer and cannot give your opinion.
    e. Decline to answer the question, and instead refer it to one of the senior officers.

21. Select the MOST EFFECTIVE action in response to the situation.
22. Select the LEAST EFFECTIVE action in response to the situation.

Situation 12:
Another officer in your section is granted a week-long leave to visit his sick grandmother. However, during this period you discover through social media that the officer is actually on a beach vacation with his girlfriend, hundreds of miles from the hospital where his grandmother was supposedly being treated.

Possible actions:
    a. Tell the officer that he has been dishonest and unethical, and that you will have to report any future episodes of this nature.
    b. Ignore the situation, since it does not really involve you.
    c. Tell the officer that you will agree to keep his misbehavior a secret if he will do your weekly reports.
    d. Immediately notify your commanding officer in person.
    e. Anonymously forward the evidence of the officer's misbehavior to your commanding officer.

23. Select the MOST EFFECTIVE action in response to the situation.
24. Select the LEAST EFFECTIVE action in response to the situation.

Situation 13:
One of the soldiers in your unit has displayed a marked decline in her performance over the past month. She appears frustrated and burnt out with her normal duties. She is one of the more popular members of the unit, so her negative attitude has a bad influence on her fellow soldiers.
Possible actions:
> a. Convene a meeting of the entire unit, and use this as a chance to single out the soldier for her poor performance.
> b. Meet with the soldier and offer whatever help you can give to improve her performance, while emphasizing the effect that her behavior has on the other soldiers.
> c. Reassign the soldier to a different unit so that her bad attitude doesn't continue to affect the other soldiers.
> d. Do nothing, in the hopes that the situation will improve without your influence.
> e. Inform your senior officer about the situation.

25. Select the MOST EFFECTIVE action in response to the situation.
26. Select the LEAST EFFECTIVE action in response to the situation.

Situation 14:
You are about to transfer to a new unit, where you will have duties in areas where you have little experience. A week before you are due to make the transfer, you receive an email from the senior officer in charge of your new unit. She reminds you that you will be entering her unit at a very important time for them, because they will be leading a set of training exercises for highly-skilled soldiers. She wants to make sure that you are ready to contribute immediately to the success of the unit, and that you will not need a great deal of assistance to complete your work.
Possible actions:
> a. Do not respond to the email, and assume that you will be able to move into your new role seamlessly.
> b. Thank the senior officer for her message, and do some internet research on your new duties.
> c. Request that the transfer be canceled, and remain with your original unit.
> d. Email the senior officer back, requesting a personal meeting where you can get more information about how to make a good transition to your new role.
> e. Ask one of the other officers in your new unit if he will quietly bring you up to speed when you arrive.

27. Select the MOST EFFECTIVE action in response to the situation.
28. Select the LEAST EFFECTIVE action in response to the situation.

Situation 15:
One of your responsibilities is to keep a set of officers briefed on some confidential activities that are taking place at your base. One day, you accidentally send an email containing some information about these confidential activities to an officer who has not received the security clearance.
Possible actions:
    a. Notify your supervisor of your mistake and let him resolve the situation.
    b. Immediately send a second email to the improper recipient, requesting that he or she destroy the email. Inform your supervisors of your mistake.
    c. Send a second email to the improper recipient, claiming that your email account has been hacked and that he should disregard any earlier messages.
    d. Wait to see if there will be any negative consequences of your mistake.
    e. Ask your supervisor if the improper recipient could be given the security clearance retroactively.

29. Select the MOST EFFECTIVE action in response to the situation.
30. Select the LEAST EFFECTIVE action in response to the situation.

Situation 16:
After completing your work at the base, you return to your living quarters for the evening. An hour later, you realize that you failed to affix your signature to a set of papers that are to be forwarded on to a different unit for completion. Without your signature, the papers cannot be sent. The content of the papers is not particularly urgent, but the delay will require the other unit officer to stay later than normal at his post.
Possible actions:
    a. Wait until your next shift to sign the papers, since they are not considered urgent.
    b. Call the base and ask a junior officer to forge your signature so the paperwork can be sent.
    c. Call the officer at the other base and explain that the papers will be arriving a little later than expected.
    d. Arrive for your next shift a little early and sign the papers first.
    e. Return to the base and sign the necessary papers.

31. Select the MOST EFFECTIVE action in response to the situation.
32. Select the LEAST EFFECTIVE action in response to the situation.

Situation 17:
Several of the soldiers in your unit have yet to complete a basic training module at a nearby base. They have asked to participate in the next training session, but instead a group of soldiers from another unit have been selected. The selected soldiers have not been waiting nearly as long as your soldiers to complete this training module. You suspect that the director of the training session dislikes you personally, though you have no specific evidence of this.

Possible actions:
    a. Request a meeting with the training director, and ask why your soldiers have been passed over, emphasizing the importance of this module for their development.
    b. Do nothing, in the hopes that the situation will improve on its own.
    c. Ask the officer in charge of the other unit if your soldiers can attend the training session instead of his.
    d. Write a critical letter to the training director and your senior officer, outlining what you perceive as the injustice of the situation.
    e. Ask the training director if your soldiers can attend the next training session along with the selected soldiers.

33. Select the MOST EFFECTIVE action in response to the situation.
34. Select the LEAST EFFECTIVE action in response to the situation.

Situation 18:
You have been assigned to draft an important report along with another officer in your unit. It is expected that the report will take approximately one month to complete. However, after about a week of work, the other officer falls ill and is required to go on leave. He is only supposed to be gone for about ten days, but at the end of this period he has not returned and there is no definitive word on when he will. You need the expertise of this officer in order to finish the report.

Possible actions:
    a. Wait until the other officer returns from leave, even if it means delaying the report.
    b. Ask for an extension, without making excuses for your failure to complete the report on time.
    c. Work extra hours to complete the report as best as you can.
    d. Tell your commanding officer the situation, and request assistance in completing the report.
    e. Order a junior officer to assist you in the completion of the report.

35. Select the MOST EFFECTIVE action in response to the situation.
36. Select the LEAST EFFECTIVE action in response to the situation.

Situation 19:
Two months ago, you joined a new unit. The leader of this unit was very welcoming to you, and made sure to give you as much assistance as you needed in learning your new duties and responsibilities. However, you are now feeling more comfortable in your role and would like more independence in your work.
Possible actions:
    a. Request a meeting with the unit leader, thank him for his assistance, and indicate that you would like to work on your own a bit more.
    b. Request that another officer be transferred to your unit, so that the unit leader will divert his attention to training this new arrival.
    c. Without saying anything directly, try to avoid the unit leader as much as possible.
    d. Tell the unit leader's commanding officer that you need the unit leader to give you more space.
    e. Keep extensive records of your work, so that you can demonstrate your competence to the unit leader.

37. Select the MOST EFFECTIVE action in response to the situation.
38. Select the LEAST EFFECTIVE action in response to the situation.

Situation 20:
One of the other officers in your division is going to make an important presentation at the end of the week. He is nervous, but has done good work in the past and has spent a great deal of time preparing the report. He asks you to look over his work in advance. You notice a few things that need to be changed, but the other officer disagrees with your corrections. You are certain that you are right and that the other officer will be sorry he did not listen to you.
Possible actions:
    a. Let your coworker go ahead with the uncorrected presentation, but make yourself look better by mentioning the errors to a senior officer before the presentation.
    b. Allow your coworker to go ahead with the presentation as is, without trying to convince him to make the corrections.
    c. Discuss the matter with your senior officer, and ask him or her to mandate the corrections.
    d. Contrive an excuse to be absent from the presentation.
    e. Make every effort to convince your colleague to make the necessary corrections.

39. Select the MOST EFFECTIVE action in response to the situation.
40. Select the LEAST EFFECTIVE action in response to the situation.

Situation 21:
You have been working with a new unit for the past six weeks. During that time, you have noticed some inefficiencies in the unit's operations, and you have developed a set of proposals for eliminating them. The majority of your coworkers agree with your proposals, but the senior officer in charge of the unit does not. The senior officer believes that implementing your proposals would be too risky and would undermine the stability of the unit.
Possible actions:
   a. Confront your senior officer, using the support of your coworkers as a reason to implement your proposals.
   b. Create a detailed and comprehensive report outlining the potential benefits of your proposals. Deliver the report and then obey the senior officer's final decision.
   c. Implement your proposals anyway, on the assumption that your senior officer will change his mind once he sees their success.
   d. Accept the senior officer's decision and try to succeed within the agreed-upon structure.
   e. Accept the senior officer's decision, but keep a running list of the ways your proposals could have improved performance, had they been implemented.

41. Select the MOST EFFECTIVE action in response to the situation.
42. Select the LEAST EFFECTIVE action in response to the situation.

Situation 22:
Six months ago, you were assigned a new assistant. Although you have been able to work successfully together, you have developed a personal dislike for this person. In your opinion, he is arrogant and too critical of the other officers. During a meeting with your senior officer, she mentions that she is considering transferring your assistant to another unit. This would represent a step up for him, and would make it possible for him to achieve even more promotions in a relatively short time.
Possible actions:
   a. Recommend that your assistant receive the promotion, if only to get him away from you.
   b. Strongly discourage the senior officer from choosing your assistant, so that he will not get a professional reward.
   c. Write an anonymous letter to the senior officer, outlining your complaints about your assistant.
   d. Strongly discourage the senior officer from choosing your assistant, with an emphasis on his personality flaws.
   e. Avoid interfering in the senior officer's decision, but make sure that she is aware of your opinion of your assistant.

43. Select the MOST EFFECTIVE action in response to the situation.
44. Select the LEAST EFFECTIVE action in response to the situation.

Situation 23:
Your base uses a computer program to determine the logistics related to supply deliveries. One day, while you are coordinating the arrival and unloading of several concurrent deliveries, the computer system crashes. The computer technician tells you that it could be an hour before the system is up and running again. You can see that there is a long line of trucks waiting to deliver their goods, and that the drivers are becoming impatient.
Possible actions:
   a. Ask one of your assistant to inform the drivers of the situation and the likely wait time. Offer whatever accommodations you can in the interim.
   b. Receive the deliveries despite the computer problems, and keep paper records so you can update the system later.
   c. Call your senior officer and ask for advice.
   d. Use this opportunity to take your lunch break, somewhere you are unlikely to meet any of the drivers.
   e. Encourage the drivers to make any other deliveries they have and then come back later.

45. Select the MOST EFFECTIVE action in response to the situation.
46. Select the LEAST EFFECTIVE action in response to the situation.

Situation 24:
You have developed an idea that you believe will improve the performance of your unit. However, some of the soldiers in your unit disagree with this idea, and one has gone so far as to write a letter of complaint to your senior officer without notifying you. You have not yet implemented your idea.
Possible actions:
   a. Meet with the letter-writer and other critics, emphasizing that going above your head will not be tolerated in the future.
   b. Abandon your idea and ask for suggestions from the soldiers who were critical of it.
   c. Ignore the critics in you unit, and implement your idea anyway.
   d. Harshly punish the letter writer, as a warning to the other soldiers.
   e. Implement your idea without acknowledging the letter or the criticism of other soldiers in the unit.

47. Select the MOST EFFECTIVE action in response to the situation.
48. Select the LEAST EFFECTIVE action in response to the situation.

Situation 25:
You received a promotion six months ago, and have been excelling in your new job. However, due to forces beyond your control, the quality of your work has decreased over the past few weeks. In part, you have been undermined by recent budget cuts. Unfortunately, your commanding officer does not fully understand the consequences of the budget cuts, and has expressed her displeasure with your recent work. She has even suggested that problems in your unit may be a result of poor management on your part.

Possible actions:
  a. Ask the commanding officer if you can take a brief leave to refocus.
  b. Do not make any excuses, but ask the advice of other officers who are dealing with the same budget constraints.
  c. Ask your commanding officer for a list of her specific complaints.
  d. Shift the blame for your unit's performance to your subordinates.
  e. Remind the commanding officer of the budget constraints, and defend your management style to her.

49. Select the MOST EFFECTIVE action in response to the situation.
50. Select the LEAST EFFECTIVE action in response to the situation.

Physical Science

1. A long nail is heated at one end. After a few seconds, the other end of the nail becomes equally hot. What type of heat transfer does this represent?
    a. Advection
    b. Conduction
    c. Convection
    d. Entropy
    e. Radiation

2. The measure of energy within a system is called:
    a. temperature
    b. convection
    c. entropy
    d. thermodynamics
    e. heat

3. How do two isotopes of the same element differ?
    a. They have different numbers of protons
    b. They have different numbers of neutrons
    c. They have different numbers of electrons
    d. They have different charges
    e. They have different atomic numbers

4. Which type of nuclear process features atomic nuclei splitting apart to form smaller nuclei?
    a. Fission
    b. Fusion
    c. Decay
    d. Ionization
    e. Chain reaction

5. The process whereby a radioactive element releases energy slowly over a long period of time to lower its energy and become more stable is best described as:
    a. combustion
    b. fission
    c. fusion
    d. decay
    e. radioactivity

6. What property of light explains why a pencil in a glass of water appears to be bent?
    a. reflection
    b. refraction
    c. angle of incidence = angle of reflection
    d. constructive interference
    e. destructive interference

7. What unit describes the frequency of a wave?
    a. hertz (Hz)
    b. decibels (dB)
    c. meters (m)
    d. meters per second (m/s)
    e. meters per second squared (m/s^2)

8. Which of the following is an example of kinetic energy being converted to potential energy?
    a. A child sliding down a slide.
    b. A cyclist coasting on his way up a hill.
    c. A pilot deploying airbrakes on approach to land.
    d. A motorist swerving to avoid a deer.
    e. A pair of billiard balls colliding and rebounding off each other.

9. The boiling of water is an example of:
    a. sublimation.
    b. condensation.
    c. neutralization.
    d. chemical change
    e. physical change

10. The center of an atom is called the:
    a. nucleus
    b. nuclide
    c. neutrino
    d. electron cloud
    e. electrolyte

11. When a solid is heated and transforms directly to the gaseous phases, this process is called:
    a. sublimation
    b. fusion
    c. diffusion
    d. condensation
    e. fission

12. Which scientist was responsible for developing the format of the modern periodic table?
    a. Faraday
    b. Einstein
    c. Hess
    d. Mendeleev
    e. Oppenheimer

13. The density of a material refers to its:
    a. Mass per unit volume
    b. Mass per unit length
    c. Mass per unit surface area
    d. Volume per unit surface area
    e. Volume per unit length

14. The precision of a set of experimentally obtained data points refers to:
    a. How accurate the data points are
    b. How many errors the data points contain
    c. How close the data points are to the mean of the data
    d. How close the data points are to the predicted result
    e. How close the set of data is to a normal distribution

15. Current, or the amount of electricity that is flowing, is measured in:
    a. volts
    b. watts
    c. ohms
    d. farads
    e. amperes

16. A solar eclipse can only occur if:
    a. the earth and the sun are on the same side of the moon
    b. the earth is between the sun and the moon
    c. the moon is between the earth and the sun
    d. the sun is between the earth and the moon
    e. the moon is full

17. What property of motion explains why passengers in a turning car feel pulled toward the outside of the turn?
    a. centripetal force
    b. inertia
    c. normal force
    d. impulse
    e. torque

18. According to the Ideal Gas Law, if a certain amount of gas is being held at a constant volume, and the temperature is increased, what will happen?
    a. the mass of the gas will increase
    b. the pressure of the gas will increase
    c. the density of the gas will decrease
    d. the mass of the gas will decrease
    e. the pressure of the gas will decrease

19. What wave characteristic is related to the loudness of a sound?
    a. frequency
    b. amplitude
    c. wavelength
    d. velocity
    e. period

20. Which of the following scenarios is NOT an example of a person applying work to a book?
    a. A book is picked up from the floor and put on a shelf.
    b. A book is pushed across a table top.
    c. A backpack holding a book is carried across the room.
    d. A book is held and then released so that it falls to the ground.
    e. A book is thrown vertically into the air.

Table Reading

For each question, select the number that appears in the table at the given coordinates. Recall that the first number in the ordered pair gives the column number, and the second gives the row number. For instance, the ordered pair (2, -1) refers to the number in column 2, row -1, which is 76 in the first table.

Use the table on the right to answer questions 1-5.

1. (2, 2)
    a. 35
    b. 39
    c. 59
    d. 45
    e. 52

2. (3, -2)
    a. 96
    b. 34
    c. 49
    d. 67
    e. 23

3. (1, 1)
    a. 92
    b. 99
    c. 76
    d. 91
    e. 32

4. (0, -2)
    a. 60
    b. 55
    c. 67
    d. 49
    e. 68

5. (3, -3)
    a. 59
    b. 92
    c. 39
    d. 51
    e. 93

	-3	-2	-1	0	1	2	3
3	29	11	43	26	35	31	39
2	67	82	86	99	35	52	63
1	34	37	91	73	92	49	32
0	45	55	95	52	92	94	22
-1	51	96	39	96	79	76	34
-2	23	54	55	60	68	43	67
-3	49	59	21	68	97	82	93

Use the table on the right to answer questions 6-10.

6. (-3, 1)
    a. 55
    b. 68
    c. 23
    d. 70
    e. 52

7. (3, 2)
    a. 58
    b. 67
    c. 91
    d. 11
    e. 52

8. (-2, -3)
    a. 31
    b. 39
    c. 63
    d. 79
    e. 43

9. (2, -1)
    a. 53
    b. 96
    c. 32
    d. 31
    e. 35

10. (-1, 0)
    a. 55
    b. 63
    c. 56
    d. 31
    e. 95

	-3	-2	-1	0	1	2	3
3	68	11	16	37	26	78	77
2	25	97	79	97	79	19	58
1	70	38	15	22	37	63	90
0	29	99	56	80	96	39	44
-1	29	13	18	53	44	53	47
-2	90	13	86	31	12	11	26
-3	18	63	61	78	64	86	61

Use the table on the right to answer questions 11-15.

11. (1, 2)
    a. 32
    b. 73
    c. 67
    d. 55
    e. 25

12. (0, 2)
    a. 22
    b. 67
    c. 49
    d. 93
    e. 77

13. (-2, -2)
    a. 17
    b. 21
    c. 96
    d. 82
    e. 23

14. (0, 0)
    a. 68
    b. 96
    c. 73
    d. 72
    e. 34

15. (1, 3)
    a. 20
    b. 22
    c. 54
    d. 45
    e. 49

	-3	-2	-1	0	1	2	3
3	22	14	57	62	20	16	87
2	63	90	60	77	25	86	42
1	69	78	35	54	57	32	29
0	81	97	37	72	84	39	48
-1	59	95	74	61	52	40	44
-2	99	17	93	89	25	75	19
-3	10	16	59	26	84	10	39

Use the table on the right to answer questions 16-20.

16. (1, 1)
    a. 99
    b. 74
    c. 63
    d. 49
    e. 52

17. (0, 0)
    a. 68
    b. 45
    c. 66
    d. 35
    e. 93

18. (2, 2)
    a. 55
    b. 73
    c. 35
    d. 97
    e. 21

19. (0, 1)
    a. 86
    b. 54
    c. 42
    d. 35
    e. 94

20. (1, -1)
    a. 76
    b. 58
    c. 86
    d. 54
    e. 82

	-3	-2	-1	0	1	2	3
3	38	70	36	61	88	20	86
2	42	74	20	79	38	55	58
1	64	18	31	42	74	76	54
0	39	44	66	66	84	95	40
-1	23	73	98	14	58	86	80
-2	41	84	91	49	60	97	52
-3	85	11	51	50	14	21	99

Use the table on the right to answer questions 21-25.

21. (3, -3)
     a. 94
     b. 68
     c. 25
     d. 73
     e. 79

22. (-3, -3)
     a. 26
     b. 42
     c. 52
     d. 68
     e. 31

23. (2, -2)
     a. 31
     b. 68
     c. 35
     d. 95
     e. 39

24. (-2, 0)
     a. 72
     b. 34
     c. 68
     d. 22
     e. 11

25. (-2, -1)
     a. 82
     b. 39
     c. 49
     d. 99
     e. 55

	-3	-2	-1	0	1	2	3
3	64	53	66	11	83	67	87
2	31	28	87	79	21	73	19
1	27	11	20	82	22	52	68
0	97	72	73	16	59	37	13
-1	96	49	68	96	97	42	53
-2	74	48	49	35	66	39	36
-3	42	69	71	95	60	31	25

Use the table on the right to answer questions 26-30.

26. (0, 0)
    a. 35
    b. 76
    c. 11
    d. 86
    e. 29

27. (-1, -1)
    a. 51
    b. 96
    c. 97
    d. 60
    e. 20

28. (2, 0)
    a. 37
    b. 54
    c. 82
    d. 55
    e. 43

29. (0, -3)
    a. 96
    b. 26
    c. 40
    d. 39
    e. 54

30. (1, 0)
    a. 52
    b. 86
    c. 55
    d. 92
    e. 99

	-3	-2	-1	0	1	2	3
3	89	29	33	74	87	21	61
2	42	77	18	45	16	64	97
1	22	24	28	57	67	92	78
0	49	23	14	11	92	54	24
-1	96	10	20	40	54	66	68
-2	35	79	42	95	60	34	44
-3	91	85	67	40	60	44	58

Use the table on the right to answer questions 31-35.

31. (1, -3)
    a. 68
    b. 82
    c. 60
    d. 26
    e. 34

32. (3, 0)
    a. 94
    b. 93
    c. 21
    d. 49
    e. 92

33. (-3, 0)
    a. 11
    b. 35
    c. 20
    d. 34
    e. 49

34. (3, -1)
    a. 67
    b. 43
    c. 96
    d. 29
    e. 81

35. (1, -1)
    a. 67
    b. 79
    c. 55
    d. 23
    e. 35

	-3	-2	-1	0	1	2	3
3	97	75	86	89	22	31	43
2	62	30	48	35	34	82	78
1	59	25	97	99	57	89	54
0	20	88	57	70	67	71	21
-1	84	65	50	35	67	32	81
-2	19	21	74	30	84	75	69
-3	98	19	37	47	68	14	24

Use the table on the right to answer questions 36-40.

36. (1, -3)
    a. 43
    b. 95
    c. 22
    d. 37
    e. 91

37. (-3, 0)
    a. 97
    b. 37
    c. 43
    d. 99
    e. 35

38. (-2, -1)
    a. 59
    b. 51
    c. 92
    d. 26
    e. 23

39. (-1, 3)
    a. 96
    b. 76
    c. 87
    d. 54
    e. 34

40. (2, 2)
    a. 67
    b. 52
    c. 60
    d. 92
    e. 35

	-3	-2	-1	0	1	2	3
3	54	54	87	36	15	61	94
2	71	79	75	70	49	92	59
1	18	43	92	70	98	66	63
0	97	57	35	58	25	67	79
-1	34	59	18	62	87	44	31
-2	95	97	65	45	88	51	10
-3	38	39	88	95	22	19	49

Instrument Comprehension

1. Which of the answer choices represents the orientation of the plane?

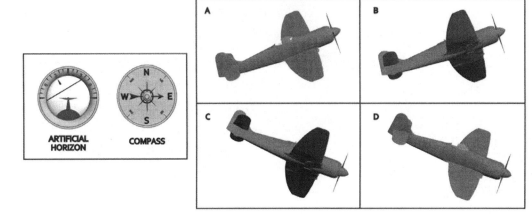

2. Which of the answer choices represents the orientation of the plane?

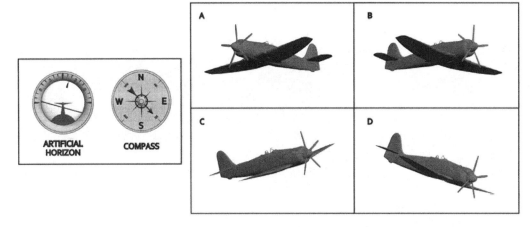

3. Which of the answer choices represents the orientation of the plane?

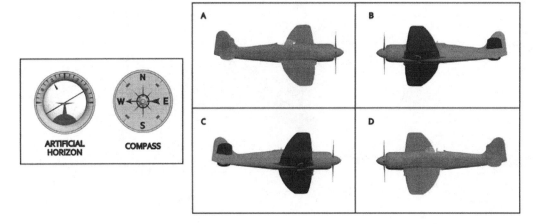

4. Which of the answer choices represents the orientation of the plane?

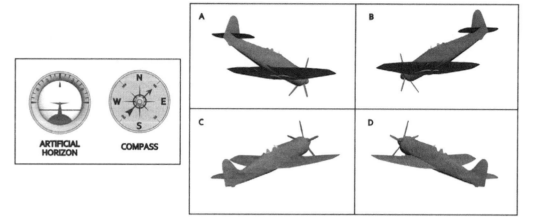

5. Which of the answer choices represents the orientation of the plane?

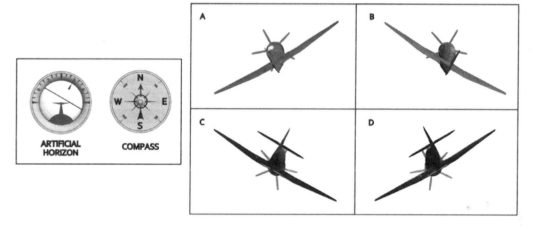

6. Which of the answer choices represents the orientation of the plane?

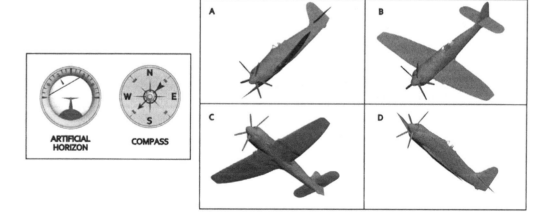

7. Which of the answer choices represents the orientation of the plane?

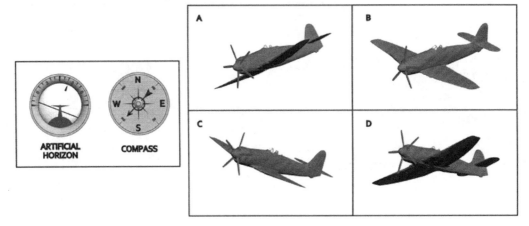

8. Which of the answer choices represents the orientation of the plane?

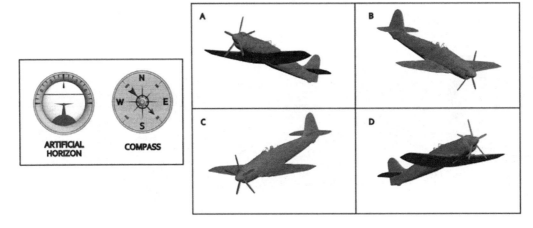

9. Which of the answer choices represents the orientation of the plane?

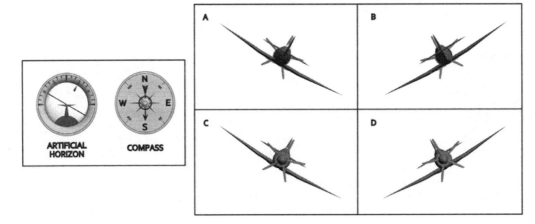

10. Which of the answer choices represents the orientation of the plane?

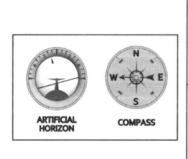
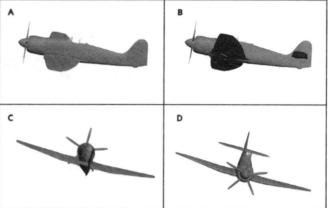

11. Which of the answer choices represents the orientation of the plane?

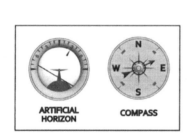

12. Which of the answer choices represents the orientation of the plane?

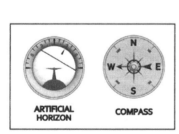

Block Counting

1. How many blocks are touching block 5 in the figure above?

2. How many blocks are touching block 1 in the figure above?

3. How many blocks are touching block 3 in the figure above?

4. How many blocks are touching block 6 in the figure above?

5. How many blocks are touching block 7 in the figure above?

6. How many blocks are touching block 9 in the figure above?

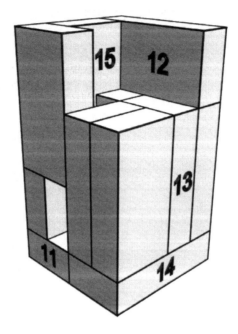

7. How many blocks are touching block 15 in the figure above?

8. How many blocks are touching block 11 in the figure above?

9. How many blocks are touching block 13 in the figure above?

10. How many blocks are touching block 16 in the figure above?

11. How many blocks are touching block 19 in the figure above?

12. How many blocks are touching block 20 in the figure above?

Aviation Information

1. From approximately how far away should a Visual Approach Slope Indicator be visible at night?
    a. Five miles
    b. Ten miles
    c. Twenty miles
    d. Thirty miles
    e. Fifty miles

2. Which of the following is NOT one of the forces a pilot must manage during flight?
    a. thrust
    b. gravity
    c. drag
    d. lift
    e. torque

3. Which control affects the angle of the main rotor blades of a helicopter?
    a. collective
    b. throttle
    c. cyclic
    d. directional control system
    e. None of the above

4. What is the flight attitude?
    a. the environment immediately around the plane
    b. the morale of the flight crew
    c. the position of the plane in motion
    d. the inclination of the elevators
    e. the positions of the ailerons

5. The curvature of an airfoil is known as the
    a. bank.
    b. position.
    c. camber.
    d. angle.
    e. attitude.

6. Which of the following is considered one of the primary flight controls?
    a. elevators
    b. leading edge devices
    c. flaps
    d. spoilers
    e. trim tabs

7. What is the term in aviation for movement around the plane's longitudinal axis?
    a. stalling
    b. rolling
    c. pitching
    d. leaning
    e. yawing

8. What is the primary determinant of air pressure in the flight envelope?
    a. the altitude at which the plane is flying
    b. the camber of the wings
    c. the amount of lift that the airfoils can create
    d. the pitch of the elevators
    e. the humidity of the air

9. A plane is said to have conventional landing gear when the third wheel is
    a. aligned with the second wheel.
    b. under the tail.
    c. directly behind the first wheel.
    d. under the nose.
    e. underneath the cockpit.

10. Which of the following is NOT part of the empennage?
    a. rudder
    b. trim tab
    c. elevator
    d. aileron
    e. horizontal stabilizer

11. A glider would be most likely to have a
    a. delta wing.
    b. triangular wing.
    c. forward swept wing.
    d. backward swept wing.
    e. straight wing.

12. Which of the following is NOT considered part of a plane's basic weight?
    a. crew
    b. fuel
    c. external equipment
    d. internal equipment
    e. fuselage

13. The support structure that runs the length of the fuselage in a monocoque plane is called a
    a. counter.
    b. former.
    c. truss.
    d. stringer.
    e. bulkhead.

14. Which control is used to manipulate the elevators on an airplane?
    a. throttle
    b. rudder
    c. joystick
    d. pedals
    e. collective

15. Which maneuver is appropriate when a pilot needs to descend quickly onto a shorter-than-normal runway?
    a. descent at minimum safe airspeed
    b. idle
    c. partial power descent
    d. glide
    e. stall

16. In aviation, which of the following is NOT one of the possible functions of a spoiler?
    a. diminishing lift
    b. raising the nose
    c. increasing drag
    d. reducing adverse yaw
    e. enabling descent without speed reduction

17. The vertical axis of a plane extends upward through the plane's
    a. cockpit.
    b. tail.
    c. landing gear.
    d. center of mass.
    e. geometric center.

18. When is trimming necessary?
    a. when the plane is ascending or descending
    b. after the elevators have been deflected upwards
    c. after the ailerons have been adjusted
    d. after the elevators have been deflected downwards
    e. after any change in the flight condition

19. How far away is an approaching plane when the Runway Centerline Lighting System lights become solid red?
    a. Five hundred feet
    b. One thousand feet
    c. Three thousand feet
    d. Five thousand feet
    e. One mile

20. What is the Coriolis force?
	a. the extra lift generated by a helicopter once it has exited its own downwash
	b. the force that spins the rotors of a helicopter even when there is no power from the engine
	c. the greater downwash at the rear half of the rotor disc, as compared with the front
	d. the phenomenon in which the effects of a force applied to a spinning disc occur ninety degrees later
	e. the change in rotational speed caused by the shift of the weight towards or away from the center of the spinning object

Answer Key

Verbal Analogies

1. B: Rash. Chastise and reprimand are synonyms. The answer choice synonym for impetuous is rash.

2. C: Femur. The humerus is a bone in the arm; the femur is a bone in the leg.

3. A: Quotient. A quotient is the result of division as a product is the result of multiplication.

4. C: Asparagus. As a sweater is something one wears, asparagus is something one eats.

5. A: Famished. As a person who is impecunious needs money, so a person who is famished needs food.

6. A: Protest. Denigrate and malign are synonyms. The answer choice synonym for demur is protest.

7. D: Generous. Obeisance and deference are synonyms. The answer choice synonym for munificent is generous.

8. D: Sow. A female goat is a nanny and a female pig is a sow.

9. B: Paucity. Cache and reserve are synonyms. The answer choice synonym for dearth is paucity.

10. E: Refuge. Arable and farmable are synonyms. The answer choice synonym for asylum is refuge.

11. A: Peripatetic. Myriad and few are antonyms. The answer choice antonym for stationary is peripatetic.

12. A: Flagon. As a mansion is a large house, so a flagon is a large bottle.

13. B: Synonyms. As a dictionary is a collection of definitions, so a thesaurus is a collection of synonyms.

14. B: Stubborn. Abstruse and esoteric are synonyms. The answer choice synonym for adamant is stubborn.

15. C: Herd. A group of bees is called a hive and a group of cattle is called a herd.

16. E: Meridian. A parallel is a line of latitude, while a meridian is a line of longitude.

17. B: Provocation. Prevention and deterrence are synonyms. The answer choice synonym for incitement is provocation.

18. A: Gauge. Value and worth are synonyms. The answer choice synonym for measure is gauge.

19. A: Oppose. Enervate and energize are antonyms. The answer choice antonym for espouse is oppose.

20. D: The analogy is tool to worker. A carpenter uses a hammer just as a doctor uses a stethoscope.

21. B: The analogy describes characteristic location. An armoire is usually kept in a bedroom, just as a desk is usually kept in an office.

22. C: This analogy describes a food source to animal relationship. Just as plankton is a food source for whales, so is bamboo a food source for pandas.

23. C: The analogy is geographic location. Just as the tundra is located in the Arctic regions, so are savannas located in tropic regions.

24. A: This is an analogy of relative degree. *Parched* is a more intense degree of *thirsty*, just as *famished* is a more intense degree of *hungry*.

25. E: This is an analogy based on antonyms. *Felicity*, or happiness, is the opposite of *sadness*, just as *ignominy*, or disgrace, is the opposite of *honor*.

Arithmetic Reasoning

1. B: First, find the total before taxes: $7.50 + $3.00 = $10.50
Then, calculate 6% of the total: $10.50 * .06 = $0.63
Finally, add the tax to find the total cost: $10.50 + $0.63 = $11.13

2. D: There are 7 days in a week. Knowing that the chef can make 25 pastries in a day, the weekly number can be calculated:
25 * 7 = 175

3. B: The woman has four days to earn $250. To find the amount she must earn each day, divide the amount she must earn ($250) by 4:
$250 / 4 = $62.50

4. A: To find the number of cars remaining, subtract the number of cars that were sold from the original number: 476 – 36 = 440

5. B: Calculate 0.5% of $450: $450 * 0.005 = $2.25
This is the amount of interest she will earn.

6. C: First, figure out how much the second child contributed: $24.00 - $15.00 = $9.00
Then, calculate how much the first two children contributed in total: 24 + 9 = $33.00
Finally, figure out how much the third child will have to contribute:
$78.00 - $33.00 = $45.00

7. C: First, figure out how many points the first woman will earn: 3 * 5 = 15
Then, figure out how many points the second woman will earn: 6 * 5 = 30
Then, add these two values together: 30 + 15 = 45 points total.

8. E: First, calculate 13% of 540 = 70
Then, add this value onto the original number of workers: 540 + 70 = 610
610 is the number of people that the company will employ after the expansion.

9. D: To find the number of apartments on each floor, divide the total number of apartments by the number of floors:
65 / 13 = 5

10. A: First, find the total number of pens: 5 * 3 = 15
Then, find the total number of pencils: 3 * 7 = 21
Finally, express it as a ratio 15 : 21

11. C: To calculate his new salary, add his raise to his original salary:
$15.23 + $2.34 = $17.57

12. A: To find the total number of passengers, multiply the number of planes by the number of passengers each can hold: 6 * 300 = 1800

13. B: Currently, there are two men for every woman. If the number of women is doubled (1 * 2 = 2), then the new ratio is 2:2. This is equivalent to 1:1.

14. C: First, calculate 3% of 250 pounds: 250 * 0.03 = 7.5 pounds
Calculate how much she weighs at the end of the first week: 250 − 7.5 = 242.5 pounds
Calculate 2% of 242.5: 242.5 * 0.02 = 4.85 pounds
Add the two values together to get the total: 7.5 + 4.85 = 12.35

15. D: Divide the total distance she must travel (583km) by the number of kilometers she drives each hour (78km) to figure out how many hours it will take to reach her destination:
583 km / 78 km = 7.47 hours

16. D: One gallon of paint can paint three rooms, so to find out how many 28 gallons can do, that number must be multiplied by 3: 28 * 3 = 84 rooms

17. A: Each earns $135, so to find the total earned, that amount must be multiplied by the number of workers: $135 * 5 = $675

18. C: First, calculate her score on the second test: 99 − 15 = 84
Then, calculate her score on the third test: 84 + 5 = 89

19. A: To find out how much he has remaining, both numbers must be subtracted from the original amount ($50.00): $50.00 - $15.64 - $7.12 = $27.24

20. E: Divide the number of students (600) by the number of classrooms they will share (20):
600 / 20 = 30

21. C: To calculate this value, divide the number of dogs (48) by the number of workers that are available to care for them (4):
48 / 4 = 12

22. C: First, calculate the length of the second office: 20 + 6 = 26 feet
Then, add both values together to get a combined length: 26 + 20 = 46 feet

23. C: Find the total cost of the items: $6.66 + $159.23 = $165.89
Then, calculate how much each individual will owe: $165.89 / 4 = $41.47

24. A: To answer this question, simply calculate half of 140 acres: 140 / 2 = 70 acres

25. C: First, calculate how many he has after selling 45: 360 − 45 = 315
Then, calculate how many he has after buying 85: 315 + 85 = 400

Word Knowledge

1. A: Spoiled has a number of meanings, and one of them is ruined. If you said somebody spoiled your fun, it would convey the same meaning as saying somebody ruined your fun.

2. B: An oath is a promise. For example, if you make an oath to keep a secret, you are promising to keep that secret.

3. B: When you inquire about something, you are asking about it or requesting more information. For example, if you told somebody you inquired about a job, it would mean you asked about it.

4. C: If you say that you comprehend something, it is the same as saying you understand it. For example, saying you comprehend what another person is saying is the same as saying you understand them.

5. A: To say that something is apparent implies that it is clear or obvious. For example, saying that it is apparent that somebody wants a job is the same as saying it is clear they want the job.

6. C: Silent or silence indicates quiet and calm. To enjoy the silence of the night is to enjoy the complete quiet of the night.

7. E: Absolutely, when used to describe a feeling or state of mind, means completely or totally. For example, saying you are absolutely certain that you made the right decision or saying you are completely certain you made the right decision conveys the same meaning.

8. D: Something that has been modified has been changed. Saying you modified your plans or saying that you changed them conveys the same meaning.

9. A: Something that is delicate can also be described as fragile. Saying that a crystal figurine is delicate or saying it is fragile conveys the same meaning.

10. B: Festivities are often commonly known as celebrations. Attending festivities implies that you are attending a celebration or party.

11. B: To say that someone is exhausted or to say that they are tired conveys a similar meaning. Usually, exhausted is a word used to describe extreme tiredness.

12. B: To cleanse something is to clean or wash it. Saying you cleansed your face or clothes is the same as saying you washed them.

13. A: To battle something is to fight it. To say that two armies battled each other and to say they fought each other conveys the same meaning.

14. C: To wander is to roam. To say someone wandered around a mall is to say they roamed or walked around aimlessly, without a specific goal or destination in mind.

15. E: Something that is done abruptly is done suddenly and without warning. For example, saying the car stopped abruptly and saying it stopped suddenly conveys the same meaning.

16. A: Somebody who has been tricked has been conned. To trick somebody is to con them, which implies that dishonest methods are used to convince another to do something they wouldn't normally do.

17. C: When used as an adjective extremely has the same meaning as very. Saying somebody is extremely happy and saying they are very happy conveys the same meaning.

18. A: To have doubts is to have uncertainties or hesitations. To say that someone is doubtful about something means that they are uncertain.

19. D: To describe something as peculiar is to say it is strange or out of the ordinary. For example, saying you are in a strange situation or saying you are in a peculiar situation conveys the same meaning.

20. B: Describing somebody as courteous implies that they are polite and well-mannered. Polite and courteous both convey the same meaning.

21. C: When somebody says they are troubled by something, it means that they are bothered by it.

22. A: Perspiration is another word for sweat. Saying somebody is perspiring is the same as saying they are sweating.

23. B: Tremble is another word for shake. To say somebody or something trembled means that it shook or shuddered.

24. A: Adhered is often used as another word for stuck. For example, to say a piece of tape adhered to the wall conveys the same meaning as saying the piece of tape stuck to the wall.

25. D: When something is described as tidy, it usually means that it is neat and that things are in their proper place. Saying a house is tidy and saying it is neat conveys the same meaning.

Math Knowledge

1. B: The perimeter of a figure is the sum of all of its sides. Since a rectangle's width and length will be the same on opposite sides, the perimeter of a rectangle can be calculated by using the following formula: perimeter = 2(width) + 2(length)
Using the numbers given in the question:
perimeter = 2(7cm) + 2(9cm)
perimeter = 14cm + 18cm
perimeter = 32cm

2. D: First, gather the like terms on opposite sides of the equation to make it easier to solve:
-3q – 4q ≥ -30 – 12
-7q ≥ -42
Then, divide both sides by -7 to solve for q:
-7q/-7 ≥ -42/-7
q ≥ 6
Finally, when both sides are divided by a negative number, the direction of the sign must be reversed:
q ≤ 6

3. C: To solve for x, it is necessary to add 6 to both sides to isolate the variable:
x – 6 + 6 = 0 + 6
x = 6

4. A: To calculate the value of this expression, substitute -3 for x each time it appears in the expression: $3(-3)^3 + (3(-3)+ 4) – 2(-3)^2$
According to the order of operations, any operations inside of brackets must be done first:
$3(-3)^3 + (-9+ 4) – 2(-3)^2$
$3(-3)^3 + -5 – 2(-3)^2$
Then, the value of the expression can be calculated:
3(-27) + -5 – 2(9)
-81 + -5 – 18
-104

5. C: First, combine like terms to make the equation easier to solve:
3x + 2x = 45 + 30
5x = 75
Then, divide both sides by 5 to solve for x:
5x/5 = 75/5
x = 15

6. A: First, add 25 to both sides to isolate x:
1/4x – 25 + 25 ≥ 75 + 25
1/4x ≥ 100
Then, multiply both sides by 4 to solve for x:
1/4x * 4 ≥ 100 * 4
x ≥ 400

7. A: First, add 5 to both sides to isolate x:
$x^2 - 5 + 5 = 20 + 5$
$x^2 = 25$
Then, take the square root of both sides to solve for x
$\sqrt{x^2} = \sqrt{25}$
x = 5

8. B: First, we must calculate the length of one side of the square. Since we know the perimeter is 8cm, and that a square has 4 equal sides, the length of each side can be calculated by dividing the perimeter (8cm) by 4: 8cm / 4 = 2cm
The formula for the area of a square is length^2
Therefore, to calculate the area of this square: 2cm^2 or 2cm * 2cm
Area = 4cm^2

9. D: To find the value of this expression, substitute the given values for x and y into the expression:
3(4)(2) – 12(2) + 5(4)
Then, calculate the value of the expression:
3*8 – 12*2 + 5*4
24 – 24 + 20
20

10. D: First, subtract 10 from both sides to isolate x:
0.65x + 10 – 10 = 15 – 10
0.65x = 5
Then, divide both sides by 0.65 to solve for x:
0.65x/0.65 = 5/0.65
x = 7.69

11. B: Use the FOIL method (first, outside, inside, and last) to get rid of the brackets:
12x^2 -18x + 20x -30
Then, combine like terms to simplify the expression:
12x^2 -18x + 20x -30
12x^2 + 2x -30

12. B: To simplify this expression, it is necessary to follow the law of exponents that states: $x^n/x^m = x^{n-m}$
First, the 50 can be divided by 5: 50/5 = 10
Then, it is simply a matter of using the law of exponents described above to simplify the expression:
$10x^{18-5}t^{6-2}w^{3-2}z^{20-19}$
$10x^{13}t^4wz$

13. E: To calculate the value of this permutation, it is necessary to multiply each number between one and 4: 1 * 2 * 3 * 4 = 24

14. D: Because it is a cube, it is known that the width and the height of the cube is also 5cm. Therefore, to find the volume of the cube, we must cube 5cm: 5cm^3
This is the same as: 5 * 5 * 5 = 125
The volume of the cube is 125cm^3.

15. A: First, factor this equation to make solving for x easier:
(x − 6) (x − 7) = 0
Then, solve for both values of x:
1) x − 6 = 0
x = 6
2) x − 7 = 0
x = 7

16. C: The area of a triangle can be calculated by using the following formula: A = 1/2b*h
Therefore, by using the values given in the question:
A = 1/2(12cm) * 12cm
A = 6cm * 12cm
A = 72cm²

17. D: To simplify this expression, it is necessary to observe the law of exponents that states:
$x^n * x^m = x^{n+m}$
Therefore: $3*7x^{7+2} + 2*9y^{12+3}$
$21x^9 + 18y^{15}$

18. C: First, subtract 27 from both sides to isolate x:
x/3 + 27 − 27 = 30 − 27
x/3 = 3
Then, both sides must be multiplied by 3 to solve for x:
3(x/3) = 3 * 3
x = 9

19. B: To find the slope of a line, it is necessary to calculate the change in y and the change in x:
Change in y: 1 − 8 = -7
Change in x: 4 − (-13) = 17
The slope of a line is expressed as change in y over change in x: -7/17

20. A: To solve for x, it is necessary to calculate the value of 20% of 200:
200 * 0.20 = 40
Therefore, x = 40

21. B: First, calculate the total number of balloons in the bag: 47 + 5 + 10 = 62
Ten of these are black, so divide this number by 62, then multiply by 100 to express the probability as a percentage:
10 / 62 = 0.16
0.16 * 100 = 16%

22. B: First, it is easier to find out how many tickets are sold for one winner.
If there are 2 winners for every 100 tickets, there is 1 winner for every 50 tickets.
If ten tickets are bought, the chances of winning are 10 in 50.
This can also be expressed as 1 in 5.

23. E: To find the volume of a rectangular prism, the formula is length * width * height.
Therefore, for this rectangular prism, volume = 10cm * 5cm * 6cm
The volume of this rectangular prism is 300cm³

24. C: To calculate the midpoint of a line, find the sum of the points and divide by two.
For x, the midpoint can be calculated as follows: 6 + 10 = 16; 16/2 = 8
For y, the midpoint can be calculated as follows: 40 + 20 = 60; 60/2 = 30
Therefore, the midpoint is (8, 30)

25. A: First, subtract 60 from both sides to isolate x:
5x + 60 – 60 = 75 – 60
5x = 15
Then, divide both sides by 5 to solve for x:
5x/5 = 15/5
x = 3

Reading Comprehension

1. A. Explaining the qualities of air that may affect flight is the primary purpose of the passage.

2. C. The best definition for *inversely* as it is used in the second paragraph is *in the opposite direction*. The author is indicating that the density of air decreases as the temperature rises, and increases as the temperature falls.

3. D. A pilot can expect the air density to decrease as the plane gains altitude. At the end of the second paragraph, the author states that a gain in altitude will usually lead to a decrease in air density, no matter what changes there may be in the temperature.

4. B. The author would most likely agree that air density is more important than relative humidity. Though the passage states that humidity can have an effect on aircraft performance, the author admits that it is not considered an essential factor. Indeed, humidity is just one of three factors (along with pressure and temperature) that affect air density.

5. C. The most likely reason why there is no chart for assessing the effects of humidity on density altitude is that humidity does not affect flight performance very much. The author mentions several times that flight performance is not significantly affected by humidity, and so it is seems that a special chart for this purpose would be unnecessary.

6. D. *Influences on Climb Performance* would be the best title for this passage. The passage surveys the various factors that affect climb performance and the choices made by pilots as they gain altitude, both during take-off and while the aircraft is already in flight.

7. E. The best definition for *pronounced* as it is used in the second paragraph is *noticeable*. The author is stating that the weight of an aircraft has a significant impact on aircraft performance. In this sentence *pronounced* is being used as an adjective, but it can also be used as a verb, meaning *spoke* or *said*.

8. C. An increase in weight means that the angle of attack must be higher in order to maintain altitude. Greater weight diminishes reserve power, lowers the climb rate, diminishes the maximum rate of climb, and increases drag.

9. B. The author would most likely agree that at the end of a long journey a plane will have a higher maximum rate of climb. The maximum rate of climb of a plane increases as the weight decreases, and the weight of a plane will decrease as it burns off fuel over the course of a long journey.

10. A. A helicopter that weighs two tons and has rotor blades that cover five hundred square feet would have a disc loading measure of four pounds per square foot. Pounds per square foot is the standard measure for disc loading. There are two thousand pounds in a ton. Disc loading is calculated by dividing the weight of the helicopter by the area covered by the rotor blades.

11. B. Discussing the interrelationships of airspeed, power, and pitch attitude is the primary purpose of the passage. Answer choices *C* and *D* are partially correct, but they leave out large sections of the passage, and therefore do not comprehensively describe the purpose or the content of the passage.

12. A. *Inadvertently*, as it is used in the fifth paragraph, most nearly means *unintentionally*. The author is indicating that a pilot can enter the region of reversed command without meaning to, if he or she attempts to climb out of ground effect without first attaining normal climb pitch attitude and airspeed.

13. B. Most flight occurs in the region of normal command. In the region of normal command, increasing power increases the airspeed, and decreasing power decreases the airspeed. This information is given in the last sentence of the second paragraph.

14. E. The author would most likely agree that as the speed of flight decreases, the power required to maintain altitude increases. This inverse relationship between required power and airspeed is expressed in the last sentence of the first paragraph.

15. D. The best title for this passage would be *Power Requirements During Flight*. The passage discusses how the need for and effects of changes in power are influenced by factors such as airspeed, pitch attitude, and altitude.

16. E. The primary purpose of the passage is to describe the factors that influence landing distance. The passage begins by discussing the minimum and normal landing distances, and goes on to cover influences on landing distance, as for instance gross weight, wind, and density altitude.

17. B. When making a normal landing, a pilot will rely on aerodynamic drag in order to avoid wearing down the brakes and tires. This point is made several times in the third paragraph of the passage.

18. A. The best definition for *principal* as it is used in the fourth paragraph is *most important*. The author is trying to make the point that gross weight has an enormous effect on landing distance.

19. C. A heavier plane must be landed at a higher airspeed to avoid hitting the runway with too much force. This idea is explored in the fourth paragraph. In order to generate an amount of lift sufficient for a smooth landing, a higher airspeed must be maintained.

20. C. The author would most likely agree that gross weight and minimum landing distance are positively correlated. The information in the passage makes it clear that as gross weight rises, minimum landing distance increases.

21. D. The primary purpose of the article is to discuss aeronautical decision making, or ADM. The article does give some examples of decision-making strategies, but these are given in the context of a description of ADM, not as the main body of the article.

22. D. The passage explains that aviation safety is distinct from other areas of safety because there is a much smaller margin for error. In other words, even small accidents in aviation can be catastrophic, because of the risks inherent in flight.

23. A. The closest definition for *conjunction* as it is used in the second paragraph is *combination*. The author is stating that the FAA manuals worked well when combined with the usual flight training.

24. A. The author would most likely agree that the body of knowledge about ADM is increasing, and this will have a positive effect on flight safety. The article details the efforts to improve ADM, and suggests that these have already improved flight safety a great deal.

25. E. For a pilot, reading the account of a recent aviation accident is an opportunity for an indirect learning experience. The passage distinguishes between direct learning experiences, which are events in one's own life, and indirect learning experiences, which are things that happen to others.

Situational Judgment

1. C (Most effective)
2. A (Least effective)
3. A (Most effective)
4. C (Least effective)
5. E (Most effective)
6. B (Least effective)
7. D (Most effective)
8. A (Least effective)
9. E (Most effective)
10. C (Least effective)
11. B (Most effective)
12. C (Least effective)
13. A (Most effective)
14. B (Least effective)
15. E (Most effective)
16. B (Least effective)
17. C (Most effective)
18. A (Least effective)
19. A (Most effective)
20. B (Least effective)
21. B (Most effective)
22. D (Least effective)
23. A (Most effective)
24. C (Least effective)
25. B (Most effective)
26. E (Least effective)
27. D (Most effective)
28. A (Least effective)
29. B (Most effective)
30. C (Least effective)
31. E (Most effective)
32. B (Least effective)
33. A (Most effective)
34. D (Least effective)
35. D (Most effective)
36. C (Least effective)
37. A (Most effective)
38. D (Least effective)
39. E (Most effective)
40. A (Least effective)
41. B (Most effective)
42. C (Least effective)
43. E (Most effective)
44. A or C (Least effective)
45. A (Most effective)
46. D (Least effective)
47. A (Most effective)
48. B (Least effective)
49. B (Most effective)
50. D (Least effective)

Physical Science

1. B: A long nail or other type of metal, substance or matter that is heated at one end and then the other end becomes equally hot is an example of conduction. Conduction is energy transfer by neighboring molecules from an area of hotter temperature to cooler temperature.

2. E: The measure of energy within a system is called heat.

3. B: They have a different number of neutrons. The distinguishing feature of an isotope is its number of neutrons. Two different isotopes of the same element will have the same number of protons but different numbers of neutrons.

4. A: Fission is a nuclear process where atomic nuclei split apart to form smaller nuclei. Nuclear fission can release large amounts of energy, emit gamma rays and form daughter products. It is used in nuclear power plants and bombs.

5. D: The process whereby a radioactive element releases energy slowly over a long period of time to lower its energy and become more stable is best described as decay. The nucleus undergoing decay spontaneously releases energy, most commonly through the emission of an alpha particle, a beta particle or a gamma ray.

6. B: Light within a single medium travels in a straight line. When it changes to a different medium, however, the light rays bend according to the refractive index of each substance. Light coming from the submerged portion of the pencil is refracted as it passes through the air-water barrier, giving the perception of a bent pencil.

7. A: Hertz (Hz) is a unit of measure used for frequency, often described as 1 cycle/second. In the context of wave motion, it is the number of complete waves that pass a given point in one second.

8. B: A cyclist coasting up a hill is trading his speed for increased altitude. This is an example of kinetic energy being converted to potential energy. The other options are examples of potential energy being converted to kinetic, kinetic energy being dissipated, and conservation of kinetic energy.

9. E: Phase changes such as boiling, melting, and freezing are physical changes. No chemical reaction takes place when water is boiled.

10. A: The center of an atom is known as the nucleus. It is composed of protons and neutrons.

11. A: Sublimation is the process of a solid changing directly into a gas without entering the liquid phase.

12. D: Mendeleev was able to connect the trends of the different elements behaviors and develop a table that showed the periodicity of the elements and their relationship to each other.

13. A: Density is mass per unit volume, typically expressed in units such as g/cm^3, or kg/m^3.

14. C: The closer the data points are to each other, the more precise the data. This does not mean the data is accurate, but rather that the results are reproducible.

15. E: Current is measured in units of amperes or amps.

16. C: In order for a solar eclipse to occur, the moon must come directly between the earth and the sun, blocking the sun's light from the earth.

17. B: Inertia is the tendency of objects that are in motion to continue moving in the same direction. The turning car initiates a change in direction, but the passengers' mass wants to continue going straight, causing them to feel a pull in that direction. This phenomenon is sometimes referred to as centrifugal force.

18. B: According to the ideal gas law, when volume is held constant, the temperature of a gas is directly proportional to the pressure of the gas. Thus, when the temperature is increased, the pressure will also increase.

19. B: The amplitude of a sound wave is what determines how loud the sound is perceived by the ear.

20. D: In all of the other examples, there is a person applying work to the book, either directly, by being picked up, pushed, or thrown, or indirectly, by being carried in a backpack. In the example of a book being released so that it falls, the only work being applied to the book is being done by gravity.

Table Reading

1. E
2. D
3. A
4. A
5. E
6. D
7. A
8. C
9. A
10. C
11. E
12. E
13. A
14. D
15. A
16. B
17. C
18. A
19. C
20. B
21. C
22. B
23. E
24. A
25. C
26. C
27. E
28. B
29. C
30. D
31. A
32. C
33. C
34. E
35. A
36. C
37. A
38. A
39. C
40. D

Instrument Comprehension

1. D

2. B

3. B

4. C

5. D

6. A

7. C

8. B

9. C

10. B

11. C

12. C

Block Counting

1. 6: 1 on the front, 2 on the back, 2 on the top, 1 on the bottom.

2. 3: 2 on the right, 1 on the top.

3. 6: 1 on the front, 3 on the left, 2 on the top.

4. 5: 1 on the back, 1 on the right, 3 on the top.

5. 5: 3 on the front, 2 on the top.

6. 9: 1 on the front, 3 on the back, 3 on the top, 2 on the bottom.

7. 4: 1 on the front, 1 on the back, 1 on the right, 1 on the bottom.

8. 3: 1 on the right, 2 on the top.

9. 5: 2 on the front, 1 on the back, 1 on the left, 1 on the bottom.

10. 9: 4 on the front, 4 on the back, 1 on the top.

11. 6: 2 on the back, 3 on the left, 1 on the bottom.

12. 4: 1 on the right, 3 on the top.

Aviation Information

1. C: A Visual Approach Slope Indicator should be visible from approximately twenty miles away at night. A Visual Approach Slope Indicator (VASI) is a common feature at large airports. This system helps guide the approaching pilot to the runway. The pilot will see white lights at the lower border of the glide path, and red lights at the upper border. In normal conditions, the lights of the VASI system should be visible for three to five miles during the day, and for twenty miles at night. If the VASI system is working properly, the plane will be safe so long as it stays within ten degrees of the extended runway centerline and four nautical miles of the runway threshold.

2. E: Torque is not one of the forces a pilot must manage during flight. The four forces a pilot must manage are lift, gravity, thrust, and drag. Lift pushes the plane up, gravity pulls the plane down, thrust propels the plane forward, and drag holds the plane back. The overall admixture of these forces as they operate on the plane is called the flight envelope.

3. A: The collective affects the angle of the main rotor blades of a helicopter. It is a long tube that extends from the floor of the cockpit. In most helicopters, it is situated on the pilot's left. The collective has two parts: a handle that can be raised or lowered, to control the pitch of the blades; and a throttle, to control the torque of the engine. The handle is the part that affects the angle of the main rotor blades. When it is raised, the leading edge of the blade is raised higher than the trailing edge.

4. C: The flight attitude is the position of a plane in motion. The flight attitude is described in terms of its position with respect to three axes: vertical, lateral, and longitudinal. The vertical axis runs up through the plane's center of gravity. A plane's position with respect to this axis is known as its yaw. The lateral axis of a plane runs from wingtip to wingtip, and the motion of the plane around this axis is known as pitch. The longitudinal axis, finally, is an imaginary line extending from the nose of the plane to its tail. The position of the plane in relation to the longitudinal axis is called roll. The attitude of the plane is controlled with the joystick, rudder pedals, and throttle.

5. C: The curvature of an airfoil is known as the camber. An airfoil (wing) is considered to have a high camber if it is very curved. A related piece of wing terminology is the mean camber line, which runs along the inside of the wing, such that the upper and lower wings are equal in thickness.

6. A: The elevators are considered one of the primary flight controls. The elevators are responsible for the plane's pitch, or movement around the lateral axis. The other primary flight controls are the ailerons and the rudder. The ailerons control the roll, or movement around the longitudinal axis, while the rudder controls the yaw, or movement around the vertical axis. The secondary flight controls are the flaps, spoilers, leading edge devices, and trim systems.

7. B: In aviation, the term for movement around the plane's longitudinal axis is rolling. The longitudinal axis runs from the nose of the plane to its tail. For the most part, the plane's roll is controlled by the joystick. By moving the stick to the right or left, the pilot dips the wings.

8. A: The altitude at which the plane is flying is the primary determinant of air pressure in the flight envelope. The most important elements of the flight envelope are the temperature, air pressure, and humidity. The conditions in the atmosphere have a great deal of influence over the amount of lift created by the airfoils. Greater air pressure is the same as greater air density. A plane will generate greater lift when it is in cool air, because cool air is less dense than warm air.

9. B: A plane is said to have conventional landing gear when the third wheel is under the tail. Typically, a plane's landing gear will consist of three wheels or sets of wheels. Two of these are under either wing or on opposing sides of the fuselage. In the conventional arrangement, the third wheel or wheel set is under the tail, while in the tricycle arrangement it is under the nose. This third wheel can rotate, which will make it possible for the plane to turn while moving on the ground.

10. D: An aileron is not part of the empennage. The empennage, otherwise known as the tail assembly, includes the elevators, vertical and horizontal stabilizers, rudders, and trim tabs. A fixed wing aircraft will typically have both vertical and horizontal stabilizers, which are immobile surfaces that extend from the back of the fuselage. The horizontal stabilizers have mobile surfaces along their trailing edges; these surfaces are called the elevators. The elevators deflect up and down to raise or lower the nose of the plane. The rudder is a single, large flap connected by a hinge to the vertical stabilizer. The rudders back-and-forth motion controls the motion of the plane with respect to its vertical axis. The trim tabs, finally, are connected to the trailing edges of one or more of the primary flight controls (i.e., ailerons, elevators, rudder).

11. E: A glider would be most likely to have a straight wing. A straight wing may be tapered, elliptical, or rectangular. This planform (wing shape) is common in aircraft that move at extremely low speed. A straight wing is often found on sailplanes and gliders. The swept wing, on the other hand, is appropriate for high-speed aircraft. The wing may be swept forward or back. This will make the plane unstable at low speeds, but will produce much less drag. A swept wing requires high-speed takeoff and landing. The delta wing, which is also known as the triangular wing, has a straight trailing edge and a high angle of sweep. This allows the plane to take off and land at high speeds.

12. A: The crew is not considered part of a plane's basic weight. The basic weight is the aircraft plus whatever internal or external equipment will remain on the plane during its journey. The crew is not included in the basic weight, though it is a part of the operating weight (basic weight plus crew), gross weight (total weight of the aircraft at any particular time), landing gross weight (weight of the plane and its contents upon touchdown), and zero fuel weight (weight of the airplane when it has no usable fuel).

13. D: The support structure that runs the length of the fuselage in a monocoque plane is called a stringer. In a monocoque plane, the fuselage is supported by stringers, formers, and bulkheads. Stringers and formers are generally made out of the same material, though they run perpendicular to one another (that is, formers run in circles around the width of the fuselage). Bulkheads are the walls that divide the sections of the fuselage. The other style of fuselage, known as a truss, is composed of triangular groupings of aluminum or steel tubing.

14. C: The joystick is used to manipulate the elevators. When the stick is pulled back, the elevators deflect upwards. This decreases the camber of the horizontal tail surface, which moves the nose up and pushes the tail down. Pushing the joystick forward, on the other hand, pushes the elevators down, which creates an upward force on the tail by increasing the camber of the horizontal tail surface.

15. A: When a pilot needs to descend quickly onto a shorter-than-normal runway, he or she will descend at the minimum safe airspeed. A descent at the minimum safe airspeed is achieved by slightly lifting the nose and moving the plane into the landing configuration. During such a descent, the plane should not exceed 1.3 times the stall speed. This technique is appropriate for landing

quickly on a short runway because the rate of descent is much faster. However, if the rate of descent should become too great, the pilot should be ready to increase power.

16. B: Raising the nose is not one of the possible functions of a spoiler. Spoilers can diminish lift and reduce drag, which enables the plane to descend without reducing its speed. However, spoilers also control the plane's roll, partly by reducing any adverse yaw. A pilot uses the spoiler in this way by raising the spoiler on the side of the turn. That side will thereby have less lift and more drag, making it drop. The plane will then bank and yaw in the intended direction. When the pilot raises both of the spoilers at the same time, the plane will descend without losing any speed. Another incidental benefit of spoilers is improved brake performance, which occurs because the plane has lift and is pushed down towards the ground.

17. D: The vertical axis of a plane extends up through the plane's center of gravity. Movement around this axis is called yawing. The position of a plane is also described with respect to the lateral and longitudinal axes. The lateral axis extends from wingtip to wingtip. The longitudinal axis runs from the nose of the plane to its tail.

18. E: Trimming is necessary after any change in the flight condition. Trimming is the adjustment of the trim tabs, which are small flaps that extend from the trailing edges of the elevators, rudder, and ailerons. Trimming generally occurs after the pilot has achieved the desired pitch, power, attitude, and configuration. The trim tabs are then used to resolve the remaining control pressures. A small plane may only have a single tab, which is controlled with a small wheel or crank.

19. D: When the Runway Centerline Lighting System lights become solid red, an approaching plane is one thousand feet away. A Runway Centerline Lighting System is a line of white lights every fifty feet or so along the centerline. The lights change their color and pattern as the plane nears the runway. When the plane gets within 3000 feet of the runway, the lights blink red and white. Within a thousand feet, the lights will turn solid red.

20. E: The Coriolis force is the change in rotational speed caused by the shift of the weight towards or away from the center of the spinning object. This phenomenon has an important application for helicopters, in which the rotor will move faster or will require less power to maintain its speed when the weight is closer to the base of the blade. The other answer choices are similarly related to helicopters. The extra lift generated by a helicopter once it has exited its own downwash is known as translational lift. The force that spins the rotors of a helicopter even when there is no power from the engine is autorotation. Greater downwash at the rear half of the rotor disc, as compared to the front half, is the result of applying the lateral cyclic. The phenomenon in which the effects of a force applied to a spinning disc occur ninety degrees later is called gyroscopic precession.

Secret Key #1 - Time is Your Greatest Enemy

Pace Yourself

Wear a watch. At the beginning of the test, check the time (or start a chronometer on your watch to count the minutes), and check the time after every few questions to make sure you are "on schedule."

If you are forced to speed up, do it efficiently. Usually one or more answer choices can be eliminated without too much difficulty. Above all, don't panic. Don't speed up and just begin guessing at random choices. By pacing yourself, and continually monitoring your progress against your watch, you will always know exactly how far ahead or behind you are with your available time. If you find that you are one minute behind on the test, don't skip one question without spending any time on it, just to catch back up. Take 15 fewer seconds on the next four questions, and after four questions you'll have caught back up. Once you catch back up, you can continue working each problem at your normal pace.

Furthermore, don't dwell on the problems that you were rushed on. If a problem was taking up too much time and you made a hurried guess, it must be difficult. The difficult questions are the ones you are most likely to miss anyway, so it isn't a big loss. It is better to end with more time than you need than to run out of time.

Lastly, sometimes it is beneficial to slow down if you are constantly getting ahead of time. You are always more likely to catch a careless mistake by working more slowly than quickly, and among very high-scoring test takers (those who are likely to have lots of time left over), careless errors affect the score more than mastery of material.

Secret Key #2 - Guessing is not Guesswork

You probably know that guessing is a good idea - unlike other standardized tests, there is no penalty for getting a wrong answer. Even if you have no idea about a question, you still have a 20-25% chance of getting it right.

Most test takers do not understand the impact that proper guessing can have on their score. Unless you score extremely high, guessing will significantly contribute to your final score.

Monkeys Take the Test

What most test takers don't realize is that to insure that 20-25% chance, you have to guess randomly. If you put 20 monkeys in a room to take this test, assuming they answered once per question and behaved themselves, on average they would get 20-25% of the questions correct. Put 20 test takers in the room, and the average will be much lower among guessed questions. Why?
1. The test writers intentionally write deceptive answer choices that "look" right. A test taker has no idea about a question, so picks the "best looking" answer, which is often wrong. The monkey has no idea what looks good and what doesn't, so will consistently be lucky about 20-25% of the time.
2. Test takers will eliminate answer choices from the guessing pool based on a hunch or intuition. Simple but correct answers often get excluded, leaving a 0% chance of being correct. The monkey has no clue, and often gets lucky with the best choice.

This is why the process of elimination endorsed by most test courses is flawed and detrimental to your performance- test takers don't guess, they make an ignorant stab in the dark that is usually worse than random.

$5 Challenge

Let me introduce one of the most valuable ideas of this course- the $5 challenge:

You only mark your "best guess" if you are willing to bet $5 on it.
You only eliminate choices from guessing if you are willing to bet $5 on it.

Why $5? Five dollars is an amount of money that is small yet not insignificant, and can really add up fast (20 questions could cost you $100). Likewise, each answer choice on one question of the test will have a small impact on your overall score, but it can really add up to a lot of points in the end.

The process of elimination IS valuable. The following shows your chance of guessing it right:

If you eliminate wrong answer choices until only this many answer choices remain:	1	2	3
Chance of getting it correct:	100%	50%	33%

However, if you accidentally eliminate the right answer or go on a hunch for an incorrect answer, your chances drop dramatically: to 0%. By guessing among all the answer choices, you are GUARANTEED to have a shot at the right answer.

That's why the $5 test is so valuable- if you give up the advantage and safety of a pure guess, it had better be worth the risk.

What we still haven't covered is how to be sure that whatever guess you make is truly random. Here's the easiest way:

Always pick the first answer choice among those remaining.

Such a technique means that you have decided, **before you see a single test question**, exactly how you are going to guess- and since the order of choices tells you nothing about which one is correct, this guessing technique is perfectly random.

This section is not meant to scare you away from making educated guesses or eliminating choices- you just need to define when a choice is worth eliminating. The $5 test, along with a pre-defined random guessing strategy, is the best way to make sure you reap all of the benefits of guessing.

Secret Key #3 - Practice Smarter, Not Harder

Many test takers delay the test preparation process because they dread the awful amounts of practice time they think necessary to succeed on the test. We have refined an effective method that will take you only a fraction of the time.

There are a number of "obstacles" in your way to succeed. Among these are answering questions, finishing in time, and mastering test-taking strategies. All must be executed on the day of the test at peak performance, or your score will suffer. The test is a mental marathon that has a large impact on your future.

Just like a marathon runner, it is important to work your way up to the full challenge. So first you just worry about questions, and then time, and finally strategy:

Success Strategy

1. Find a good source for practice tests.
2. If you are willing to make a larger time investment, consider using more than one study guide- often the different approaches of multiple authors will help you "get" difficult concepts.
3. Take a practice test with no time constraints, with all study helps "open book." Take your time with questions and focus on applying strategies.
4. Take a practice test with time constraints, with all guides "open book."
5. Take a final practice test with no open material and time limits

If you have time to take more practice tests, just repeat step 5. By gradually exposing yourself to the full rigors of the test environment, you will condition your mind to the stress of test day and maximize your success.

Secret Key #4 - Prepare, Don't Procrastinate

Let me state an obvious fact: if you take the test three times, you will get three different scores. This is due to the way you feel on test day, the level of preparedness you have, and, despite the test writers' claims to the contrary, some tests WILL be easier for you than others.

Since your future depends so much on your score, you should maximize your chances of success. In order to maximize the likelihood of success, you've got to prepare in advance. This means taking practice tests and spending time learning the information and test taking strategies you will need to succeed.

Never take the test as a "practice" test, expecting that you can just take it again if you need to. Feel free to take sample tests on your own, but when you go to take the official test, be prepared, be focused, and do your best the first time!

Secret Key #5 - Test Yourself

Everyone knows that time is money. There is no need to spend too much of your time or too little of your time preparing for the test. You should only spend as much of your precious time preparing as is necessary for you to get the score you need.

Once you have taken a practice test under real conditions of time constraints, then you will know if you are ready for the test or not.

If you have scored extremely high the first time that you take the practice test, then there is not much point in spending countless hours studying. You are already there.

Benchmark your abilities by retaking practice tests and seeing how much you have improved. Once you score high enough to guarantee success, then you are ready.

If you have scored well below where you need, then knuckle down and begin studying in earnest. Check your improvement regularly through the use of practice tests under real conditions. Above all, don't worry, panic, or give up. The key is perseverance!

Then, when you go to take the test, remain confident and remember how well you did on the practice tests. If you can score high enough on a practice test, then you can do the same on the real thing.

General Strategies

The most important thing you can do is to ignore your fears and jump into the test immediately- do not be overwhelmed by any strange-sounding terms. You have to jump into the test like jumping into a pool- all at once is the easiest way.

Make Predictions

As you read and understand the question, try to guess what the answer will be. Remember that several of the answer choices are wrong, and once you begin reading them, your mind will immediately become cluttered with answer choices designed to throw you off. Your mind is typically the most focused immediately after you have read the question and digested its contents. If you can, try to predict what the correct answer will be. You may be surprised at what you can predict.

Quickly scan the choices and see if your prediction is in the listed answer choices. If it is, then you can be quite confident that you have the right answer. It still won't hurt to check the other answer choices, but most of the time, you've got it!

Answer the Question

It may seem obvious to only pick answer choices that answer the question, but the test writers can create some excellent answer choices that are wrong. Don't pick an answer just because it sounds right, or you believe it to be true. It MUST answer the question. Once you've made your selection, always go back and check it against the question and make sure that you didn't misread the question, and the answer choice does answer the question posed.

Benchmark

After you read the first answer choice, decide if you think it sounds correct or not. If it doesn't, move on to the next answer choice. If it does, mentally mark that answer choice. This doesn't mean that you've definitely selected it as your answer choice, it just means that it's the best you've seen thus far. Go ahead and read the next choice. If the next choice is worse than the one you've already selected, keep going to the next answer choice. If the next choice is better than the choice you've already selected, mentally mark the new answer choice as your best guess.

The first answer choice that you select becomes your standard. Every other answer choice must be benchmarked against that standard. That choice is correct until proven otherwise by another answer choice beating it out. Once you've decided that no other answer choice seems as good, do one final check to ensure that your answer choice answers the question posed.

Valid Information

Don't discount any of the information provided in the question. Every piece of information may be necessary to determine the correct answer. None of the information in the question is there to throw you off (while the answer choices will certainly have information to throw you off). If two seemingly unrelated topics are discussed, don't ignore either. You can be confident there is a relationship, or it wouldn't be included in the question, and you are probably going to have to determine what is that relationship to find the answer.

Avoid "Fact Traps"

Don't get distracted by a choice that is factually true. Your search is for the answer that answers the question. Stay focused and don't fall for an answer that is true but incorrect. Always go back to the question and make sure you're choosing an answer that actually answers the question and is not just a true statement. An answer can be factually correct, but it MUST answer the question asked. Additionally, two answers can both be seemingly correct, so be sure to read all of the answer choices, and make sure that you get the one that BEST answers the question.

Milk the Question

Some of the questions may throw you completely off. They might deal with a subject you have not been exposed to, or one that you haven't reviewed in years. While your lack of knowledge about the subject will be a hindrance, the question itself can give you many clues that will help you find the correct answer. Read the question carefully and look for clues. Watch particularly for adjectives and nouns describing difficult terms or words that you don't recognize. Regardless of if you completely understand a word or not, replacing it with a synonym either provided or one you more familiar with may help you to understand what the questions are asking. Rather than wracking your mind about specific detailed information concerning a difficult term or word, try to use mental substitutes that are easier to understand.

The Trap of Familiarity

Don't just choose a word because you recognize it. On difficult questions, you may not recognize a number of words in the answer choices. The test writers don't put "make-believe" words on the test; so don't think that just because you only recognize all the words in one answer choice means that answer choice must be correct. If you only recognize words in one answer choice, then focus on that one. Is it correct? Try your best to determine if it is correct. If it is, that is great, but if it doesn't, eliminate it. Each word and answer choice you eliminate increases your chances of getting the question correct, even if you then have to guess among the unfamiliar choices.

Eliminate Answers

Eliminate choices as soon as you realize they are wrong. But be careful! Make sure you consider all of the possible answer choices. Just because one appears right, doesn't mean that the next one won't be even better! The test writers will usually put more than one good answer choice for every question, so read all of them. Don't worry if you are stuck between two that seem right. By getting down to just two remaining possible choices, your odds are now 50/50. Rather than wasting too much time, play the odds. You are guessing, but guessing wisely, because you've been able to knock out some of the answer choices that you know are wrong. If you are eliminating choices and realize that the last answer choice you are left with is also obviously wrong, don't panic. Start over and consider each choice again. There may easily be something that you missed the first time and will realize on the second pass.

Tough Questions

If you are stumped on a problem or it appears too hard or too difficult, don't waste time. Move on! Remember though, if you can quickly check for obviously incorrect answer choices, your chances of guessing correctly are greatly improved. Before you completely give up, at least try to knock out a couple of possible answers. Eliminate what you can and then guess at the remaining answer choices before moving on.

Brainstorm

If you get stuck on a difficult question, spend a few seconds quickly brainstorming. Run through the complete list of possible answer choices. Look at each choice and ask yourself, "Could this answer the question satisfactorily?" Go through each answer choice and consider it independently of the other. By systematically going through all possibilities, you may find something that you would otherwise overlook. Remember that when you get stuck, it's important to try to keep moving.

Read Carefully

Understand the problem. Read the question and answer choices carefully. Don't miss the question because you misread the terms. You have plenty of time to read each question thoroughly and make sure you understand what is being asked. Yet a happy medium must be attained, so don't waste too much time. You must read carefully, but efficiently.

Face Value

When in doubt, use common sense. Always accept the situation in the problem at face value. Don't read too much into it. These problems will not require you to make huge leaps of logic. The test writers aren't trying to throw you off with a cheap trick. If you have to go beyond creativity and make a leap of logic in order to have an answer choice answer the question, then you should look at the other answer choices. Don't overcomplicate the problem by creating theoretical relationships or explanations that will warp time or space. These are normal problems rooted in reality. It's just that the applicable relationship or explanation may not be readily apparent and you have to figure things out. Use your common sense to interpret anything that isn't clear.

Prefixes

If you're having trouble with a word in the question or answer choices, try dissecting it. Take advantage of every clue that the word might include. Prefixes and suffixes can be a huge help. Usually they allow you to determine a basic meaning. Pre- means before, post- means after, pro - is positive, de- is negative. From these prefixes and suffixes, you can get an idea of the general meaning of the word and try to put it into context. Beware though of any traps. Just because con is the opposite of pro, doesn't necessarily mean congress is the opposite of progress!

Hedge Phrases

Watch out for critical "hedge" phrases, such as likely, may, can, will often, sometimes, often, almost, mostly, usually, generally, rarely, sometimes. Question writers insert these hedge phrases to cover every possibility. Often an answer choice will be wrong simply because it leaves no room for exception. Avoid answer choices that have definitive words like "exactly," and "always".

Switchback Words

Stay alert for "switchbacks". These are the words and phrases frequently used to alert you to shifts in thought. The most common switchback word is "but". Others include although, however, nevertheless, on the other hand, even though, while, in spite of, despite, regardless of.

New Information

Correct answer choices will rarely have completely new information included. Answer choices typically are straightforward reflections of the material asked about and will directly relate to the question. If a new piece of information is included in an answer choice that doesn't even seem to relate to the topic being asked about, then that answer choice is likely incorrect. All of the information needed to answer the question is usually provided for you, and so you should not have to make guesses that are unsupported or choose answer choices that require unknown information that cannot be reasoned on its own.

Time Management

On technical questions, don't get lost on the technical terms. Don't spend too much time on any one question. If you don't know what a term means, then since you don't have a dictionary, odds are you aren't going to get much further. You should immediately recognize terms as whether or not you know them. If you don't, work with the other clues that you have, the other answer choices and terms provided, but don't waste too much time trying to figure out a difficult term.

Contextual Clues

Look for contextual clues. An answer can be right but not correct. The contextual clues will help you find the answer that is most right and is correct. Understand the context in which a phrase or statement is made. This will help you make important distinctions.

Don't Panic

Panicking will not answer any questions for you. Therefore, it isn't helpful. When you first see the question, if your mind goes blank, take a deep breath. Force yourself to mechanically go through the steps of solving the problem and using the strategies you've learned.

Pace Yourself

Don't get clock fever. It's easy to be overwhelmed when you're looking at a page full of questions, your mind is full of random thoughts and feeling confused, and the clock is ticking down faster than you would like. Calm down and maintain the pace that you have set for yourself. As long as you are on track by monitoring your pace, you are guaranteed to have enough time for yourself. When you get to the last few minutes of the test, it may seem like you won't have enough time left, but if you only have as many questions as you should have left at that point, then you're right on track!

Answer Selection

The best way to pick an answer choice is to eliminate all of those that are wrong, until only one is left and confirm that is the correct answer. Sometimes though, an answer choice may immediately look right. Be careful! Take a second to make sure that the other choices are not equally obvious. Don't make a hasty mistake. There are only two times that you should stop before checking other answers. First is when you are positive that the answer choice you have selected is correct. Second is when time is almost out and you have to make a quick guess!

Check Your Work

Since you will probably not know every term listed and the answer to every question, it is important that you get credit for the ones that you do know. Don't miss any questions through careless mistakes. If at all possible, try to take a second to look back over your answer selection and make sure you've selected the correct answer choice and haven't made a costly careless mistake (such as marking an answer choice that you didn't mean to mark). This quick double check should more than pay for itself in caught mistakes for the time it costs.

Beware of Directly Quoted Answers

Sometimes an answer choice will repeat word for word a portion of the question or reference section. However, beware of such exact duplication – it may be a trap! More than likely, the correct choice will paraphrase or summarize a point, rather than being exactly the same wording.

Slang

Scientific sounding answers are better than slang ones. An answer choice that begins "To compare the outcomes..." is much more likely to be correct than one that begins "Because some people insisted..."

Extreme Statements

Avoid wild answers that throw out highly controversial ideas that are proclaimed as established fact. An answer choice that states the "process should be used in certain situations, if..." is much more likely to be correct than one that states the "process should be discontinued completely." The first is a calm rational statement and doesn't even make a definitive, uncompromising stance, using a hedge word "if" to provide wiggle room, whereas the second choice is a radical idea and far more extreme.

Answer Choice Families

When you have two or more answer choices that are direct opposites or parallels, one of them is usually the correct answer. For instance, if one answer choice states "x increases" and another answer choice states "x decreases" or "y increases," then those two or three answer choices are very similar in construction and fall into the same family of answer choices. A family of answer choices is when two or three answer choices are very similar in construction, and yet often have a directly opposite meaning. Usually the correct answer choice will be in that family of answer choices. The "odd man out" or answer choice that doesn't seem to fit the parallel construction of the other answer choices is more likely to be incorrect.

Appendix: Area, Volume, Surface Area Formulas

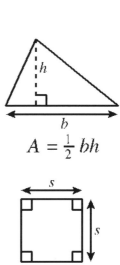
$A = \frac{1}{2} bh$

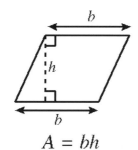
$A = bh$

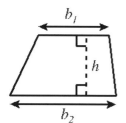
$A = \frac{1}{2} h(b_1 + b_2)$

$p = 4s$
$A = s^2$

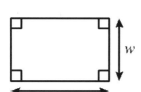
$p = 2l + 2w$
$A = lw$

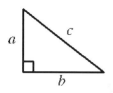
$c^2 = a^2 + b^2$

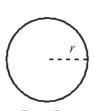
$C = 2\pi r$
$A = \pi r^2$

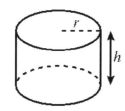
$V = \pi r^2 h$
$S.A. = 2\pi rh + 2\pi r^2$

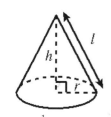
$V = \frac{1}{3} \pi r^2 h$
$S.A. = \pi rl + \pi r^2$

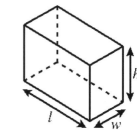
$V = lwh$
$S.A. = 2lw + 2lh + 2wh$

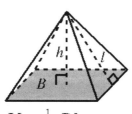
$V = \frac{1}{3} Bh$
$S.A. = \frac{1}{2} lp + B$

Pi

$\pi \approx 3.14$

Additional Bonus Material

Due to our efforts to try to keep this book to a manageable length, we've created a link that will give you access to all of your additional bonus material.

Please visit http://www.mometrix.com/bonus948/officercand to access the information.